Nalluri & Featherstone's

Civil Engineering Hydraulics

Nalluri & Featherstone's

Civil Engineering Hydraulics

Essential Theory with Worked Examples

6th Edition

Martin Marriott
University of East London

WILEY

This edition first published 2016
Fifth edition first published 2009
Fourth edition published 2001
Third edition published 1995
Second edition published 1988
First edition published 1982
First, second, third, fourth and fifth editions © 1982, 1988, 1995, 2001 and 2009 by R.E. Featherstone & C. Nalluri
This edition © 2016 by John Wiley & Sons, Ltd

Blackwell Publishing was acquired by John Wiley & Sons in February 2007. Blackwell's publishing programme has been merged with Wiley's global Scientific, Technical, and Medical business to form Wiley-Blackwell.

Registered office
John Wiley & Sons, Ltd, The Atrium, Southern Gate, Chichester, West Sussex, PO19 8SQ, United Kingdom

Editorial offices
9600 Garsington Road, Oxford, OX4 2DQ, United Kingdom
The Atrium, Southern Gate, Chichester, West Sussex, PO19 8SQ,
United Kingdom

For details of our global editorial offices, for customer services and for information about how to apply for permission to reuse the copyright material in this book please see our website at www.wiley.com/wiley-blackwell.

The right of the author to be identified as the author of this work has been asserted in accordance with the UK Copyright, Designs and Patents Act 1988.

Designations used by companies to distinguish their products are often claimed as trademarks. All brand names and product names used in this book are trade names, service marks, trademarks or registered trademarks of their respective owners. The publisher is not associated with any product or vendor mentioned in this book.

Limit of Liability/Disclaimer of Warranty: While the publisher and author(s) have used their best efforts in preparing this book, they make no representations or warranties with respect to the accuracy or completeness of the contents of this book and specifically disclaim any implied warranties of merchantability or fitness for a particular purpose. It is sold on the understanding that the publisher is not engaged in rendering professional services and neither the publisher nor the author shall be liable for damages arising herefrom. If professional advice or other expert assistance is required, the services of a competent professional should be sought.

Library of Congress Cataloging-in-Publication Data

Names: Marriott, Martin, author. | Featherstone, R. E., author. | Nalluri, C., author.
Title: Nalluri & Featherstone's civil engineering hydraulics : essential theory with worked examples.
Other titles: Civil engineering hydraulics
Description: 6th edition/ Martin Marriott, University of East London. | Chichester, West Sussex,
    United Kingdom : John Wiley & Sons, Inc., 2016. | Includes bibliographical references and index.
Identifiers: LCCN 2015044914 (print) | LCCN 2015045968 (ebook) | ISBN 9781118915639 (pbk.) |
    ISBN 9781118915806 (pdf) | ISBN 9781118915660 (epub)
Subjects: LCSH: Hydraulic engineering. | Hydraulics.
Classification: LCC TC145 .N35 2016 (print) | LCC TC145 (ebook) | DDC 627–dc23
LC record available at http://lccn.loc.gov/2015044914

A catalogue record for this book is available from the British Library.

Wiley also publishes its books in a variety of electronic formats. Some content that appears in print may not be available in electronic books.

Cover image: Itoshiro Dam/structurae.de

Set in 9.5/11.5pt Sabon by Aptara Inc., New Delhi, India

Printed in Singapore by C.O.S. Printers Pte Ltd

6   2016

Contents

Preface to Sixth Edition   xi
About the Author   xiii
Symbols   xv

1   Properties of Fluids   1

    1.1   Introduction   1
    1.2   Engineering units   1
    1.3   Mass density and specific weight   2
    1.4   Relative density   2
    1.5   Viscosity of fluids   2
    1.6   Compressibility and elasticity of fluids   2
    1.7   Vapour pressure of liquids   2
    1.8   Surface tension and capillarity   3
    Worked examples   3
    References and recommended reading   5
    Problems   5

2   Fluid Statics   7

    2.1   Introduction   7
    2.2   Pascal's law   7
    2.3   Pressure variation with depth in a static incompressible fluid   8
    2.4   Pressure measurement   9
    2.5   Hydrostatic thrust on plane surfaces   11
    2.6   Pressure diagrams   14
    2.7   Hydrostatic thrust on curved surfaces   15
    2.8   Hydrostatic buoyant thrust   17
    2.9   Stability of floating bodies   17
    2.10   Determination of metacentre   18
    2.11   Periodic time of rolling (or oscillation) of a floating body   20
    2.12   Liquid ballast and the effective metacentric height   20
    2.13   Relative equilibrium   22
    Worked examples   24

Reference and recommended reading   41
Problems   41

3   Fluid Flow Concepts and Measurements   47

    3.1   Kinematics of fluids   47
    3.2   Steady and unsteady flows   48
    3.3   Uniform and non-uniform flows   48
    3.4   Rotational and irrotational flows   49
    3.5   One-, two- and three-dimensional flows   49
    3.6   Streamtube and continuity equation   49
    3.7   Accelerations of fluid particles   50
    3.8   Two kinds of fluid flow   51
    3.9   Dynamics of fluid flow   52
    3.10   Energy equation for an ideal fluid flow   52
    3.11   Modified energy equation for real fluid flows   54
    3.12   Separation and cavitation in fluid flow   55
    3.13   Impulse–momentum equation   56
    3.14   Energy losses in sudden transitions   57
    3.15   Flow measurement through pipes   58
    3.16   Flow measurement through orifices and mouthpieces   60
    3.17   Flow measurement in channels   64
    Worked examples   69
    References and recommended reading   85
    Problems   85

4   Flow of Incompressible Fluids in Pipelines   89

    4.1   Resistance in circular pipelines flowing full   89
    4.2   Resistance to flow in non-circular sections   94
    4.3   Local losses   94
    Worked examples   95
    References and recommended reading   115
    Problems   115

5   Pipe Network Analysis   119

    5.1   Introduction   119
    5.2   The head balance method ('loop' method)   120
    5.3   The quantity balance method ('nodal' method)   121
    5.4   The gradient method   123
    Worked examples   125
    References and recommended reading   142
    Problems   143

6   Pump–Pipeline System Analysis and Design   149

    6.1   Introduction   149
    6.2   Hydraulic gradient in pump–pipeline systems   150
    6.3   Multiple pump systems   151
    6.4   Variable-speed pump operation   153

6.5   Suction lift limitations   153
Worked examples   154
References and recommended reading   168
Problems   168

7   Boundary Layers on Flat Plates and in Ducts   171

7.1   Introduction   171
7.2   The laminar boundary layer   171
7.3   The turbulent boundary layer   172
7.4   Combined drag due to both laminar and turbulent boundary layers   173
7.5   The displacement thickness   173
7.6   Boundary layers in turbulent pipe flow   174
7.7   The laminar sub-layer   176
Worked examples   178
References and recommended reading   185
Problems   185

8   Steady Flow in Open Channels   187

8.1   Introduction   187
8.2   Uniform flow resistance   188
8.3   Channels of composite roughness   189
8.4   Channels of compound section   190
8.5   Channel design   191
8.6   Uniform flow in part-full circular pipes   194
8.7   Steady, rapidly varied channel flow energy principles   195
8.8   The momentum equation and the hydraulic jump   196
8.9   Steady, gradually varied open channel flow   198
8.10   Computations of gradually varied flow   199
8.11   The direct step method   199
8.12   The standard step method   200
8.13   Canal delivery problems   201
8.14   Culvert flow   202
8.15   Spatially varied flow in open channels   203
Worked examples   205
References and recommended reading   241
Problems   241

9   Dimensional Analysis, Similitude and Hydraulic Models   247

9.1   Introduction   247
9.2   Dimensional analysis   248
9.3   Physical significance of non-dimensional groups   248
9.4   The Buckingham π theorem   249
9.5   Similitude and model studies   249
Worked examples   250
References and recommended reading   263
Problems   263

10 Ideal Fluid Flow and Curvilinear Flow — 265

10.1 Ideal fluid flow — 265
10.2 Streamlines, the stream function — 265
10.3 Relationship between discharge and stream function — 266
10.4 Circulation and the velocity potential function — 267
10.5 Stream functions for basic flow patterns — 267
10.6 Combinations of basic flow patterns — 269
10.7 Pressure at points in the flow field — 269
10.8 The use of flow nets and numerical methods — 270
10.9 Curvilinear flow of real fluids — 273
10.10 Free and forced vortices — 274
Worked examples — 274
References and recommended reading — 285
Problems — 285

11 Gradually Varied Unsteady Flow from Reservoirs — 289

11.1 Discharge between reservoirs under varying head — 289
11.2 Unsteady flow over a spillway — 291
11.3 Flow establishment — 292
Worked examples — 293
References and recommended reading — 302
Problems — 302

12 Mass Oscillations and Pressure Transients in Pipelines — 305

12.1 Mass oscillation in pipe systems – surge chamber operation — 305
12.2 Solution neglecting tunnel friction and throttle losses for sudden discharge stoppage — 306
12.3 Solution including tunnel and surge chamber losses for sudden discharge stoppage — 307
12.4 Finite difference methods in the solution of the surge chamber equations — 308
12.5 Pressure transients in pipelines (waterhammer) — 309
12.6 The basic differential equations of waterhammer — 311
12.7 Solutions of the waterhammer equations — 312
12.8 The Allievi equations — 312
12.9 Alternative formulation — 315
Worked examples — 316
References and recommended reading — 322
Problems — 322

13 Unsteady Flow in Channels — 323

13.1 Introduction — 323
13.2 Gradually varied unsteady flow — 323
13.3 Surges in open channels — 324
13.4 The upstream positive surge — 325
13.5 The downstream positive surge — 326
13.6 Negative surge waves — 327

13.7   The dam break   329
Worked examples   330
References and recommended reading   333
Problems   333

14 Uniform Flow in Loose-Boundary Channels   335

14.1   Introduction   335
14.2   Flow regimes   335
14.3   Incipient (threshold) motion   335
14.4   Resistance to flow in alluvial (loose-bed) channels   337
14.5   Velocity distributions in loose-boundary channels   339
14.6   Sediment transport   339
14.7   Bed load transport   340
14.8   Suspended load transport   343
14.9   Total load transport   345
14.10 Regime channel design   346
14.11 Rigid-bed channels with sediment transport   350
Worked examples   352
References and recommended reading   367
Problems   368

15 Hydraulic Structures   371

15.1   Introduction   371
15.2   Spillways   371
15.3   Energy dissipators and downstream scour protection   376
Worked examples   379
References and recommended reading   389
Problems   390

16 Environmental Hydraulics and Engineering Hydrology   393

16.1   Introduction   393
16.2   Analysis of gauged river flow data   393
16.3   River Thames discharge data   395
16.4   Flood alleviation, sustainability and environmental channels   396
16.5   Project appraisal   397
Worked examples   398
References and recommended reading   405
Problems   406

17 Introduction to Coastal Engineering   409

17.1   Introduction   409
17.2   Waves and wave theories   409
17.3   Wave processes   420
17.4   Wave set-down and set-up   428
17.5   Wave impact, run-up and overtopping   429
17.6   Tides, surges and mean sea level   430
17.7   Tsunami waves   432

x   Contents

Worked examples   433
References and recommended reading   438
Problems   439

Answers   441
Index   447

Preface to Sixth Edition

This book has regularly been on reading lists for hydraulics and water engineering modules for university civil engineering degree students. The concise summary of theory and the worked examples have been useful to me both as a practising engineer and as an academic.

The fifth edition aimed to retain all the good qualities of Nalluri and Featherstone's previous editions, with updating as necessary and with an additional chapter on environmental hydraulics and hydrology.

The latest sixth edition now adds a new chapter on coastal engineering prepared by my colleague Dr Ravindra Jayaratne based on original material and advice from Dr Dominic Hames of HR Wallingford. As before, each chapter contains theory sections, after which there are worked examples followed by a list of references and recommended reading. Then there are further problems as a useful resource for students to tackle. The numerical answers to these are at the back of the book, and solutions are available to download from the publisher's website: **http://www.wiley.com/go/Marriott.**

I am grateful to all those who have helped me in many ways, either through their advice in person or through their published work, and of course to the many students with whom I have enjoyed studying this material.

Martin Marriott
University of East London
2016

About the Author

This well-established text draws on Nalluri and Featherstone's extensive teaching experience at Newcastle University, including material provided by Professor J. Saldarriaga of the University of Los Andes, Colombia. The text has been updated and extended by Dr Martin Marriott with input from Dr Ravindra Jayaratne of the University of East London and Dr Dominic Hames of HR Wallingford.

Martin Marriott is a chartered civil engineer, with degrees from the Universities of Cambridge, Imperial College London and Hertfordshire. He has wide professional experience in the UK and overseas with major firms of consulting engineers, followed by many years of experience as a lecturer in higher education, currently at the University of East London.

Symbols

The following is a list of the main symbols used in this book (with their SI units, where appropriate). Various subscripts have also been used, for example to denote particular locations. Note that some symbols are inevitably used with different meanings in different contexts, and so a number of alternatives are listed below. Readers should be aware of this, and check the context for clarification.

a area (m^2); distance (m); acceleration (m/s^2)
b width (m); probability weighted moment of flows (m^3/s)
c wave celerity (m/s)
d diameter (m); water depth (m)
f force (N); function; silt factor; frequency
g gravitational acceleration (≈ 9.81 m/s^2)
h height (m); pressure head difference (m); head loss (m)
i rank in descending order
j rank in ascending order
k radius of gyration (m); roughness height (m); constant; coefficient
m metacentric height (m); mass (kg)
n Manning's coefficient; exponent; number; wave steepness; group velocity parameter
p pressure (N/m^2)
q discharge per unit width (m^2/s)
r radius (m); discount rate
s relative density; distance (m); sinuosity; standard deviation of sample
t time (s); L-moment ratios
u velocity (m/s); parameter
v velocity (m/s)
w velocity (m/s)
x distance (m); variable
y distance (m); reduced variate; depth (m)
z elevation (m); vertical distance (m)

A area (m^2)
B width (m); centre of buoyancy; benefit
C constant; centre of pressure; coefficient; cost

D diameter (m)
E specifie energy (J/N = m); elastic modulus (N/m^2); wave energy (J/m^2)
F force (N); head loss coefficient (s^2/m); annual probability of non-exceedance
Fr Froude number
G centroid
H height (m); head (m); wave height (m)
I second moment of area (m^4); inflow (m^3/s)
J junction or node
K bulk modulus of elasticity (N/m^2); coefficient; conveyance (m^3/s); circulation (m^2/s)
L length (m); L-moment of flows (m^3/s); wavelength (m)
M metacentre; mass (kg)
N number; rotational speed (rev/min)
P height of weir (m); wetted perimeter (m); power (W); annual exceedance
        probability; wave power (W/m)
Q discharge (m^3/s)
R resultant force (N); hydraulic radius (m); radius (m)
Re Reynolds number
S slope; energy gradient; storage volume (m^3); wave spectrum (m^{2}s)
T thrust (N); time period (s); return period (years); surface width (m); thickness (m);
        wave period (s)
U velocity (m/s)
V volume (m^3); velocity (m/s)
W weight (N); fall velocity (m/s)
We Weber number
Z elevation (m); section factor (m$^{5/2}$)

α angular acceleration (rad/s^2); angle (rad); Coriolis coefficient; parameter
β momentum correction factor (Boussinesq coefficient); parameter; slope
γ specific weight (N/m^3)
δ boundary layer thickness (m)
ζ factor
η efficiency; wave profile (m)
θ angle (radian or degree); slope; wave direction
κ constant
λ Darcy–Weisbach friction factor; scale
μ dynamic or absolute viscosity (Ns/m^2); ripple factor; mean
υ kinematic viscosity (m^2/s)
ξ spillway loss coefficient; displacement (m)
π circle circumference-to-diameter ratio ($\approx$ 3.142); Buckingham dimensionless group
ρ mass density (kg/m^3)
σ surface tension (N/m); safety factor
τ shear stress (N/m^2)
ϕ function; potential (m^2/s); transport parameter; angle of repose (degree)
ψ stream function (m^2/s); flow parameter
ω angular velocity (rad/s)
Δ increment; submerged relative density

Chapter 1
Properties of Fluids

1.1 Introduction

A **fluid** is a substance which deforms continuously, or flows, when subjected to shear stresses. The term fluid embraces both gases and liquids; a given mass of liquid will occupy a definite volume whereas a gas will fill its container. Gases are readily compressible; the low compressibility, or elastic volumetric deformation, of liquids is generally neglected in computations except those relating to large depths in the oceans and in pressure transients in pipelines.

This text, however, deals exclusively with liquids and more particularly with Newtonian liquids (i.e. those having a linear relationship between shear stress and rate of deformation).

Typical values of different properties are quoted in the text as needed for the various worked examples. For more comprehensive details of physical properties, refer to tables such as Kaye and Laby (1995) or internet versions of such information.

1.2 Engineering units

The **metre–kilogram–second (mks) system** is the agreed version of the international system (SI) of units that is used in this text. The physical quantities in this text can be described by a set of three primary dimensions (units): mass (kg), length (m) and time (s). Further discussion is contained in Chapter 9 regarding dimensional analysis. The present chapter refers to the relevant units that will be used.

The unit of force is called newton (N) and 1 N is the force which accelerates a mass of 1 kg at a rate of 1 m/s^2 (1 N = 1 kg m/s^2).

The unit of work is called joule (J) and it is the energy needed to move a force of 1 N over a distance of 1 m. Power is the energy or work done per unit time and its unit is watt (W) (1 W = 1 J/s = 1 N m/s).

Nalluri & Featherstone's Civil Engineering Hydraulics: Essential Theory with Worked Examples,
Sixth Edition. Martin Marriott.
© 2016 John Wiley & Sons, Ltd. Published 2016 by John Wiley & Sons, Ltd.
Companion Website: www.wiley.com/go/Marriott

1.3 Mass density and specific weight

Mass density (ρ) or **density** of a substance is defined as the mass of the substance per unit volume (kg/m^3) and is different from **specific weight** (γ), which is the force exerted by the earth's gravity (g) upon a unit volume of the substance ($\gamma = \rho g$: N/m^3). In a satellite where there is no gravity, an object has no specific weight but possesses the same density that it has on the earth.

1.4 Relative density

Relative density (s) of a substance is the ratio of its mass density to that of water at a standard temperature (4°C) and pressure (atmospheric) and is dimensionless.

For water, $\rho = 10^3$ kg/m^3, $\gamma = 10^3 \times 9.81 \simeq 10^4$ N/m^3 and $s = 1$.

1.5 Viscosity of fluids

Viscosity is that property of a fluid which by virtue of cohesion and interaction between fluid molecules offers resistance to shear deformation. Different fluids deform at different rates under the action of the same shear stress. Fluids with high viscosity such as syrup deform relatively more slowly than fluids with low viscosity such as water.

All fluids are viscous and 'Newtonian fluids' obey the linear relationship

$$\tau = \mu \frac{du}{dy} \quad \text{(Newton's law of viscosity)} \qquad [1.1]$$

where τ is the shear stress (N/m^2), du/dy the velocity gradient or the rate of deformation (rad/s) and μ the coefficient of dynamic (or absolute) viscosity (N s/m^2 or kg/(m s)).

Kinematic viscosity (ν) is the ratio of dynamic viscosity to mass density expressed in metres squared per second.

Water is a Newtonian fluid having a dynamic viscosity of approximately 1.0×10^{-3} N s/m^2 and kinematic viscosity of 1.0×10^{-6} m^2/s at 20°C.

1.6 Compressibility and elasticity of fluids

All fluids are compressible under the application of an external force and when the force is removed they expand back to their original volume, exhibiting the property that stress is proportional to volumetric strain.

$$\text{The bulk modulus of elasticity,} \, K = \frac{\text{pressure change}}{\text{volumetric strain}}$$

$$= -\frac{dp}{(dV/V)} \qquad [1.2]$$

The negative sign indicates that an increase in pressure causes a decrease in volume.

Water with a bulk modulus of 2.1×10^9 N/m^2 at 20°C is 100 times more compressible than steel, but it is ordinarily considered incompressible.

1.7 Vapour pressure of liquids

A liquid in a closed container is subjected to partial vapour pressure due to the escaping molecules from the surface; it reaches a stage of equilibrium when this pressure reaches saturated vapour pressure. Since this depends upon molecular activity, which is a function of temperature, the vapour pressure of a fluid also depends upon its temperature and increases with it. If the pressure above a liquid reaches the vapour pressure of the liquid, boiling occurs; for example, if the pressure is reduced sufficiently, boiling may occur at room temperature.

The saturated vapour pressure for water at 20°C is 2.45×10^3 N/m².

1.8 Surface tension and capillarity

Liquids possess the properties of cohesion and adhesion due to molecular attraction. Due to the property of cohesion, liquids can resist small tensile forces at the interface between the liquid and air, known as **surface tension** (σ: N/m). If the liquid molecules have greater adhesion than cohesion, then the liquid sticks to the surface of the container with which it is in contact, resulting in a capillary rise of the liquid surface; a predominating cohesion, in contrast, causes capillary depression. The surface tension of water is 73×10^{-3} N/m at 20°C.

The capillary rise or depression h of a liquid in a tube of diameter d can be written as

$$h = \frac{4\sigma \cos \theta}{\rho g d} \qquad [1.3]$$

where θ is the angle of contact between liquid and solid.

Surface tension increases the pressure within a droplet of liquid. The internal pressure p balancing the surface tensional force of a small spherical droplet of radius r is given by

$$p = \frac{2\sigma}{r} \qquad [1.4]$$

Worked examples

Example 1.1

The density of an oil at 20°C is 850 kg/m³. Find its relative density and kinematic viscosity if the dynamic viscosity is 5×10^{-3} kg/(m s).

Solution:

$$\text{Relative density, } s = \frac{\rho \text{ of oil}}{\rho \text{ of water}}$$

$$= \frac{850}{10^3}$$

$$= 0.85$$

$$\text{Kinematic viscosity, } v = \frac{\mu}{\rho}$$

$$= \frac{5 \times 10^{-3}}{850}$$

$$= 5.88 \times 10^{-6} \text{ m}^2/\text{s}$$

Example 1.2

If the velocity distribution of a viscous liquid ($\mu = 0.9$ N s/m²) over a fixed boundary is given by $u = 0.68y - y^2$, in which u is the velocity (in metres per second) at a distance y (in metres) above the boundary surface, determine the shear stress at the surface and at $y = 0.34$ m.

Solution:

$$u = 0.68y - y^2$$
$$\Rightarrow \frac{du}{dy} = 0.68 - 2y$$

Hence, $(du/dy)_{y=0} = 0.68$ s^{-1} and $(du/dy)_{y=0.34m} = 0$.

Dynamic viscosity of the fluid, $\mu = 0.9$ N s/m²

From Equation 1.1,

$$\text{shear stress } (\tau)_{y=0} = 0.9 \times 0.68$$
$$= 0.612 \text{ N/m}^2$$

and at $y = 0.34$ m, $\tau = 0$.

Example 1.3

At a depth of 8.5 km in the ocean the pressure is 90 MN/m². The specific weight of the sea water at the surface is 10.2 kN/m³ and its average bulk modulus is 2.4×10^6 kN/m². Determine (a) the change in specific volume, (b) the specific volume and (c) the specific weight of sea water at 8.5 km depth.

Solution:

Change in pressure at a depth of 8.5 km, $dp = 90$ MN/m²
$$= 9 \times 10^4 \text{ kN/m}^2$$

Bulk modulus, $K = 2.4 \times 10^6$ kN/m²

From $K = -\dfrac{dp}{(dV/V)}$

$$\frac{dV}{V} = \frac{-9 \times 10^4}{2.4 \times 10^6} = -3.75 \times 10^{-2}$$

Defining specific volume as $1/\gamma$ (m³/kN), the specific volume of sea water at the surface = $1/10.2 = 9.8 \times 10^{-2}$ m³/kN.

Change in specific volume between that at the surface and at 8.5 km depth, dV
$$= -3.75 \times 10^{-2} \times 9.8 \times 10^{-2}$$
$$= -36.75 \times 10^{-4} \text{ m}^3/\text{kN}$$

The specific volume of sea water at 8.5 km depth $= 9.8 \times 10^{-2} - 36.75 \times 10^{-4}$
$$= 9.44 \times 10^{-2} \, \text{m}^3/\text{kN}$$

The specific weight of sea water at 8.5 km depth $= \dfrac{1}{\text{specific volume}}$

$$= \dfrac{1}{9.44 \times 10^{-2}}$$

$$= 10.6 \, \text{kN/m}^3$$

References and recommended reading

Kaye, G. W. C. and Laby, T. H. (1995) *Tables of Physical and Chemical Constants*, 16th edn, Longman, London. http://www.kayelaby.npl.co.uk

Massey, B. S. and Ward-Smith, J. (2012) *Mechanics of Fluids*, 9th edn, Taylor & Francis, Abingdon, UK.

Problems

1. **(a)** Explain why the viscosity of a liquid decreases while that of a gas increases with an increase of temperature.

   (b) The following data refer to a liquid under shearing action at a constant temperature. Determine its dynamic viscosity.

du/dy (s^{-1})	0	0.2	0.4	0.6	0.8
τ (N/m^2)	0	0	1.9	3.1	4.0

2. A 300 mm wide shaft sleeve moves along a 100 mm diameter shaft at a speed of 0.5 m/s under the application of a force of 250 N in the direction of its motion. If 1000 N of force is applied, what speed will the sleeve attain? Assume the temperature of the sleeve to be constant and determine the viscosity of the Newtonian fluid in the clearance between the shaft and its sleeve if the radial clearance is estimated to be 0.075 mm.

3. A shaft of 100 mm diameter rotates at 120 rad/s in a bearing 150 mm long. If the radial clearance is 0.2 mm and the absolute viscosity of the lubricant is 0.20 kg/(m s), find the power loss in the bearing.

4. A block of dimensions 300 mm × 300 mm × 300 mm and mass 30 kg slides down a plane inclined at 30° to the horizontal, on which there is a thin film of oil of viscosity 2.3×10^{-3} N s/m^2. Determine the speed of the block if the film thickness is estimated to be 0.03 mm.

5. Calculate the capillary effect (in millimetres) in a glass tube of 6 mm diameter when immersed in (i) water and (ii) mercury, both liquids being at 20°C. Assume σ to be 73×10^{-3} N/m for water and 0.5 N/m for mercury. The contact angles for water and mercury are 0 and 130°, respectively.

6. Calculate the internal pressure of a 25 mm diameter soap bubble if the tension in the soap film is 0.5 N/m.

Chapter 2
Fluid Statics

2.1 Introduction

Fluid statics is the study of pressures throughout a fluid at rest and the pressure forces on finite surfaces. Since the fluid is at rest there are no shear stresses in it. Hence the pressure p at a point on a plane surface (inside the fluid or on the boundaries of its container), defined as the limiting value of the ratio of normal force to surface area as the area approaches zero size, always acts normal to the surface and is measured in newtons per square metre (pascals, Pa) or in bars (1 bar = 10^5 N/m^2 or 10^5 Pa).

2.2 Pascal's law

Pascal's law states that the pressure at a point in a fluid at rest is the same in all directions. This means it is independent of the orientation of the surface around the point.

Consider a small triangular prism of unit length surrounding the point in a fluid at rest (Figure 2.1).

Since the body is in static equilibrium, we can write

$$p_1(\text{AB} \times l) - p_3(\text{BC} \times l)\cos\theta = 0 \qquad \text{(i)}$$

and

$$p_2(\text{AC} \times l) - p_3(\text{BC} \times l)\sin\theta - W = 0 \qquad \text{(ii)}$$

From Equation (i) $p_1 = p_3$, since $\cos\theta = \text{AB}/\text{BC}$, and Equation (ii) gives $p_2 = p_3$, since $\sin\theta = \text{AC}/\text{BC}$ and $W = 0$ as the prism shrinks to a point.

$$\Rightarrow p_1 = p_2 = p_3$$

Nalluri & Featherstone's Civil Engineering Hydraulics: Essential Theory with Worked Examples, Sixth Edition. Martin Marriott.
© 2016 John Wiley & Sons, Ltd. Published 2016 by John Wiley & Sons, Ltd.
Companion Website: www.wiley.com/go/Marriott

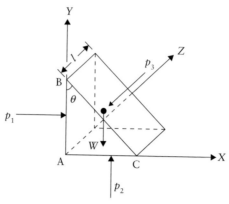

Figure 2.1   Pressure at a point.

2.3 Pressure variation with depth in a static incompressible fluid

Consider an elementary cylindrical volume of fluid (of length L and cross-sectional area dA) within the static fluid mass (Figure 2.2), p being the pressure at an elevation of y and dp being the pressure variation corresponding to an elevation variation of dy.

For equilibrium of the elementary volume,

$$p \, dA - \rho g \, dA \, L \sin \theta - (p + dp) \, dA = 0$$

or

$$dp = -\rho g \, dy \quad \left(\text{since } \sin \theta = \frac{dy}{L} \right) \qquad [2.1]$$

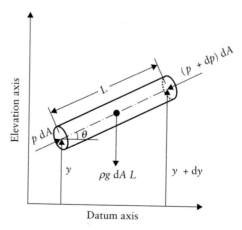

Figure 2.2   Pressure variation with elevation.

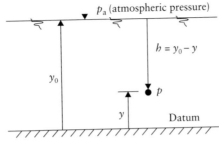

Figure 2.3   Pressure and pressure head at a point.

ρ being constant for incompressible fluids, we can write

$$\int dp = -\rho g \int dy$$

which gives

$$p = -\rho g y + C \qquad \text{(i)}$$

When $y = y_0, p = p_a$, the atmospheric pressure (Figure 2.3).
From Equation (i),

$$p - p_a = \rho g (y_0 - y)$$
$$= \rho g h$$

or the pressure at a depth $h, p = p_a + \rho g h$

$$= \rho g h \text{ above atmospheric pressure} \qquad [2.2]$$

Note:

(a) If $p = \rho g h, h = p/\rho g$ and is known as the pressure head (in metres) of fluid of density ρ.

(b) Equation (i) can be written as $p/\rho g + y = \text{constant}$, which shows that any increase in elevation is compensated by a corresponding decrease in pressure head. $(p/\rho g + y)$ is known as the piezometric head and such a variation is known as the hydrostatic pressure distribution.

If the static fluid is a compressible liquid, ρ is no longer constant and it is dependent on the degree of its compressibility. Equations 1.2 and 2.1 yield the relationship

$$\frac{1}{\rho} = \frac{1}{\rho_0} - \frac{gh}{K} \qquad [2.3]$$

where ρ is the density at a depth h below the free surface at which its density is ρ_0.

2.4 Pressure measurement

The pressure at the earth's surface depends upon the air column above it. At sea level this atmospheric pressure is about 101 kN/m², equivalent to 10.3 m of water or 760 mm of mercury columns. A perfect vacuum is an empty space where the pressure is zero. **Gauge pressure** is the pressure measured above or below atmospheric pressure. The pressure

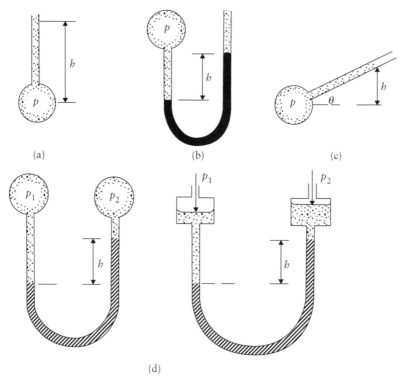

Figure 2.4 Pressure measurement devices: (a) piezometer, (b) U-tube, (c) inclined mano-meter and (d) differential manometers.

below atmospheric pressure is also called **negative or partial vacuum pressure. Absolute pressure** is the pressure measured above a perfect vacuum, the absolute zero.

(a) A simple vertical tube fixed to a system whose pressure is to be measured is called a piezometer (Figure 2.4a). The liquid rises to such a level that the liquid column's height balances the pressure inside.

(b) A bent tube in the form of a U, known as a U-tube manometer, is much more con-venient than a simple piezometer. Heavy immiscible manometer liquids are used to measure large pressures, and lighter liquids to measure small pressures (Figure 2.4b).

(c) An inclined tube or U-tube (Figure 2.4c) is used as a pressure-measuring device when the pressures are very small. The accuracy of measurement is improved by providing suitable inclination.

(d) A differential manometer (Figure 2.4d) is essentially a U-tube manometer containing a single liquid capable of measuring large pressure differences between two systems. If the pressure difference is very small, the manometer may be modified by providing enlarged ends and two different liquids in the two limbs and is called a differential micromanometer.

If the density of water is ρ, a water column of height h produces a pressure $p = \rho g h$ and this can be expressed in terms of any other liquid column h_1 as $\rho_1 g h_1$, ρ_1 being its density:

$$\Rightarrow h \text{ in water column} = \left(\frac{\rho_1}{\rho}\right) h_1 = s h_1 \qquad [2.4]$$

where s is the relative density of the liquid.

For each one of the pressure measurement devices shown in Figure 2.4, an equation can be written using the principle of hydrostatic pressure distribution, expressing the pressures (in metres) of the water column (Equation 2.4) for convenience.

2.5 Hydrostatic thrust on plane surfaces

Let the plane surface be inclined at an angle θ to the free surface of water, as shown in Figure 2.5.

If the plane area A is assumed to consist of elemental areas dA, the elemental forces dF always normal to the surface area are parallel. Therefore the system is equivalent to one resultant force F, known as the **hydrostatic thrust**. Its point of application C, which would produce the same moment effects as the distributed thrust, is called the **centre of pressure**.

We can write

$$F = \int_A \mathrm{d}F = \int_A \rho g h \, \mathrm{d}A = \rho g \sin\theta \int_A \mathrm{d}A \, x$$
$$= \rho g \sin\theta \, A\bar{x}$$
$$= \rho g \bar{h} A \qquad [2.5]$$

where $\bar{h}$ is the vertical depth of the centroid G.

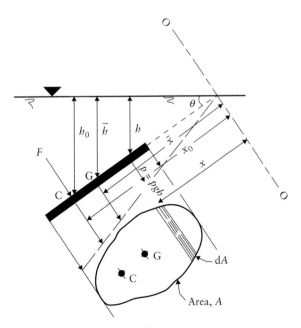

Figure 2.5 Hydrostatic thrust on a plane surface.

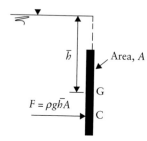

Figure 2.6   Vertical plane surface.

Taking moments of these forces about the axis O–O shown in Figure 2.5,

$$Fx_0 = \rho g \sin \theta \int_A dA \, x^2$$

The distance to the centre of pressure, C, is therefore

$$x_0 = \frac{\int_A dA \, x^2}{\int_A dA \, x}$$

$$= \frac{\text{second moment of the area about axis O–O}}{\text{first moment of the area about axis O–O}} \qquad [2.6]$$

$$= \frac{I_0}{A\bar{x}}$$

But $I_0 = I_g + A\bar{x}^2$ (parallel-axis rule), where I_g is the second moment of area of the surface about an axis through its centroid and parallel to axis O–O.

$$\Rightarrow x_0 = \bar{x} + \frac{I_g}{A\bar{x}} \qquad [2.7]$$

which shows that the centre of pressure is always below the centroid of the area.

Depth of centre of pressure below free surface, $h_0 = x_0 \sin \theta$

$$\Rightarrow h_0 = \bar{h} + \frac{I_g \sin^2 \theta}{A\bar{h}} \qquad [2.8]$$

For a vertical surface, $\theta = 90°$.

$$\Rightarrow h_0 = \bar{h} + \frac{I_g}{A\bar{h}} \qquad [2.9]$$

The distance between the centroid and centre of pressure is

$$GC = \frac{I_g}{A\bar{h}} \quad \text{(see Figure 2.6)} \qquad [2.10]$$

The moment of F about the centroid is written as

$$F \times GC = \rho g \bar{h} A \times \frac{I_g}{A\bar{h}}$$

$$= \rho g I_g$$

which is independent of the depth of submergence.

Table 2.1   Second moments of plane areas

Shape	Size	Second moment of area I_g about an axis GG through centroid
Rectangle		$I_g = \int_{-d/2}^{+d/2} x^2 b \, dx = 2b \int_{0}^{d/2} x^2 \, dx = \frac{1}{12} bd^3$
Triangle		$I_g = \frac{1}{36} bh^3$
Circle		$I_g = \frac{\pi d^4}{64}$

Note:

(a)   Radius of gyration of the area about G is

$$k_g = \sqrt{\frac{I_g}{A}} \qquad [2.11]$$

giving

$$h_0 = \bar{h} + \frac{k_g^2}{\bar{h}} \qquad [2.12]$$

(b)   When the surface area is symmetrical about its vertical centroidal axis, the centre of pressure always lies on this symmetrical axis but below the centroid of the area.

If the area is not symmetrical, an additional coordinate y_0 must be fixed to locate the centre of pressure completely.

By moments (Figure 2.7),

$$y_0 \int_A dF = \int_A dF \, y$$

$$\Rightarrow y_0 \, \rho g \bar{x} \sin \theta \, A = \int_A \rho g x \sin \theta \, dA \, y$$

$$\Rightarrow y_0 = \frac{1}{A\bar{x}} \int_A xy \, dA \qquad [2.13]$$

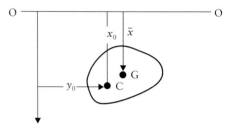

Figure 2.7   Centre of pressure of an asymmetrical plane surface.

2.6 Pressure diagrams

Another approach to determine hydrostatic thrust and its location is by the concept of pressure distribution over the surface (Figure 2.8).

Total thrust on a rectangular vertical surface subjected to water pressure on one side (Figure 2.9) by a pressure diagram:

$$\text{Average pressure on the surface} = \frac{\rho g H}{2}$$

$$\text{Total thrust, } F = \text{average pressure} \times \text{area of surface}$$

$$= \left(\frac{\rho g H}{2} \right) \times H \times B$$

$$= \frac{1}{2} \rho g H^2 \times B$$

$$= \text{volume of the pressure prism} \qquad [2.14]$$

or

$$\frac{\text{total thrust}}{\text{unit width}} = \frac{1}{2} \rho g H^2$$

$$= \text{area of the pressure diagram} \qquad [2.15]$$

and the centre of pressure is the centroid of the pressure prism.

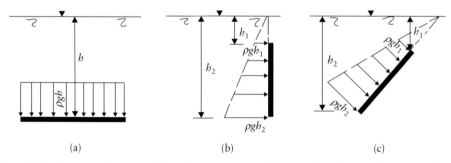

(a)   (b)   (c)

Figure 2.8   Pressure diagrams: (a) horizontal surface, (b) vertical surface and (c) inclined surface.

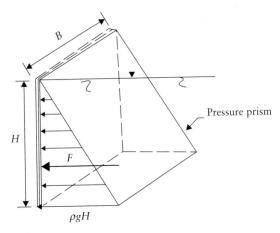

Figure 2.9   Pressure prism.

2.7 Hydrostatic thrust on curved surfaces

Consider a curved gate surface subjected to water pressure as in Figure 2.10. The pressure at any point h below the free water surface is $\rho g h$ and is normal to the gate surface, and the nature of its distribution over the entire surface makes the analytical integration difficult.

However, the total thrust acting normally on the surface can be split into two components and the problem of determining the thrust approached indirectly by combining these two components.

Considering an elementary area of the surface dA (Figure 2.11) at an angle θ to the vertical, pressure intensity on this elementary area is $\rho g h$.

$$\text{Total thrust on this area, } dF = \rho g h\, dA$$
$$\text{Horizontal component of } dF,\, dF_x = \rho g h\, dA\, \cos\theta$$
$$\text{Vertical component of } dF,\, dF_y = \rho g h\, dA\, \sin\theta$$

Horizontal component of the total thrust on the curved area A:

$$F_x = \int_A \rho g h\, dA\, \cos\theta = \rho g \bar{h} A_v$$

where A_v = the vertically projected area of the curved surface

F_x = pressure intensity at the centroid of a vertically projected area

(BD) × vertically projected area   [2.16]

and vertical component, $F_y = \int_A \rho g h\, dA\, \sin\theta$

$$= \rho g \int_A dV$$

dV being the volume of the water prism (real or virtual) over the area dA.

$$\Rightarrow F_y = \rho g V$$

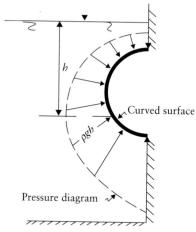

Figure 2.10   Hydrostatic thrust on curved surface.

= the weight of water (real or virtual) above the curved surface
BC bounded by the vertical BD and the free water surface CD   [2.17]

The resultant thrust, $F = \sqrt{F_x^2 + F_y^2}$ [2.18]

acting normally to the surface at an angle α to the horizontal

$$\alpha = \tan^{-1}\left(\frac{F_y}{F_x}\right)$$ [2.19]

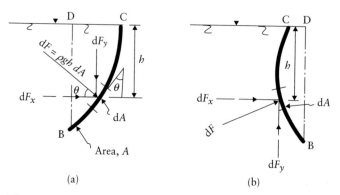

Figure 2.11   Thrust components on curved surfaces: (a) surface containing liquid and (b) surface displacing liquid.

2.8 Hydrostatic buoyant thrust

When a body is submerged or floating in a static fluid, various parts of the surface of the body are exposed to pressures dependent on the depths of submergence.

Consider two elemental cylindrical volumes (one vertical and one horizontal) of the body shown (Figure 2.12) submerged in a fluid, the cross-sectional area of each cylinder being dA.

$$\text{Vertical upthrust on the cylinder BC} = (p_c - p_b)\, dA$$

$$\text{Total upthrust on the body} = \int_A (p_c - p_b)\, dA$$

$$= \int_A \rho g h\, dA$$

$$= \int_A \rho g\, dV$$

$$= \rho g V = \text{weight of fluid displaced} \quad [2.20]$$

where V is the volume of the submerged body displacing the fluid.

$$\text{Horizontal thrust on the cylinder BD} = (p_b - p_d)\, dA$$

$$\text{Total horizontal thrust on the body} = \int_A (p_b - p_d)\, dA$$

$$= 0 \quad (\text{since } p_b = p_d)$$

Hence it can be concluded that the only force acting on the body is the vertical upthrust known as the buoyant thrust or force, which is equal to the weight of the fluid displaced by the body (Archimedes' principle). This buoyant thrust acts through the centroid of the displaced fluid volume.

2.9 Stability of floating bodies

The buoyant thrust on a body of weight W and centroid G acts through the centroid of the displaced fluid volume, and this point or application of the buoyant force is

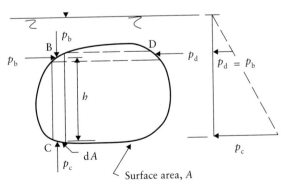

Figure 2.12 Submerged body and buoyant thrust.

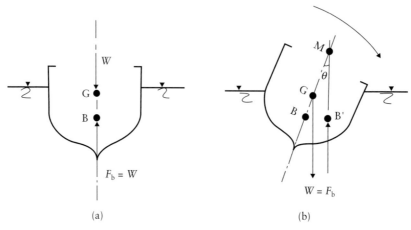

Figure 2.13   Centre of buoyancy and metacentre: (a) equilibrium condition and (b) disturbed condition.

called the **centre of buoyancy**, B, of the body. For the body to be in equilibrium, the weight W must equal the buoyant thrust F_b, both acting along the same vertical line (Figure 2.13).

For small angles of heel, the intersection point of the vertical through the new centre of buoyancy B' and the line BG produced is known as the metacentre M, and the body thus disturbed tends to oscillate about, M. The distance between G and M is the metacentric height.

Conditions of equilibrium:

(a) *Stable equilibrium* (Figure 2.14a): If M lies above G (i.e. positive metacentric height), the couple so produced sets in a restoring moment equal to $W\,GM\,\sin\theta$ opposing the disturbing moment, thereby bringing the body back to its original position, and the body is said to be in stable equilibrium; this is achieved when BM > BG.

(b) *Unstable equilibrium* (Figure 2.14b): If M is below G (i.e. negative metacentric height), the moment of the couple further disturbs the displacement and the body is then in unstable equilibrium. This condition therefore exists when BM < BG.

(c) *Neutral equilibrium* (Figure 2.14c): If G and M coincide (i.e. zero metacentric height), the body floats stably in its displaced position. This condition of neutral equilibrium exists when BM = BG.

2.10 Determination of metacentre

In Figure 2.15, AA is the waterline; and when the body is given a small tilt, $\theta°$, two wedge forces, due to the submergence and the emergence of the wedge areas AOA' on either side of the axis of rolling, are imposed on the body, forming a couple which tends to restore the body to its undisturbed condition. The effect of this couple is the same as the moment caused by the shift of the total buoyant force F_b from B to B', the new centre of buoyancy.

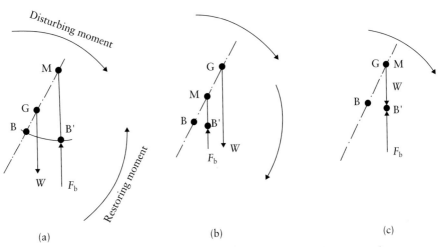

Figure 2.14 Conditions of equilibrium: (a) stable, (b) unstable and (c) neutral.

The buoyant force acting through B′,

$$F'_b = W + df - df = W = F_b$$

By moments about B, $F'_b \times BB' = df \times \overline{gg}$

$$\Rightarrow BB' = \frac{df \times \overline{gg}}{F'_b} = \frac{df \times \overline{gg}}{W}$$

$$= \frac{df \times \overline{gg}}{\rho g V} \tag{i}$$

where V is the volume of the displaced fluid.

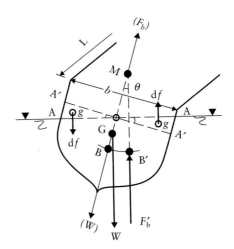

Figure 2.15 Determination of metacentre.

The wedge force $df = \rho g \times \frac{1}{2}AA' \times \frac{1}{2}b \times L$ (for small angles), where L is the length of the body.

$$AA' = \frac{1}{2}b\theta \quad \text{and} \quad \overline{gg} = \frac{2}{3}\left(\frac{1}{2}b\right) + \frac{2}{3}\left(\frac{1}{2}b\right) = \frac{2}{3}b$$

$$\Rightarrow BB' = BM\theta = \frac{\rho g \times \frac{1}{4}b\theta \times \frac{1}{2}b \times L \times \frac{2}{3}b}{\rho g V} \quad \text{(from Equation (i))}$$

or

$$BM = \frac{1}{12}\frac{Lb^3}{V} = \frac{I}{V} \qquad [2.21]$$

where I is the second moment of the plan area of the body at water level about its longitudinal axis.

Hence the metacentric height, $GM = BM - BG$

$$= \frac{I}{V} - BG \qquad [2.22]$$

2.11 Periodic time of rolling (or oscillation) of a floating body

For a small displacement θ, the restoring moment is $Wm\theta$, where W is the weight of the body and m is the distance GM or metacentric height. Angular acceleration α is found by dividing this restoring moment by the mass moment of inertia of the body:

$$\alpha = \frac{Wm\,\theta}{(W/g)k^2} = \frac{mg\theta}{k^2}$$

which shows that α is proportional to θ.

Hence the motion is simple harmonic and its periodic time is:

$$T = 2\pi\sqrt{\frac{\text{displacement}}{\text{acceleration}}}$$

$$= 2\pi\sqrt{\frac{\theta}{mg\theta/k^2}} = 2\pi\sqrt{\frac{k^2}{gm}} \qquad [2.23]$$

For larger values of m, the floating body will no doubt be stable (BM > BG) but the period of oscillation will decrease, thereby increasing the frequency of rolling which may be uncomfortable to passengers and also the body may be subjected to damage. Hence the metacentric height must be fixed, by experience, according to the type of vessel.

2.12 Liquid ballast and the effective metacentric height

For a tilt angle θ, the fluid in the tank (Figure 2.16a) is displaced, thereby shifting its centroid from G to G'. This is analogous to the case of a floating vessel, the centre of buoyancy of which shifts from B to B' through a small heel angle θ.

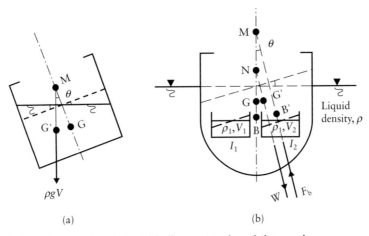

Figure 2.16 Floating vessel with liquid ballast: (a) tank and (b) vessel.

Chapter 2

Hence we can write

$$GG' = GM\theta = \frac{\theta I}{V} \quad \left(\text{since in the case of floating vessel}\right.$$

$$\left. BB' = BM\theta = \frac{\theta I}{V}\right)$$

When the vessel heels, the centroids of the volumes V_1 and V_2 in the compartments (Figure 2.16b) will move by $\theta I_1/V_1$ and $\theta I_2/V_2$, thus shifting the centroid of the vessel from G to G'.

By taking moments,

$$WGG' = \frac{W_1\theta I_1}{V_1} + \frac{W_2\theta I_2}{V_2}$$

or

$$\rho g V GG' = \frac{\rho_1 g V_1 \theta I_1}{V_1} + \frac{\rho_1 g V_2 \theta I_2}{V_2}$$

Hence

$$GG' = \frac{\rho_1 \theta (I_1 + I_2)}{\rho V}$$

V being the volume of the displaced fluid by the vessel whose second moment of area at floating level about its longitudinal axis is I. B' is the new centroid of the displaced liquid through which the buoyant force F_b $(=\rho g V)$ acts, thereby setting a restoring moment equal to $\rho g V\, NM\, \theta$, NM being the effective metacentric height.

$$NM = BM - GN - BG$$

$$BM = \frac{I}{V}; \quad GN = \frac{GG'}{\theta} = \frac{(\rho_1/\rho)(I_1 + I_2)}{V}$$

$$\Rightarrow NM = \frac{I - (\rho_1/\rho)(I_1 + I_2)}{V} - BG \qquad [2.24]$$

and if $\rho_1 = \rho$,

$$NM = \frac{I - (I_1 + I_2)}{V} - BG$$

2.13 Relative equilibrium

If a body of fluid is subjected to motion such that no layer moves relative to an adjacent layer, shear stresses do not exist within the fluid. In other words, in a moving fluid mass if the fluid particles do not move relative to each other, they are said to be in static condition and a relative or dynamic equilibrium exists between them under the action of accelerating force, and fluid pressures are everywhere normal to the surfaces on which they act.

2.13.1 Uniform linear acceleration

A liquid in an open vessel subjected to a uniform acceleration adjusts to the acceleration after some time so that it moves as a solid and the whole mass of liquid will be in relative equilibrium.

A horizontal acceleration (Figure 2.17a) a_x causes the free liquid surface to slope upwards in a direction opposite to a_x and the entire mass of liquid is then under the action of gravity force, hydrostatic forces and the accelerating or inertial force ma_x, m being the liquid mass.

For equilibrium of a particle of mass m, say, on the free surface,

$$F \sin \theta = ma_x \quad \text{and} \quad F\cos \theta - mg = 0 \quad \text{or} \quad F\cos \theta = mg$$

$$\text{Slope of free surface,} \tan \theta = \frac{ma_x}{mg} = \frac{a_x}{g} \qquad [2.25]$$

and the lines of constant pressure will be parallel to the free liquid surface.

A vertical acceleration (Figure 2.17b) (positive upwards) a_y causes no disturbance to the free surface and the fluid mass is in equilibrium under gravity, hydrostatic forces and the inertial force ma_y.

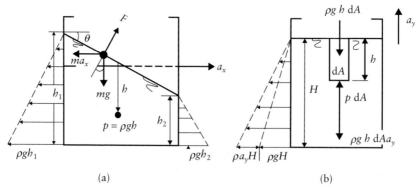

(a)   (b)

Figure 2.17   Fluid subjected to linear accelerations: (a) horizontal acceleration and (b) vertical acceleration.

For equilibrium of a small column of liquid of area dA,

$$p \, dA = \rho h \, dA \, g + \rho h \, dA \, a_y$$

The pressure intensity at a depth h below the free surface is

$$p = \rho g h \left(1 + \frac{a_y}{g} \right) \qquad [2.26]$$

2.13.2 Radial acceleration

Fluid particles moving in a curved path experience radial acceleration. When a cylindrical container partly filled with a liquid is rotated at a constant angular velocity ω about a vertical axis, the rotational motion is transmitted to different parts of the liquid and after some time the whole fluid mass assumes the same angular velocity as a solid and the fluid particles experience no relative motion.

A particle of mass m on the free surface (Figure 2.18) is in equilibrium under the action of gravity, hydrostatic force and the centrifugal accelerating force $m\omega^2 r$, $\omega^2 r$ being the centrifugal acceleration due to rotation.

The gradient of the free surface is

$$\tan\theta = \frac{dy}{dr} = \frac{m\omega^2 r}{mg} = \frac{\omega^2 r}{g}$$

$$\Rightarrow y = \frac{\omega^2 r^2}{2g} + \text{constant, } C$$

when $r = 0, y = 0$ and hence $C = 0$.

$$\Rightarrow y = \frac{\omega^2 r^2}{2g} \qquad [2.27]$$

which shows that the free liquid surface is a paraboloid of revolution and this principle is used in a hydrostatic tachometer.

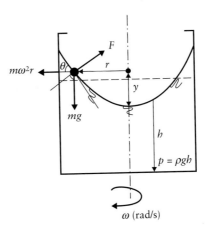

Figure 2.18 Fluid subjected to radial acceleration.

Worked examples

Example 2.1

A hydraulic jack having a ram 150 mm in diameter lifts a weight of 20 kN under the action of a 30 mm diameter plunger. The stroke length of the plunger is 250 mm and if it makes 100 strokes per minute, find by how much the load is lifted per minute and what power is required to drive the plunger.

Solution:

Since the pressure is the same in all directions and is transmitted through the fluid in the hydraulic jack (Figure 2.19),

$$\text{Pressure intensity, } p = \frac{F}{a} = \frac{W}{A}$$
$$\text{Force on the plunger, } F = W\left(\frac{a}{A}\right)$$
$$= 20 \times 10^3 \times \left(\frac{30^2}{150^2}\right)$$
$$= 800 \text{ N}$$
$$\text{Distance moved per minute by the plunger} = 100 \times 0.25$$
$$= 25 \text{ m}$$
$$\text{Distance through which the weight is lifted per minute} = 25 \times \left(\frac{30^2}{150^2}\right)$$
$$= 1 \text{ m}$$
$$\text{Power required} = 20 \times 10^3 \times \frac{1}{60} \text{ N m/s}$$
$$= 333.3 \text{ W}$$

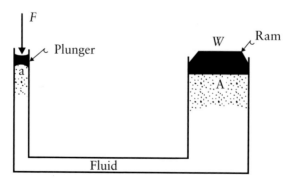

Figure 2.19   Hydraulic jack.

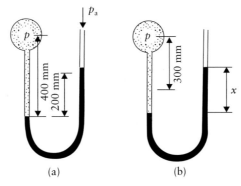

Figure 2.20   U-tube manometer: (a) initial situation and (b) if pressure falls.

Example 2.2

A U-tube containing mercury (relative density 13.6) has its right-hand limb open to atmosphere and the left-hand limb connected to a pipe conveying water under pressure, the difference in levels of mercury in the two limbs being 200 mm. If the mercury level in the left-hand limb is 400 mm below the centre line of the pipe, find the absolute pressure in the pipeline in kilopascals. Also find the new difference in levels of the mercury in the U-tube if the pressure in the pipe falls by 2 kN/m².

Solution:

Starting from the left-hand-side end (Figure 2.20a):

$$\frac{p}{\rho g} + 0.40 - 13.6 \times 0.20 = 0 \quad \text{(atmosphere)}$$

$$\Rightarrow \frac{p}{\rho g} = 2.32 \text{ m of water}$$

$$\text{or } p = 10^3 \times 9.81 \times 2.32 = 22.76 \text{ kN/m}^2$$

The corresponding absolute pressure $= 101 + 22.76$

$$= 123.76 \text{ kN/m}^2$$

$$= 123.76 \text{ kPa}$$

When the manometer is not connected to the system, the mercury levels in both the limbs equalise and are 300 mm below the centre line of the pipe, and writing the manometer equation for new conditions (Figure 2.20b):

$$20.76 \times \frac{10^3}{10^3 \times 9.81} + 0.30 + \frac{x}{2} - 13.6x = 0$$

$\Rightarrow$ The new difference in mercury levels, $x = 0.184$ m or 184 mm

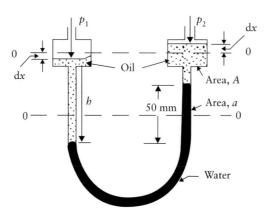

Figure 2.21 Differential micromanometer.

Example 2.3

A double-column enlarged-ends manometer is used to measure a small pressure difference between the two points of a system conveying air under pressure, the diameter of U-tube being 1/10 of the diameter of the enlarged ends. The heavy liquid used is water and the lighter liquid in both limbs is oil of relative density 0.82. Assuming the surfaces of the lighter liquid to remain in the enlarged ends, determine the difference in pressure in millimetres of water for a manometer displacement of 50 mm.

What would be the manometer reading if carbon tetrachloride (of relative density 1.6) were used in place of water, the pressure conditions remaining the same?

Solution:

Referring to Figure 2.21, the manometer equation can be written as

$$\frac{p_1}{\rho g} + 0.82h - 0.05 - (h - 0.05 + 2 \, dx)\, 0.82 = \frac{p_2}{\rho g}$$

and by volumes displaced

$$dx \, A = \left(\frac{0.05}{2}\right) a$$

or

$$2 \, dx = 0.05 \left(\frac{a}{A}\right) = 0.05 \left(\frac{1}{10}\right)^2$$

$$\Rightarrow \frac{p_1 - p_2}{\rho g} = 9.41 \times 10^{-3} \text{ m or } 9.41 \text{ mm of water}$$

For the same pressure conditions if y is the manometer reading using carbon tetrachloride, the manometer equation is

$$\frac{p_1}{\rho g} + 0.82h - 1.6y - 0.82(h - y + 2 \, dx) = \frac{p_2}{\rho g}$$

and

$$2\,dx = \frac{y}{10^2}$$

$$\Rightarrow \frac{p_1 - p_2}{\rho g} = 9.41 \times 10^{-3} = 0.788y$$

Hence the manometer displacement

$$y = \frac{9.41 \times 10^{-3}}{0.788}$$

$$= 11.94 \times 10^{-3} \text{ m or } 11.94 \text{ mm of carbon tetrachloride}$$

Example 2.4

One end of an inclined U-tube manometer is connected to a system carrying air under a very small pressure. If the other end is open to atmosphere, the angle of inclination is 3° to the horizontal and the tube contains oil of relative density 0.8, calculate (i) the air pressure in the system for a manometer reading of 500 mm along the slope and (ii) the equivalent vertical water column height.

Solution:

The manometer equation gives (Figure 2.22)

$$\frac{p}{\rho_o g} - z = 0 \quad \text{and} \quad z = h \sin \theta$$

$$p = \rho_o g h \sin \theta \quad (\rho_o \text{ being the density of oil})$$

$$= 0.8 \times 1000 \times 9.81 \times 0.5 \times \sin 3°$$

$$= 205.36 \text{ N/m}^2$$

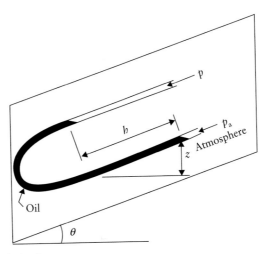

Figure 2.22   Inclined U-tube manometer.

If h' is the equivalent water column height and ρ the density of water, we can write

$$p = \rho g h' = \rho_o g h \sin\theta$$

$$\Rightarrow h' = \left(\frac{\rho_o}{\rho}\right) h \sin\theta$$

$$= s h \sin\theta$$

$$= 0.8 \times 0.5 \times \sin 3°$$

$$= 2.09 \times 10^{-2} \text{ m}$$

Example 2.5

(a) An open steel tank of base 4 m^2 has its sides sloping outwards such that its top is 7 m^2. If the tank is 2 m high and is filled with water, determine the total thrust and its location (i) on the base and (ii) on one of the sloping sides.

(b) If the four sides of the tank slope inwards so that its top is 1 m^2, find the thrust and its location on the base when it is filled with water.

Solution:

$$\text{Pressure intensity on the base, } p = \rho g \times 2 \text{ N/m}^2$$
$$\text{Hence total thrust on the base, } P = p \times A$$

Referring to Figures 2.23a and 2.23b, thrust $P = \rho g \times 2 \times 4 \times 4 = 314$ kN for both cases (Pascal's or hydrostatic paradox), and by symmetry this acts through the centroid of the base.

Total thrust on a side (Figure 2.23a):

$$\text{Length of sloping side} = \sqrt{1.5^2 + 2^2} = 2.5 \text{ m}$$

By moments,

$$\frac{(7+4)}{2} \times 2.5 \times \bar{x} = 4 \times 2.5 \times \frac{2.5}{2} + 2 \times \frac{1}{2} \times 1.5 \times 2.5 \times \frac{2.5}{3}$$

$$13.75\bar{x} = 12.5 + 3.125$$

$$\Rightarrow \bar{x} = 1.136 \text{ m}$$

$$\text{Depth of immersion, } \bar{h} = \bar{x} \sin\theta$$

$$= 1.36 \times \frac{2}{2.5}$$

$$= 0.91 \text{ m}$$

Hence the total thrust, $F = \rho g \bar{h} A$

$$= 10^3 \times 9.81 \times 0.91 \times \frac{(7+4)}{2} \times 2.5$$

$$= 122.75 \text{ kN}$$

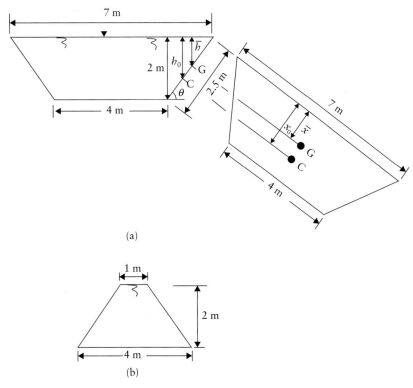

(a)

(b)

Figure 2.23 Open tank with sloping sides: (a) sides sloping outwards and (b) sloping inwards.

$$\text{Centre of pressure, } h_0 = \bar{h} + \frac{I_g \sin^2 \theta}{A\bar{h}}$$

$$I_g = \frac{1}{12} \times 4 \times 2.5^3 + 2 \times \frac{1}{36} \times 1.5 \times 2.5^3$$
$$= 5.208 + 1.302$$
$$= 6.51 \text{ m}^4$$

$$\Rightarrow h_0 = 0.91 + \frac{6.51(2/2.5)^2}{[(7+4)/2] \times 2.5 \times 0.91}$$
$$= 0.91 + 0.333$$
$$= 1.243 \text{ m}$$

Example 2.6

A 2 m × 2 m tank with vertical sides contains oil of density 900 kg/m³ to a depth of 0.8 m floating on 1.2 m depth of water. Calculate the total thrust and its location on one side of the tank (see Figure 2.24).

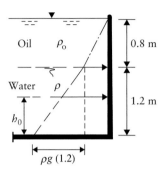

Figure 2.24   Oil and water thrusts on a side of a tank.

Solution:

Total thrust on one vertical side, $F =$ volume of the pressure prism

$$= \left[\frac{1}{2} \rho_o g (0.8)^2 + \rho_o g \times 0.8 \times 1.2 + \frac{1}{2} \rho g (1.2)^2 \right] 2$$

$$= 5.65 + 16.95 + 14.13$$

$$= 36.73 \text{ kN}$$

By moments, centre of pressure, h_0,

$$36.73 \times h_0 = 5.65 \left(1.2 + \frac{0.8}{3} \right) + 16.95 \times \frac{1.2}{2} + 14.13 \times \frac{1.2}{3}$$

$$= 8.29 + 10.17 + 5.65$$

$$= 24.11$$

$$\Rightarrow h_0 = \frac{24.11}{36.73}$$

$$= 0.656 \text{ m above the base}$$

Example 2.7

(a) A circular butterfly gate pivoted about a horizontal axis passing through its centroid is subjected to hydrostatic thrust on one side and counterbalanced by a force F applied at the bottom, as shown in Figure 2.25. If the diameter of the gate is 4 m and the water depth is 1 m above the gate, determine the force, F, required to keep the gate in position.

(b) If the gate is to retain water to its top level on the other side also, determine the net hydrostatic thrust on the gate and suggest the new conditions for the gate to be in equilibrium (see Figure 2.26).

Solution:

(a)   Water on one side only:

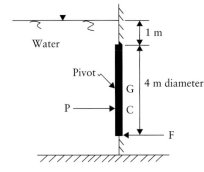

Figure 2.25   Circular gate.

Hydrostatic thrust on the gate,

$$P = \rho \bar{h} A = 10^3 \times 9.81 \times 3 \times \frac{1}{4}\pi 4^2$$

$$= 369.83 \text{ kN}$$

and

the distance, CG $= \dfrac{I_g}{A\bar{h}}$

$$= \frac{(\pi/64)4^4}{(\pi/4)4^2} \times 3$$

$$= 0.333 \text{ m}$$

Taking moments about G,

$$369.83 \times 0.333 = F \times 2$$

$$F = 61.64 \text{ kN}$$

(b)   If the gate is retaining water on the other side also, the net hydrostatic thrust is due to the resultant pressure diagram with a uniform pressure distribution of intensity equal to $\rho g h$ (Figure 2.26).

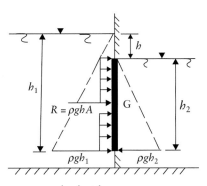

Figure 2.26   Gate retaining water on both sides.

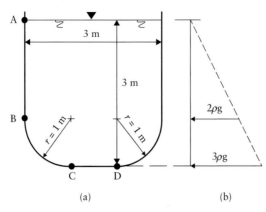

Figure 2.27 Open tank with plane and curved surfaces: (a) tank elevation and (b) pressure diameter.

$$\text{Net hydrostatic thrust, } R = \rho g h A$$

$$= 10^3 \times 9.81 \times 1 \times \frac{1}{4}\pi 4^2$$

$$= 123.28 \text{ kN}$$

This acts through the centroid of the gate, G, and since its moment about G is zero, $F = 0$ for the gate to be in equilibrium, for any depth h of water above the gate on the other side.

Example 2.8

An open tank 3 m × 1 m in cross section (Figure 2.27a) holds water to a depth of 3 m. Determine the magnitude, direction and line of action of the forces exerted upon the plane surfaces AB and CD and the curved surface BC of the tank.

Solution:

Force on face AB/m length = area of pressure diagram (see Figure 2.27b)

$$= \frac{1}{2}\rho g \times 2^2 = 19.62 \text{ kN/m}$$

acting normal to the face AB at a depth of $(2/3) \times 2 = 1.33$ m from the water surface.
   Force on curved surface BC/m length:

Horizontal component, F_x (from the pressure diagram) $= \rho g \times 2 \times 1 + \frac{1}{2}\rho g \times 1^2$

$$= 24.52 \text{ kN/m}$$

Vertical component, F_y = weight of water above the surface

$$= \rho g \times 2 \times 1 \times 1 + \rho g \times \frac{\pi \times 1^2}{4} \times 1$$

$$= 27.32 \text{ kN/m}$$

$$\text{Resultant thrust, } F = \sqrt{24.52^2 + 27.32^2}$$
$$= 36.71 \text{ kN/m}$$

acting at an angle $\alpha = \tan^{-1}(27.32/24.52) = 48°5'$ to the horizontal and passing through the centre of curvature of the surface BC.

Force on surface CD/m length:

$$\text{Uniform pressure intensity on CD} = \rho g \times 3 \text{ N/m}^2$$
$$\text{Total thrust on CD} = \text{uniform pressure} \times \text{area}$$
$$= \rho g \times 3 \times 1 \times 1$$
$$= 29.43 \text{ kN/m}$$

acting vertically downwards (normal to CD) through the mid-point of the surface CD.

Example 2.9

A 3 m diameter roller gate retains water on both sides of a spillway crest as shown in Figure 2.28. Determine (i) the magnitude, direction and location of the resultant hydrostatic thrust acting on the gate per unit length; and (ii) the horizontal water thrust on the spillway per unit length.

Solution:

Thrust on the gate:

$$\textit{Left-hand side:}\quad \text{Horizontal component} = \frac{1}{2}\rho g \times 3^2$$
$$= 44.14 \text{ kN/m}$$
$$\text{Vertical component} = \rho g \times \frac{1}{2}\frac{\pi}{4}3^2 \times 1$$
$$= 34.67 \text{ kN/m}$$

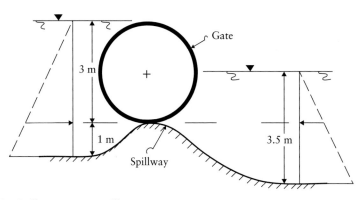

Figure 2.28   Roller gate on a spillway.

Right-hand side: Horizontal component $= \dfrac{1}{2}\rho g(1.5)^2$

$$= 11.03 \text{ kN/m}$$

Vertical component $= \rho g \times \dfrac{1}{4}\dfrac{\pi}{4}3^2 \times 1$

$$= 17.34 \ \text{kN/m}$$

Net horizontal component on the gate (left to right) $= 44.14 - 11.03$

$$= 33.11 \text{ kN/m}$$

and

net vertical component (upwards) $= 34.67 + 17.34$

$$= 50.01 \text{ kN/m}$$

Resultant hydrostatic thrust on the gate $= \sqrt{(33.11)^2 + (50.01)^2}$

$$= 60 \text{ kN/m}$$

acting at an angle $\alpha = \tan^{-1}(33.11/50.01) = 33°30'$ to the vertical and passes through the centre of the gate (normal to the surface).

Depth of centre of pressure $= r + r\cos\alpha$

$$= 1.5 \ (1 + \cos 33°30')$$

$$= 2.75 \text{ m below the free surface of the left-hand side}$$

Horizontal thrust on the spillway:
   From pressure diagrams (see Figure 2.28),

thrust from left-hand side $= \dfrac{1}{2}(\rho g \times 3 + \rho g \times 4) \times 1$

$$= 34.33 \text{ kN/m}$$

and

thrust from right-hand side $= \dfrac{1}{2}(\rho g \times 1.5 + \rho g \times 3.5) \times 2$

$$= 49.05 \text{ kN/m}$$

Resultant thrust (horizontal) on the spillway $= 49.05 - 34.33$

$$= 14.72 \text{ kN/m towards left}$$

Example 2.10

The gates of a lock (Figure 2.29) are 5 m high and when closed include an angle of 120°. The width of the lock is 6 m. Each gate is carried on two hinges placed on the top and bottom of the gate. If the water levels are 4.5 m and 3 m on the upstream and downstream

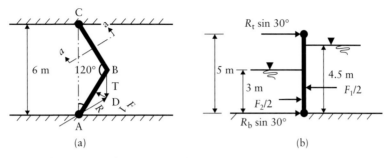

Figure 2.29 Lock gates: (a) plan view and (b) section a–a.

sides, respectively, determine the magnitudes of the forces on the hinges due to water pressure.

Solution:

Forces on any one gate (say, AB) are as follows: F is the resultant water thrust; T, the thrust of gate; BC, normal to contact surface; and R, the resultant of hinge forces. Since these three forces keep the gate in equilibrium, they should meet at a point D (Figure 2.29a).

Resolution of forces along AB and normal to AB gives (ABD = BAD = 30°)

$$T \cos 30° = R \cos 30° \quad \text{or} \quad T = R \tag{i}$$

and

$$F = R \sin 30° + T \sin 30° \quad \text{or} \quad F = R \tag{ii}$$

Length of the gate $= 3/\sin 60° = 3.464$ m

The resultant water thrusts on either side of the gate, $F = F_1 - F_2$:

$$F_1 = \frac{1}{2}\rho g (4.5)^2 \times 3.464$$
$$= 344 \text{ kN acting at } 4.5/3 = 1.5 \text{ m from the base}$$

$$\text{and } F_2 = \frac{1}{2}\rho g \times 3^2 \times 3.464$$
$$= 153 \text{ kN acting at } 3/3 = 1 \text{ m from the base}$$

Resultant water thrust, $F = 344 - 153$
$$= 191 \text{ kN} = R \quad \text{(from Equation (ii))}$$

Total hinge reaction, $R = R_t + R_b$ (sum of top and bottom hinge forces) (iii)

From Equation (ii),

$$\frac{F}{2} = R \sin 30°$$

or

$$\frac{F_1 - F_2}{2} = R_t \sin 30° + R_b \sin 30° \tag{iv}$$

Taking moments about the bottom hinge (Figure 2.29b),

$$\frac{344}{2} \times 1.5 - \frac{153}{2} \times 1 = R_t \sin 30° \times 5$$

$$\Rightarrow R_t = 72.6 \text{ kN}$$

and hence from Equation (iii),

$$R_b = 191 - 72.6$$
$$= 118.4 \text{ kN}$$

Example 2.11

A rectangular block of wood floats in water with 50 mm projecting above the water surface. When placed in glycerine of relative density 1.35, the block projects 75 mm above the surface of glycerine. Determine the relative density of the wood.

Solution:

Weight of wooden block, W = upthrust in water = upthrust in glycerine
= weight of fluid displaced

$$W = \rho_w gAh = \rho gA(h - 50 \times 10^{-3}) = \rho_G gA(h - 75 \times 10^{-3})$$

ρ, ρ_w and ρ_G being the densities of water, wood and glycerine, respectively; A the cross-sectional area of the block; and h its height.

The relative density of glycerine, $\dfrac{\rho_G}{\rho} = \dfrac{h - 50 \times 10^{-3}}{h - 75 \times 10^{-3}} = 1.35$

$\Rightarrow h = 146.43 \times 10^{-3}$ m or 146.43 mm

Hence the relative density of wood, $\dfrac{\rho_w}{\rho} = \dfrac{146.43 - 50}{146.43} = 0.658$

Example 2.12

(a) A ship of 50 MN displacement has a weight of 100 kN moved 10 m across the deck causing a heel angle of 5°. Find the metacentric height of the ship.

(b) A homogeneous circular cylinder of radius R and height H is to float stably in a liquid. Show that R must not be less than $\sqrt{2r(1-r)}H$ in order to float with its axis vertical, where r is the ratio of relative densities of the cylinder and the liquid. Hence establish the condition for R/H to be minimum.

Solution:

Referring to Figure 2.30a,

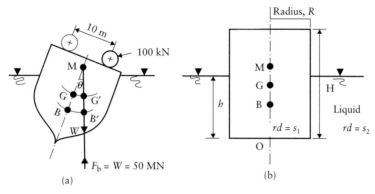

Figure 2.30 Determination of metacentric height and stability conditions: (a) ship and (b) cylinder.

Moment heeling the ship $= 100 \times 10 = 1000$ kN m

$\qquad = $ moment due to the shifting of W from G to G'

$\qquad = W \times GG'$

$\Rightarrow GG' = GM \sin\theta = \dfrac{1000}{50 \times 10^3} = \dfrac{1}{50}$ m

Hence the metacentric height, $GM = \dfrac{1}{50 \times \sin 5°} = 0.23$ m

Referring to Figure 2.30b,

$\qquad$ weight of cylinder = weight of liquid displaced

If the depth of submergence is h, we can write

$$\rho g s_1 \pi R^2 H = \rho g s_2 \pi R^2 h \quad (\rho \text{ being the density of water})$$

and hence

$$\frac{s_1}{s_2} = r = \frac{h}{H} \tag{i}$$

$$OG = \frac{H}{2} \tag{ii}$$

$$OB = \frac{h}{2} = \frac{rH}{2} \tag{iii}$$

$$\Rightarrow BG = OG - OB = \frac{H}{2} - \frac{rH}{2}$$

$$= \left(\frac{H}{2}\right)(1 - r)$$

and

$$BM = \frac{I}{V} = \frac{\pi R^4/4}{\pi R^2 h H}$$

$$= \frac{R^2}{4rH}$$

For stable condition,

$$BM > BG$$

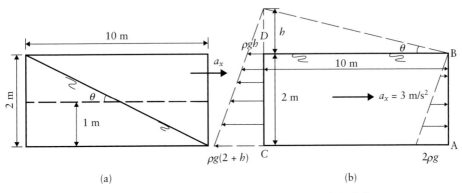

Figure 2.31 Oil tanker subjected to accelerations: (a) half full and (b) full.

$$\frac{R^2}{4rH} > \left(\frac{H}{2}\right)(1-r) \quad \text{or} \quad \frac{R^2}{H^2} > 2r(1-r)$$
$$\Rightarrow \frac{R}{H} > \sqrt{2r(1-r)}$$

and hence

$$R > \sqrt{2r(1-r)}H$$

For limiting value of R/H, $r(1-r)$ is to be minimum or

$$d\,[r(1-r)] = 0$$
$$\Rightarrow 1 - 2r = 0 \text{ and hence } r = \frac{1}{2}$$

Example 2.13

An oil tanker 3 m wide, 2 m deep and 10 m long contains oil of density 800 kg/m³ to a depth of 1 m. Determine the maximum horizontal acceleration that can be given to the tanker such that the oil just reaches its top end.

If this tanker is closed and completely filled with the oil and accelerated horizontally at 3 m/s², determine the total liquid thrust on (i) the front end, (ii) the rear end and (iii) one of its longitudinal vertical sides (see Figure 2.31).

Solution:

From Figure 2.31a,

$$\text{maximum possible surface slope} = \frac{1}{5} = \frac{a_x}{g}$$

$$\Rightarrow \text{The maximum horizontal acceleration, } a_x = \left(\frac{1}{5}\right) \times 9.81$$

$$= 1.962 \text{ m/s}^2$$

When the tanker is completely filled and closed, there will be pressure built up at the rear-end equivalent to the virtual oil column (h) that would assume a slope of a_x/g (Figure 2.31b).

(i) Total thrust on front end AB = $\frac{1}{2}\rho g \times 2^2 \times 3 = 58.86$ kN

(ii) Total thrust on rear end CD:
    Virtual rise of oil level at rear end,

$$h = 10 \times \tan\theta = 10 \times \frac{a_x}{g} = 10 \times \frac{3}{9.81} = 3.06 \text{ m}$$

$$\text{Total thrust on CD} = \frac{\rho g(3.06) + \rho g(2 + 3.06)}{2} \times 2 \times 3$$

$$= 239 \text{ kN}$$

(iii) Total thrust on side ABCD = volume of the pressure prism

$$= \frac{1}{2}\rho g \times 2^2 \times 10 + \frac{1}{2}\rho g \times 3.06 \times 10 \times 2$$

$$= \frac{1}{2}\rho g(2 + 3.06) \times 2 \times 10$$

$$= 496 \text{ kN}$$

Example 2.14

A vertical hoist carries a square tank of 2 m × 2 m containing water to the top of a construction scaffold with a varying speed of 2 m/s. If the water depth is 2 m, calculate the total hydrostatic thrust on the bottom of the tank.
    If this tank of water is lowered with an acceleration equal to that of gravity, what are the thrusts on the floor and sides of the tank?

Solution:

$$\text{Vertical upward acceleration,} \, a_y = 2 \text{ m/s}^2$$

$$\text{Pressure intensity at a depth } h = \rho g h \left(1 + \frac{a_y}{g}\right)$$

$$= \rho g h \left(1 + \frac{2}{9.81}\right)$$

$$= 1.204 \times g h \text{ kN/m}^2$$

$$\text{Total hydrostatic thrust on the floor} = \text{intensity} \times \text{area}$$

$$= 1.204 \times 9.81 \times 2 \times 2 \times 2$$

$$= 94.5 \text{ kN}$$

$$\text{Downward acceleration} = -9.81 \text{ m/s}^2$$

$$\text{Pressure intensity at a depth } h = \rho g h \left(1 - \frac{9.81}{9.81}\right)$$

$$= 0$$

Therefore, there exists no hydrostatic thrust on the floor, nor on the sides.

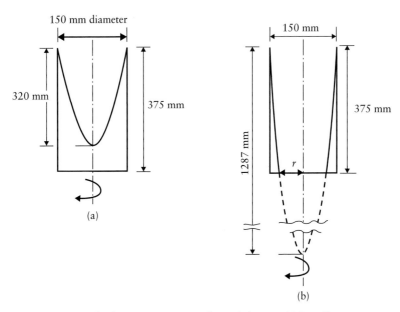

Figure 2.32   Rotating cylinder: (a) $\omega = 33.5$ rad/s and (b) $\omega = 67.0$ rad/s.

Example 2.15

A 375 mm high open cylinder, 150 mm in diameter, is filled with water and rotated about its vertical axis at an angular speed of 33.5 rad/s. Determine (i) the depth of water in the cylinder when it is brought to rest and (ii) the volume of water that remains in the cylinder if the speed is doubled (see Figure 2.32).

Solution:

$$\text{Angular velocity, } \omega = 33.5 \text{ rad/s}$$

$$\text{Height of the paraboloid (Figure 2.32a), } y = \frac{\omega^2 r^2}{2g}$$

$$= \frac{(33.5 \times 0.075)^2}{19.62}$$

$$= 0.32 \text{ m}$$

$$\text{Amount of water spilled out} = \text{volume of the paraboloid}$$

$$= \frac{1}{2} \times \text{volume of circumscribing cylinder}$$

$$= \frac{1}{2}\pi(0.075)^2 \times 0.32$$

$$= 2.83 \times 10^{-3} \text{ m}^3$$

$$\text{Original volume of water} = \pi(0.075)^2 \times 0.375$$

$$= 6.63 \times 10^{-3} \text{ m}^3$$

$$\text{Remaining volume of water} = (6.63 - 2.83) \times 10^{-3}$$
$$= 3.8 \times 10^{-3} \text{ m}^3$$
$$\text{Hence the depth of water at rest} = \frac{3.8 \times 10^{-3}}{\pi(0.075)^2}$$
$$= 0.215 \text{ m}$$

If the speed is doubled, $\omega = 67$ rad/s.

$$\text{Height of paraboloid} = \frac{(67 \times 0.075)^2}{2g}$$
$$= 1.287 \text{ m}$$

The free surface in the vessel assumes the shape as shown in Figure 2.32b, and we can write

$$1.287 - 0.375 = \frac{\omega^2 r^2}{2g}$$

$$r = \sqrt{2g \times \frac{0.912}{67^2}} = 0.063 \text{ m}$$

Therefore, the volume of water spilled out $= \frac{1}{2}\pi(0.075)^2$

$$\times 1.287 - \frac{1}{2}\pi(0.063)^2 \times 0.912$$
$$= 5.684 \times 10^{-3} \text{m}^3$$

$$\text{Hence the volume of water left} = (6.63 - 5.684) \times 10^{-3}$$
$$= 0.946 \times 10^{-3} \text{ m}^3$$

Reference and recommended reading

Zipparro, V. J. and Hasen, H. (1993) *Davis' Handbook of Applied Hydraulics*, 4th edn, McGraw-Hill, New York.

Problems

1. (a) A large storage tank contains a salt solution of variable density given by $\rho = 1050 + kh$ in kilograms per cubic metre, where $k = 50$ kg/m^4, at a depth h (in metres) below the free surface. Calculate the pressure intensity at the bottom of the tank holding 5 m of the solution.

    (b) A Bourdon-type pressure gauge is connected to a hydraulic cylinder activated by a piston of 20 mm diameter. If the gauge balances a total mass of 10 kg placed on the piston, determine the gauge reading (in metres) of water.

2. A closed cylindrical tank 4 m high is partly filled with oil of density 800 kg/m^3 to a depth of 3 m. The remaining space is filled with air under pressure. A U-tube containing mercury (of relative density 13.6) is used to measure the air pressure, with one end open to atmosphere. Find the gauge pressure at the base of the tank when the mercury deflection

in the open limb of the U-tube is (i) 100 mm above and (ii) 100 mm below the level in the other limb.

3. A manometer consists of a glass tube, inclined at 30° to the horizontal, connected to a metal cylinder standing upright. The upper end of the cylinder is connected to a gas supply under pressure. Find the pressure (in millimetres) of water when the manometer fluid of relative density 0.8 reads a deflection of 80 mm along the tube. Take the ratio, r, of the diameters of the cylinder and the tube as 64. What value of r would you suggest so that the error due to disregarding the change in level in the cylinder will not exceed 0.2%?

4. In order to measure the pressure difference between two points in a pipeline carrying water, an inverted U-tube is connected to the points, and air under atmospheric pressure is entrapped in the upper portion of the U-tube. If the manometer deflection is 0.8 m and the downstream tapping is 0.5 m below the upstream point, find the pressure difference between the two points.

5. A high-pressure gas pipeline is connected to a macromanometer consisting of four U-tubes in series with one end open to atmosphere, and a deflection of 500 mm of mercury (of relative density 13.6) has been observed. If water is entrapped between the mercury columns of the manometer and the relative density of the gas is 1.2×10^{-3}, calculate the gas pressure in newtons per millimetre square, the centre line of the pipeline being at a height of 0.50 m above the top mercury level.

6. A dock gate is to be reinforced with three identical horizontal beams. If the water stands to depths of 5 m and 3 m on either side, find the positions of the beams, measured above the floor level, so that each beam will carry an equal load, and calculate the load on each beam per unit length.

7. A storage tank of a sewage treatment plant is to discharge excess sewage into the sea through a horizontal rectangular culvert 1 m deep and 1.3 m wide. The face of the discharge end of the culvert is inclined at 40° to the vertical and the storage level is controlled by a flap gate weighing 4.5 kN, hinged at the top edge and just covering the opening. When the sea water stands to the hinge level, to what height above the top of the culvert will the sewage be stored before a discharge occurs? Take the density of the sewage as 1000 kg/m³ and of the sea water as 1025 kg/m³.

8. A radial gate, 2 m long, hinged about a horizontal axis closes the rectangular sluice of a control dam by the application of a counterweight W (see Figure 2.33). Determine (i) the total hydrostatic thrust and its location on the gate when the storage depth is 4 m and (ii) for the gate to be stable, the counterweight W. Explain what will happen if the storage increases beyond 4 m.

9. A sector gate of radius 3 m and length 4 m retains water as shown in Figure 2.34. Determine the magnitude, direction and location of the resultant hydrostatic thrust on the gate.

10. The profile of the inner face of a dam is a parabola with equation $y = 0.30x^2$ (see Figure 2.35). The dam retains water to a depth of 30 m above the base. Determine the hydrostatic thrust on the dam per unit length, its inclination to the vertical and the point at which the line of action of this thrust intersects the horizontal base of the dam.

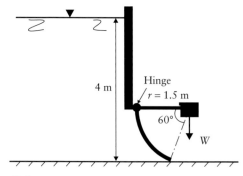

Figure 2.33 Hinged radial gate.

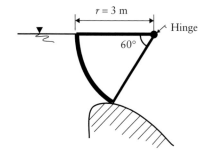

Figure 2.34 Sector or tainter gate.

11. A homogeneous wooden cylinder of circular section (of relative density 0.7) is required to float in oil of density 900 kg/m³. If d and h are the diameter and height of the cylinder, respectively, establish the upper limiting value of the ratio h/d for the cylinder to float with its axis vertical.

12. A conical buoy floating in water with its apex downwards has a diameter d and a vertical height h. If the relative density of the material of the buoy is s, prove that for

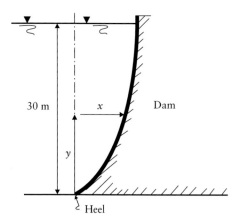

Figure 2.35 Parabolic profile of the inner face of a dam.

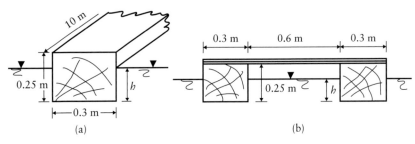

Figure 2.36   Floating platform: (a) single beam and (b) platform.

stable equilibrium,

$$\frac{h}{d} < \frac{1}{2}\sqrt{\frac{s^{1/3}}{1 - s^{1/3}}}$$

13. A cyclindrical buoy weighing 20 kN is to float in sea water whose density is 1020 kg/m³. The buoy has a diameter of 2 m and height of 2.5 m. Prove that it is unstable.
     If the buoy is anchored with a chain attached to the centre of its base, find the tension in the chain to keep the buoy in vertical position.

14. A floating platform for offshore drilling purposes is in the form of a square floor supported by four vertical cylinders at the corners. Determine the location of the centroid of the assembly in terms of the side L of the floor and the depth of submergence h of the cylinders so as to float in neutral equilibrium under a uniformly distributed loading condition.

15. A platform constructed by joining two 10 m long wooden beams as shown in Figure 2.36 is to float in water. Examine the stability of a single beam and of the platform and determine their stability moments. Neglect the weight of the connecting pieces and take the density of wood as 600 kg/m³.

16. A rectangular barge 10 m wide and 20 m long is 5 m deep and weighs 6 MN when loaded without any ballast. The barge has two compartments, each 4 m wide and 20 m long, symmetrically placed about its central axis, and each containing 1 MN of water ballast. The water surface in each compartment is free to move. The centre of gravity without ballast is 3.0 m above the bottom and on the geometrical centre of the plan. (i) Calculate the metacentric height for rolling, and (ii) if 100 kN of the deck load is shifted 5 m laterally find the approximate heel angle of the barge.

17. A U-tube acceleration meter consists of two vertical limbs connected by a horizontal tube of 400 mm long parallel to the direction of motion. Calculate the level difference of the liquid in the U-tube when it is subjected to a horizontal uniform acceleration of 6 m/s².

18. An open rectangular tank 4 m long and 3 m wide contains water up to a depth of 2 m. Calculate the slope of the free surface of water when the tank is accelerated at 2 m/s² (i) up a slope of 30° and (ii) down a slope of 30°.

19. Prove that in the forced vortex motion (fluids subjected to rotation externally) of a liquid, the rate of increase of the pressure p with respect to the radius r at a point in liquid is given by $dp/dr = \rho\omega^2 r$, in which ω is the angular velocity of the liquid and ρ is its mass density. Hence calculate the thrust of the liquid on the top of a closed vertical cylinder of 450 mm diameter, completely filled with water under a pressure of 10 N/cm^2, when the cylinder rotates about its axis at 240 rpm.

Chapter 3
Fluid Flow Concepts and Measurements

3.1 Kinematics of fluids

The kinematics of fluids deals with space–time relationships for fluids in motion. In the Lagrangian method of describing the fluid motion, one is concerned to trace the paths of the individual fluid particles (elements) and to find their velocities, pressures and so on with the passage of time. The coordinates of a particle $A(x, y, z)$ at any time t (Figure 3.1a) are dependent on its initial coordinates (a, b, c) at the instant t_0 and can be written as functions of a, b, c and t; in other words,

$$x = \phi_1(a, b, c, t)$$
$$y = \phi_2(a, b, c, t)$$
$$z = \phi_3(a, b, c, t)$$

The path traced by the particles over a period of time is known as the **pathline**. Due to the diffusivity phenomena of fluids and their flows, it is difficult to describe the motion of individual particles of a flow field with time. More appropriate for describing the fluid motion is to know the flow characteristics such as velocity and pressure of a particle or group of particles at a chosen point in the flow field at any particular time; such a description of fluid flow is known as the **Eulerian method**.

In any flow field, velocity is the most important characteristic to be identified at any point. The velocity vector at a point in the flow field is a function of s and t and can be resolved into u, v and w components, representing velocities in the x-, y- and z-directions, respectively; these components are functions of x, y, z and t and written as (Figure 3.1b)

$$u = f_1(x, y, z, t)$$
$$v = f_2(x, y, z, t)$$
$$w = f_3(x, y, z, t)$$

defining the vector V at each point in the space at any instant t. A continuous curve traced tangentially to the velocity vector at each point in the flow field is known as the **streamline**.

Nalluri & Featherstone's Civil Engineering Hydraulics: Essential Theory with Worked Examples, Sixth Edition. Martin Marriott.
© 2016 John Wiley & Sons, Ltd. Published 2016 by John Wiley & Sons, Ltd.
Companion Website: www.wiley.com/go/Marriott

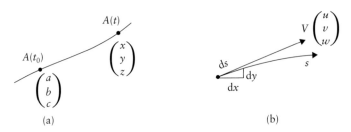

Figure 3.1   Descriptions of fluid flow: (a) pathline and (b) velocity vector.

3.2 Steady and unsteady flows

The flow parameters such as velocity, pressure and density of a fluid flow are independent of time in a steady flow, whereas they depend on time in unsteady flows. For example, this can be written as

$$\left(\frac{\partial V}{\partial t}\right)_{x_0 y_0 z_0} = 0 \quad \text{for steady flow} \tag{3.1a}$$

$$\text{and } \left(\frac{\partial V}{\partial t}\right)_{x_0 y_0 z_0} \neq 0 \quad \text{for unsteady flow} \tag{3.1b}$$

At a point, in reality these parameters are generally time dependent but often remain constant on average over a time period T. For example, the average velocity $\bar{u}$ can be written as

$$\bar{u} = \frac{1}{T} \int_t^{t+T} u \, dt \quad \text{where } u = u(t) = \bar{u} \pm u'(t)$$

where u' is the velocity fluctuation from the mean, with time t; such velocities are called temporal mean velocities.

In steady flow, the streamline has a fixed direction at every point and is therefore fixed in space. A particle always moves tangentially to the streamline and hence in steady flow the path of a particle is a streamline.

3.3 Uniform and non-uniform flows

A flow is uniform if its characteristics at any given instant remain the same at different points in the direction of flow; otherwise it is termed as non-uniform flow. Mathematically, this can be expressed as

$$\left(\frac{\partial V}{\partial s}\right)_{t_0} = 0 \quad \text{for uniform flow} \tag{3.2a}$$

$$\text{and } \left(\frac{\partial V}{\partial s}\right)_{t_0} \neq 0 \quad \text{for non-uniform flow} \tag{3.2b}$$

The flow through a long uniform pipe at a constant rate is steady uniform flow and at a varying rate is unsteady uniform flow. Flow through a diverging pipe at a constant rate is steady non-uniform flow and at a varying rate is unsteady non-uniform flow.

3.4 Rotational and irrotational flows

If the fluid particles within a flow have rotation about any axis, the flow is called rotational and if they do not suffer rotation, the flow is in irrotational motion. The non-uniform velocity distribution of real fluids close to a boundary causes particles to deform with a small degree of rotation, whereas the flow is irrotational if the velocity distribution is uniform across a section of the flow field.

3.5 One-, two- and three-dimensional flows

The velocity component transverse to the main flow direction is neglected in one-dimensional flow analysis. Flow through a pipe may usually be characterised as one-dimensional. In two-dimensional flow, the velocity vector is a function of two coordinates and the flow conditions in a straight, wide river may be considered as two-dimensional. Three-dimensional flow is the most general type of flow in which the velocity vector varies with space and is generally complex.

Thus, in terms of the velocity vector $V(s, t)$, we can write

$$V = f(x, t) \qquad \text{(one-dimensional flow)} \qquad [3.3a]$$

$$V = f(x, y, t) \qquad \text{(two-dimensional flow)} \qquad [3.3b]$$

$$V = f(x, y, z, t) \qquad \text{(three-dimensional flow)} \qquad [3.3c]$$

3.6 Streamtube and continuity equation

A streamtube consists of a group of streamlines whose bounding surface is made up of these several streamlines. Since the velocity at any point along a streamline is tangential to it, there can be no flow across the surface of a streamtube, and therefore, the streamtube surface behaves like a boundary of a pipe across which there is no flow. This concept of the streamtube is very useful in deriving the continuity equation.

Considering an elemental streamtube of the flow (Figure 3.2), we can state

$$\frac{\text{mass entering the tube}}{\text{second}} = \frac{\text{mass leaving the tube}}{\text{second}}$$

since there is no mass flow across the tube (principle of mass conservation).

$$\Rightarrow \rho_1 V_1 \, dA_1 = \rho_2 V_2 \, dA_2$$

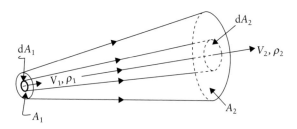

Figure 3.2   Streamtube.

Chapter 3

where V_1 and V_2 are the steady average velocities at the entrance and exit of the elementary streamtube of cross-sectional areas dA_1 and dA_2, and ρ_1 and ρ_2 are the corresponding densities of entering and leaving fluid.

Therefore, for a collection of such streamtubes along the flow,

$$\bar{\rho}_1 \bar{V}_1 A_1 = \bar{\rho}_2 \bar{V}_2 A_2 \qquad [3.4]$$

where $\bar{\rho}_1$ and $\bar{\rho}_2$ are the average densities of fluid at the entrance and exit, and $\bar{V}_1$ and $\bar{V}_2$ are the average velocities over the entire entrance and exit sections of areas A_1 and A_2 of the flow tube.

For incompressible steady flow, Equation 3.4 reduces to the one-dimensional continuity equation

$$A_1 \bar{V}_1 = A_2 \bar{V}_2 = Q \qquad [3.5]$$

and Q is the volumetric rate of flow called discharge, expressed in metres cubed per second, often referred to as cumecs.

3.7 Accelerations of fluid particles

In general, the velocity vector V of a flow field is a function of space and time, written as

$$V = f(s, t)$$

which shows that the fluid particles experience accelerations due to (a) change in velocity in space (convective acceleration) and (b) change in velocity in time (local or temporal acceleration).

3.7.1 Tangential acceleration

If V_s in the direction of motion is equal to $f(s, t)$, then

$$dV_s = \frac{\partial V_s}{\partial s} ds + \frac{\partial V_s}{\partial t} dt$$

$$\text{or} \quad \frac{dV_s}{dt} = \frac{ds}{dt} \frac{\partial V_s}{\partial s} + \frac{\partial V_s}{\partial t}$$

$$= V_s \frac{\partial V_s}{\partial s} + \frac{\partial V_s}{\partial t} \qquad [3.6]$$

where dV_s/dt is the total tangential acceleration equal to the sum of tangential convective and tangential local accelerations.

3.7.2 Normal acceleration

The velocity vectors of the particles negotiating curved paths (Figure 3.3a) may experience change in both direction and magnitude.

Along a flow line of radius of curvature r, the velocity vector V_s at A changes to $V_s + \Delta V$ at B. The vector change ΔV can be resolved into two components, one along the vector V_s and the other normal to the vector V_s. The tangential change in velocity vector ΔV_s

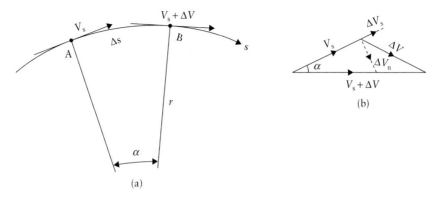

Figure 3.3   Curved motion: (a) curved path and (b) velocity vectors.

produces tangential convective acceleration, whereas the normal component ΔV_n produces normal convective acceleration:

$$\frac{\Delta V_n}{\Delta t} = \left(\frac{\Delta V_n}{\Delta s}\frac{\Delta s}{\Delta t}\right)$$

From similar triangles (Figure 3.3b),

$$\frac{\Delta V_n}{V_s} = \frac{\Delta s}{r}$$

$$\text{or } \frac{\Delta V_n}{\Delta s} = \frac{V_s}{r}$$

The total normal acceleration can now be written as

$$\frac{dV_n}{dt} = \frac{V_s^2}{r} + \frac{\partial V_n}{\partial t} \qquad [3.7]$$

where $\partial V_n/\partial t$ is the local normal acceleration.

Examples of streamline patterns and their corresponding types of acceleration in steady flows ($\partial V/\partial t = 0$) are shown in Figure 3.4.

Fluid flows between straight parallel boundaries (Figure 3.4a) do not experience any kind of accelerations, whereas between straight converging (Figure 3.4b) or diverging boundaries the flow suffers tangential convective accelerations or decelerations.

Flow in a concentric curved bend (Figure 3.4c) experiences normal convective accelerations, while in a converging (Figure 3.4d) or diverging bend both tangential and normal convective accelerations or decelerations exist.

3.8 Two kinds of fluid flow

Fluid flow may be classified as laminar or turbulent. In laminar flow, the fluid particles move along smooth layers, one layer gliding over an adjacent layer. Viscous shear stresses dominate in this kind of flow in which the shear stress and velocity distribution are governed by Newton's law of viscosity (Equation 1.1). In turbulent flows, which occur most commonly in engineering practice, the fluid particles move in erratic paths causing instantaneous fluctuations in the velocity components. These turbulent fluctuations cause an

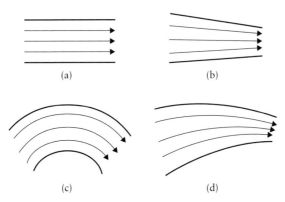

Figure 3.4   Streamline patterns and types of acceleration: (a) no accelerations exist; (b) tangential convective accelerations; (c) normal convective accelerations; and (d) tangential and normal convective accelerations.

exchange of momentum setting up additional shear stresses of large magnitudes. An equation of the form similar to Newton's law of viscosity (Equation 1.1) may be written for turbulent flow by replacing μ by η. The coefficient η, called the eddy viscosity, depends on the fluid motion and the density.

The type of a flow is identified by the Reynolds number, $Re = \rho VL/\mu$, where ρ and μ are the density and viscosity of the fluid, V is the flow velocity and L is a characteristic length such as the pipe diameter (D) in the case of a pipe flow. Reynolds number represents the ratio of inertial forces to the viscous forces that exist in the flow field and is dimensionless.

The flow through a pipe is always laminar if the corresponding Reynolds number $(Re = \rho VD/\mu)$ is less than 2000, and for all practical purposes the flow may be assumed to pass through a transition to full turbulent flow in the range of Reynolds numbers from 2000 to 4000.

3.9 Dynamics of fluid flow

The study of fluid dynamics deals with the forces responsible for fluid motion and the resulting accelerations. A fluid in motion experiences, in addition to gravity, pressure forces, viscous and turbulent shear resistances, boundary resistance and forces due to surface tension and compressibility effects of the fluid. The presence of such a complex system of forces in real fluid flow problems makes the analysis very complicated.

However, a simplifying approach to the problem may be made by assuming the fluid to be ideal or perfect (i.e. non-viscous or frictionless and incompressible). Water has a relatively low viscosity and is practically incompressible and is found to behave like an ideal fluid. The study of ideal fluid motion is a valuable background information to encounter the problems of civil engineering hydraulics.

3.10 Energy equation for an ideal fluid flow

Consider an elemental streamtube in motion along a streamline (Figure 3.5) of an ideal fluid flow.

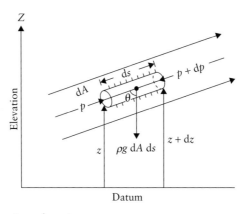

Figure 3.5   Euler's equation of motion.

The forces responsible for its motion are the pressure forces, gravity and accelerating force due to change in velocity along the streamline. All frictional forces are assumed to be zero and the flow is irrotational (i.e. uniform velocity distribution across streamlines).

By Newton's second law of motion along the streamline (force = mass × acceleration),

$$p \, dA - (p + dp) \, dA - \rho g \, dA \, ds \, \cos \theta = \rho \, dA \, ds \, \frac{dV}{dt}$$

$$\text{or} - dp - \rho g \, ds \, \cos \theta = \rho \, ds \, \frac{dV}{dt}$$

The tangential acceleration (along streamline) for steady flow can be expressed as

$$\frac{dV}{dt} = V \frac{dV}{ds} \quad \text{(Equation 3.6)}$$

$$\text{and} \ \cos \theta = \frac{dz}{ds} \quad \text{(Figure 3.5)}$$

$$\Rightarrow - dp - \rho g \, dz = \rho V \, dV$$

$$\text{or} \ \ dz + \frac{dp}{\rho g} + \frac{d(V^2)}{2g} = 0 \qquad [3.8]$$

Equation 3.8 is the Euler equation of motion applicable to steady-state irrotational flow of an ideal and incompressible fluid.

Integrating along the streamline, we get

$$z + \frac{p}{\rho g} + \frac{V^2}{2g} = \text{Constant} \qquad [3.9a]$$

$$\text{or} \ z_1 + \frac{p_1}{\rho g} + \frac{V_1^2}{2g} = z_2 + \frac{p_2}{\rho g} + \frac{V_2^2}{2g} \qquad [3.9b]$$

The three terms on the left-hand side of Equation 3.9a have the dimension of length, and the sum can be interpreted as the total energy of a fluid element of unit weight. For this reason Equation 3.9b, known as Bernoulli's equation, is sometimes called the energy equation for steady ideal fluid flow along a streamline between two sections, 1 and 2.

Bernoulli's theorem states that the total energy at all points along a steady continuous streamline of an ideal incompressible fluid flow is constant and is written as

$$z + \frac{p}{\rho g} + \frac{V^2}{2g} = \text{Constant}$$

where z is the elevation, p pressure and V velocity of the fluid at a point in the flow under consideration.

The first term, z, is the elevation or potential energy per unit weight of fluid with respect to an arbitrary datum, z N m/N (or metres), called elevation or potential head. The second term, $p/\rho g$, represents the work done in pushing a body of fluid by fluid pressure. This pressure energy per unit weight is $p/\rho g$ N m/N (or metres), called pressure head. The third term, $V^2/2g$, is the kinetic energy per unit weight of fluid (KE $= \frac{1}{2}mV^2$ and mass $m = W/g$) in N m/N (or metres), known as velocity head. The sum of these three terms is known as total head.

3.11 Modified energy equation for real fluid flows

Bernoulli's equation can be modified in the case of real incompressible fluid flow (i) by introducing a loss term in Equation 3.9b which would take into account the energy expended in overcoming the frictional resistances caused by viscous and turbulent shear stresses and other resistances due to changes of section, valves, fittings and so on; and (ii) by correcting the velocity energy term for true velocity distribution. The frictional losses depend on the type of flow; in a laminar pipe flow they vary directly with the viscosity, the length and the velocity and inversely with the square of the diameter, whereas in turbulent flow they vary directly with the length and square of the velocity and inversely with the diameter. The turbulent losses also depend on the roughness of the interior surface of the pipe wall and the fluid properties of density and viscosity.

Therefore, for real incompressible fluid flow, we can write

$$z_1 + \frac{p_1}{\rho g} + \frac{\alpha_1 V_1^2}{2g} = z_1 + \frac{p_2}{\rho g} + \frac{\alpha_2 V_2^2}{2g} + \text{losses} \qquad [3.10]$$

where α is the velocity (kinetic) energy correction factor.

Total kinetic energy over the section =

$\sum$ kinetic energies of individual particles of mass m

In the case of uniform velocity distribution (Figure 3.6a), each particle moves with a velocity V and its kinetic energy is $\frac{1}{2}mV^2$.

Total kinetic energy at the section $= \frac{1}{2}(m + m + m + \cdots)V^2$

$$= \frac{1}{2}\left(\frac{W}{g}\right)V^2$$

$$= \frac{V^2}{2g} \text{ per unit weight of fluid}$$

In the case of non-uniform velocity distribution (Figure 3.6b), the particles move with different velocities.

Mass of individual elements passing through an elementary area $dA = \rho \, dA \, v$

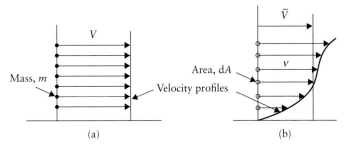

Figure 3.6 Velocity (kinetic) energy correction factor: (a) uniform distribution and (b) non-uniform distribution.

$$\text{Kinetic energy of individual mass element} = \frac{1}{2}\rho\,dA\,v\,v^2$$

and hence

$$\text{total kinetic energy over the section} = \int_A \frac{1}{2}\rho v^3\,dA = \alpha\,\frac{1}{2}\rho A \bar{V} \bar{V}^2 \text{ or } \alpha\,\frac{1}{2}\rho A \bar{V}^3$$

$\bar{V}$ being the average velocity at the section.

$$\Rightarrow \alpha = \frac{1}{A}\int_A \left(\frac{v}{\bar{V}}\right)^3 dA \qquad\qquad [3.11]$$

(For turbulent flows α lies between 1.03 and 1.3, and for laminar flows it is 2.0.)
α is commonly referred to as the Coriolis coefficient.

3.12 Separation and cavitation in fluid flow

Consider a rising main (Figure 3.7a) of uniform pipeline. At any point, by Bernoulli's equation

$$\text{Total energy} = z + \frac{p}{\rho g} + \frac{V^2}{2g} = \text{Constant}$$

For a given discharge the velocity is the same at all sections (uniform diameter) and hence we have $z + p/\rho g = $ Constant.

As the elevation z increases, the pressure p in the system decreases and if p becomes vapour pressure of the fluid, the fluid tends to boil, liberating dissolved gases and air bubbles. With further liberation of gases the bubbles tend to grow in size eventually blocking the pipe section, thus allowing the discharge to take place intermittently. This phenomenon is known as **separation** and greatly reduces the efficiency of the system.

If the tiny air bubbles formed at the separation point are carried to a high-pressure region (Figure 3.7b) by the flowing fluid, they collapse extremely abruptly or implode, producing a violent hammering action on any boundary surface on which the imploding bubbles come in contact and cause pitting and vibration to the system, which is highly undesirable. The whole phenomenon is called **cavitation** and should be avoided while designing any hydraulic system.

Chapter 3

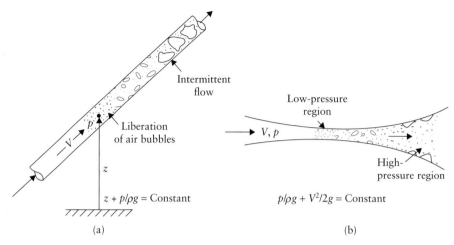

Figure 3.7   Separation and cavitation phenomena: (a) rising main of uniform diameter and (b) horizontal converging–diverging pipe.

3.13 Impulse–momentum equation

Momentum of a body is the product of its mass and velocity (kg m/s), and Newton's second law of motion states that the resultant external force acting on any body in any direction is equal to the rate of change of momentum of the body in that direction.

In the x-direction, this can be expressed as

$$F_x = \frac{d}{dt}(mv_x)$$
$$\text{or } F_x \, dt = d(mv_x) \tag{3.12}$$

Equation 3.12 is known as the implulse–momentum equation and can be written as

$$F_x \, dt = m \, dv_x \tag{3.13}$$

where m is the mass of the body and dv is the change in velocity in the direction considered; $F \, dt$ is called the impulse of applied force F.

For a fluid, the sum of external forces on a control volume may be equated to the net change in the rate of momentum flow $\rho Q v$, as shown in Example 3.7.

3.13.1 Momentum correction factor (β)

In the case of non-uniform velocity distribution (Figure 3.6b), the particles move with different velocities across a section of the flow field.

Total momentum flow = Σ momentum flow of individual particles

and can be written as

$$\int_A \rho \, dA \, v \, v = \beta \rho A \bar{V} \bar{V} \quad \text{or} \quad \beta \rho Q \bar{V}$$

where $\bar{V}$ is the average velocity at the section.

$$\Rightarrow \beta = \frac{1}{A} \int_A \left(\frac{v}{\bar{V}}\right)^2 dA \tag{3.14}$$

(For turbulent flows β is seldom greater than 1.1, and for laminar flows it is 1.33.) β is commonly referred to as the Boussinesq coefficient.

3.14 Energy losses in sudden transitions

Flow through a sudden expansion experiences separation from the boundary to some length downstream of the flow. In these regions of separation, turbulent eddies form resulting from pressure loss dissipating in the form of heat energy.

Referring to Figure 3.8a, the pressure against the annular area $A_2 - A_1$ is experimentally found to be the same as the pressure p_1, just before the entrance.

$$\text{Energy equation:} \quad \frac{p_1}{\rho g} + \frac{V_1^2}{2g} = \frac{p_2}{\rho g} + \frac{V_2^2}{2g} + \text{loss} \tag{i}$$

Momentum equation: Net force on the control volume between 1 and 2

= rate of change of momentum

$$p_1 A_1 + p_1(A_2 - A_1) - p_2 A_2 = \rho Q(V_2 - V_1) \tag{ii}$$

$$\text{Continuity equation:} \quad A_1 V_1 = A_2 V_2 = Q \tag{iii}$$

The head or energy loss between 1 and 2 (from Equations (i), (ii) and (iii))

$$h_L = \frac{(V_1 - V_2)^2}{2g} \tag{3.15}$$

Referring to Figure 3.8b, the head loss is mainly due to sudden enlargement of flow from vena contracta to section 2, and therefore, the contraction loss can be written as (from Equation 3.15)

$$h_L = \frac{(V_c - V_2)^2}{2g} \tag{iv}$$

where V_c is velocity at vena contracta v–c.

By continuity,

$$A_c V_c = A_2 V_2 = Q$$

$$\Rightarrow V_c = \left(\frac{A_2}{A_c}\right) V_2 = \frac{V_2}{C_c}$$

where C_c is the coefficient of contraction $(= A_c / A_2)$.

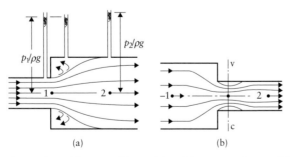

(a)   (b)

Figure 3.8   Energy losses in sudden transitions: (a) sudden expansion and (b) sudden contraction.

Equation (iv) reduces to

$$h_L = \left(\frac{1}{C_c} - 1\right)^2 \frac{V_2^2}{2g} = \frac{kV_2^2}{2g}$$ [3.16]

where k is a function of the contraction ratio A_2/A_1.

3.15 Flow measurement through pipes

Application of continuity, energy and momentum equations to a given system of fluid flow makes velocity and volume measurements possible.

3.15.1 Venturi meter and orifice meter

A pressure differential is created along the flow by providing either a gradual (venturi meter) or sudden (orifice plate meter) constriction in the pipeline, and is related to flow velocities and discharge by the energy and continuity principles (see Figure 3.9).

Bernoulli's equation between inlet section and constriction can be written as

$$\frac{p_1}{\rho g} + \frac{v_1^2}{2g} = \frac{p_2}{\rho g} + \frac{v_2^2}{2g} \quad \text{neglecting losses}$$

$$\Rightarrow \frac{p_1 - p_2}{\rho g} = h = \frac{v_1^2 - v_2^2}{2g}$$ (i)

The continuity equation gives

$$a_1 v_1 = a_2 v_2 = Q$$ (ii)

From Equations (i) and (ii),

$$v_1 = a_2 \sqrt{\frac{2gh}{a_1^2 - a_2^2}}$$

$$\text{and } Q = a_1 v_1 = \frac{a_1 a_2 \sqrt{2gh}}{\sqrt{a_1^2 - a_2^2}}$$ [3.17]

$$\text{or } Q = a_1 \sqrt{\frac{2gh}{k^2 - 1}} \quad \text{where } k = \frac{a_1}{a_2}$$ [3.18]

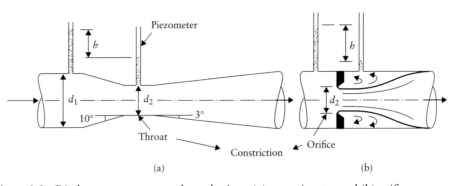

Figure 3.9   Discharge measurement through pipes: (a) venturi meter and (b) orifice meter.

Equation 3.18 is an ideal equation obtained by neglecting all losses.

The actual discharge is, therefore, written by introducing a coefficient C_d in Equation 3.18:

$$\text{Discharge, } Q = C_d a_1 \sqrt{\frac{2gh}{k^2 - 1}} \qquad [3.19]$$

The numerical value of C_d, the coefficient of discharge, will depend on the ratio a_1/a_2, type of transition, velocity and viscosity of the flowing fluid.

The gradual transitions of the venturi meter (Figure 3.9a) between its inlet and outlet induce the least amount of losses, and the value of its C_d lies between 0.96 and 0.99 for turbulent flows.

The transition in the case of an orifice plate meter (Figure 3.9b) is sudden and hence the flow within the meter experiences greater losses due to contraction and expansion of the jet through the orifice. Its discharge coefficient has a much lower value (0.6–0.63), as the area a_2 in Equation 3.17 refers to the orifice and not to the contracted jet.

The reduction in the constriction diameter causes velocity to increase, and correspondingly a large pressure differential is created between the inlet and constriction, thus enabling greater accuracy in its measurement. High velocities at the constriction cause low pressures in the system, and if these fall below the vapour pressure limit of the fluid, cavitation sets in, which is highly undesirable. Therefore, the selection of the ratio d_2/d_1 is to be considered carefully. This ratio may be kept between $1/3$ and $3/4$, and a more common value is $1/2$. The orifice plate and venturi meters provide a good illustration of hydraulic principles, showing the interchange of energy between the terms in Bernoulli's equation. An increasingly widely used alternative method of flow measurement in pipes is the electromagnetic flowmeter (see e.g. Baker, 2000 and 2006).

3.15.2 Pitot tube

A Pitot tube in its simplest form is an L-shaped tube held against the flow as shown in Figure 3.10, creating a stagnation point in the flow.

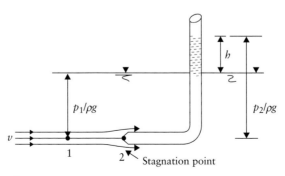

Figure 3.10   Pitot tube.

The stagnation pressure at point 2 (velocity is zero) can be expressed as

$$\frac{p_2}{\rho g} = \frac{p_1}{\rho g} + \frac{v^2}{2g} \quad \text{(by Bernoulli's equation)}$$

The rise in water level or pressure differential between 1 and 2 can be written as

$$h = \frac{(p_2 - p_1)}{\rho g} = \frac{v^2}{2g}$$

$$\Rightarrow \text{Velocity}, v = \sqrt{2gh} \qquad [3.20]$$

The actual velocity will be slightly less than the velocity given by Equation 3.20, and it is modified by introducing a coefficient K (usually between 0.95 and 1.0) as

$$v = K\sqrt{2gh} \qquad [3.21]$$

3.16 Flow measurement through orifices and mouthpieces

3.16.1 Small orifice

If the head h, causing flow through an orifice of diameter d, is constant (small orifice: $h \gg d$) as shown in Figure 3.11, by Bernoulli's equation

$$h + \frac{p_1}{\rho g} + \frac{v_1^2}{2g} = 0 + \frac{p_2}{\rho g} + \frac{v_2^2}{2g} + \text{losses}$$

With $p_1 = p_2$ (both atmospheric), assuming $v_1 \simeq 0$ and ignoring losses, we get

$$\frac{v_2^2}{2g} = h$$

or the velocity through the orifice,

$$v_2 = \sqrt{2gh} \qquad [3.22]$$

Equation 3.22 is called Torricelli's theorem and the velocity is called the theoretical velocity.

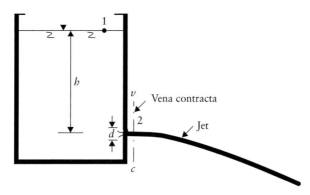

Figure 3.11 Small orifice ($h \gg d$).

The actual velocity is $C_v\sqrt{2gh}$, where C_v is the coefficient of velocity defined as

$$C_v = \frac{\text{actual velocity}}{\text{theoretical velocity}} \qquad [3.23]$$

The area of jet is much less than the area of the orifice due to contraction, and the corresponding coefficient of contraction is defined as

$$C_c = \frac{\text{area of jet}}{\text{area } a \text{ of orifice}} \qquad [3.24]$$

At a section very close to the orifice, known as the vena contracta, the velocity is normal to the cross section of the jet and hence the discharge can be written as

$$Q = \text{area of jet} \times \text{velocity of jet (at vena contracta)}$$

$$= C_c\, a \times C_v\sqrt{2gh}$$

$$= C_d\, a\sqrt{2gh} \qquad [3.25]$$

where C_d is called the coefficient of discharge and defined as

$$C_d = \frac{\text{actual discharge}}{\text{theoretical discharge}}$$

$$= C_c C_v \qquad [3.26]$$

Some typical orifices and mouthpieces (short pipe lengths attached to orifice) and their coefficients, C_c, C_v and C_d, are shown in Figure 3.12.

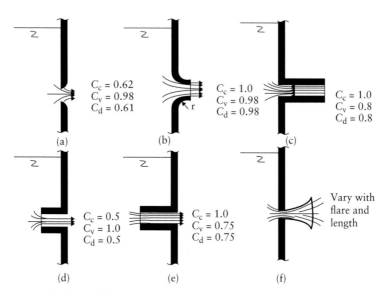

Figure 3.12 Hydraulic coefficients for some typical orifices and mouthpieces: (a) sharp-edged orifice, (b) bell-mouthed orifice, (c) mouthpiece, (d) and (e) Borda's (re-entrant) mouthpieces and (f) divergent tube.

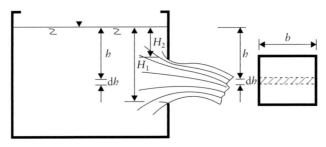

Figure 3.13   Large rectangular orifice.

3.16.2 Large rectangular orifice (see Figure 3.13)

As the orifice is large, the velocity across the jet is no longer constant; however, if we consider a small area $b\,dh$ at a depth h, then

$$\text{velocity through this area} = \sqrt{2gh} \quad \text{(Equation 3.22)}$$

The actual discharge through the strip area,

$$dQ = C_d \times \text{area of strip} \times \text{velocity through the strip}$$

$$= C_d b\,dh\sqrt{2gh}$$

The total discharge through the entire opening (h from H_2 to H_1),

$$Q = \int dQ = C_d b\sqrt{2g}\int_{H_2}^{H_1} h^{1/2}\,dh$$

$$= \frac{2}{3}C_d\sqrt{2g}\,b\left(H_1^{3/2} - H_2^{3/2}\right) \qquad [3.27]$$

Modification of Equation 3.27

(a) If V_a is the velocity of approach, the head responsible for the strip velocity is $h + \alpha V_a^2/2g$ (Figure 3.14a) and hence the strip velocity is $\sqrt{2g(h + \alpha V_a^2/2g)}$, α being the kinetic energy correction factor (Coriolis coefficient).

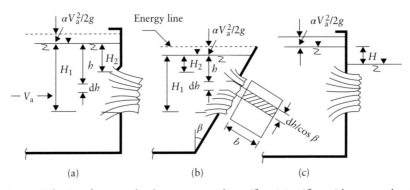

Figure 3.14   Velocity of approach – large rectangular orifice: (a) orifice with approach velocity, (b) orifice in inclined wall and (c) submerged orifice.

Discharge through the strip,

$$dQ = C_d b\, dh \sqrt{2g\left(h + \frac{\alpha V_a^2}{2g}\right)}$$

and the total discharge,

$$Q = \int dQ = \frac{2}{3} C_d \sqrt{2g} b \left[\left(H_1 + \frac{\alpha V_a^2}{2g}\right)^{3/2} - \left(H_2 + \frac{\alpha V_a^2}{2g}\right)^{3/2}\right] \qquad [3.28]$$

(b) Side wall of the tank inclined at an angle β (see Figure 3.14b):

$$\text{Effective strip area} = \frac{b\, dh}{\cos \beta}$$

$$\text{Discharge, } Q = \int dQ = \int_{H_2}^{H_1} C_d \frac{b\, dh}{\cos \beta} \sqrt{2g\left(h + \frac{\alpha V_a^2}{2g}\right)}$$

$$= \frac{2}{3} C_d \sqrt{2g}\, \frac{b}{\cos \beta} \left[\left(H_1 + \frac{\alpha V_a^2}{2g}\right)^{3/2} - \left(H_2 + \frac{\alpha V_a^2}{2g}\right)^{3/2}\right] \qquad [3.29]$$

(c) Submerged orifice (see Figure 3.14c):
It can be shown by Bernoulli's equation that the velocity across the jet is constant and equal to $\sqrt{2gH}$ or $\sqrt{2g(H + \alpha V_a^2/2g)}$ if V_a is considered.
The discharge through a submerged orifice,

$$Q = C_d \times \text{area of orifice} \times \text{velocity}$$

$$= C_d A \sqrt{2g\left(H + \frac{\alpha V_a^2}{2g}\right)} \qquad [3.30]$$

Note: Since the head causing flow is varying, the discharge through the orifice varies with time.

If h is the head at any instant t (see Figure 3.15), the velocity through the orifice at that instant is $\sqrt{2gh}$.

Let the water level drop down by a small amount dh in a time dt.

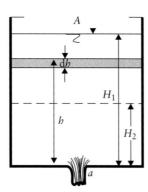

Figure 3.15   Time taken to empty a tank.

We can write

$$\text{volume reduced} = \text{volume escaped through the orifice}$$

$$-A\,dh = C_d a \sqrt{2gh}\,dt \quad (dh \text{ is negative})$$

$$\Rightarrow dt = -\frac{A}{C_d a \sqrt{2g}}\frac{dh}{h^{1/2}} \tag{3.31}$$

Time taken to lower the water level from H_1 to H_2,

$$T = \int dt = \frac{2A\left(H_1^{1/2} - H_2^{1/2}\right)}{C_d a \sqrt{2g}} \tag{3.32}$$

3.17 Flow measurement in channels

Notches and weirs are regular obstructions placed across open streams over which the flow takes place. The head over the sill of such an obstruction is related to the discharge through energy principles. A weir or a notch may be regarded as a special form of large orifice with the free water surface below its upper edge. Thus Equation 3.27 with $H^2 = 0$, for example, gives the discharge through a rectangular notch. In general, the discharge over such structures can be written as

$$Q = KH^n \tag{3.33}$$

where K and n depend on the geometry of notch.

3.17.1 Rectangular notch

Considering a small strip area of the notch at a depth h below the free water surface (see Figure 3.16), the total head responsible for the flow is written as $h + \alpha V_a^2/2g$, α being the energy correction factor.

$$\text{Velocity through the strip} = \sqrt{2g\left(h + \frac{\alpha V_a^2}{2g}\right)}$$

$$\text{and discharge, } dQ = C_d b\,dh\sqrt{2g\left(h + \frac{\alpha V_a^2}{2g}\right)}$$

$$\text{Total discharge, } Q = \int dQ = C_d \sqrt{2g}\,b \int_0^H \left(h + \frac{\alpha V_a^2}{2g}\right)^{1/2} dh$$

$$= \frac{2}{3}C_d \sqrt{2g}\,b\left[\left(H + \frac{\alpha V_a^2}{2g}\right)^{3/2} - \left(\frac{\alpha V_a^2}{2g}\right)^{3/2}\right] \tag{3.34}$$

The discharge coefficient C_d largely depends on the shape, contraction of the nappe, sill height, head causing flow, sill thickness and so on.

As the effective width of the notch for the flow is reduced by the presence of end contractions, each contraction being one-tenth of the total head (experimental result),

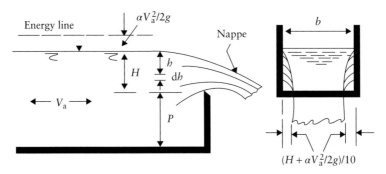

Figure 3.16   Rectangular notch with end contractions.

Equation 3.34 is modified (in SI units) as

$$Q = 1.84 \left[b - 0.1n \left(H + \frac{\alpha V_a^2}{2g} \right) \right] \left[\left(H + \frac{\alpha V_a^2}{2g} \right)^{3/2} - \left(\frac{\alpha V_a^2}{2g} \right)^{3/2} \right] \qquad [3.35]$$

taking an average value of $C_d = 0.623$. This is known as the Francis formula, n being the number of end contractions.

Note: (a) Bazin formula:

$$Q = \left(0.405 + \frac{0.003}{H_1} \right) \sqrt{2g} b (H_1)^{3/2} \qquad [3.36]$$

where $H_1 = H + 1.6(V_a^2/2g)$.
   (b) Rehbock formula:

$$Q = \left[1.78 + 0.245 \left(\frac{H_e}{P} \right) \right] b (H_e)^{3/2} \qquad [3.37]$$

where $H_e = H + 0.0012$ m, and P is the height of the sill, the coefficient of discharge C_d being

$$C_d = 0.602 + 0.083 \left(\frac{H_e}{P} \right) \qquad [3.38]$$

3.17.2 Triangular or V-notch

A similar approach to determine the discharge over a triangular notch of an included angle θ results in the equation

$$Q = \frac{8}{15} C_d \sqrt{2g} \tan \frac{\theta}{2} \left[\left(H + \frac{\alpha V_a^2}{2g} \right)^{5/2} - \left(\frac{\alpha V_a^2}{2g} \right)^{5/2} \right] \qquad [3.39a]$$

If the approach velocity V_a is neglected, Equation 3.39a reduces to

$$Q = \frac{8}{15} C_d \sqrt{2g} \tan \frac{\theta}{2} H^{5/2} \qquad [3.39b]$$

3.17.3 Cipolletti weir

This is a trapezoidal weir with $14°$ side slopes (1 horizontal : 4 vertical). The discharge over such a weir may be computed by using the formula for a suppressed (no end contractions) rectangular weir with equal sill width.

A trapezoidal notch may be considered as one rectangular notch of width b and two half V-notches (apex angle $\frac{1}{2}\theta$), and the discharge equation written as

$$Q = \frac{2}{3}C_d\sqrt{2g}(b - 0.2H)H^{3/2} + \frac{8}{15}C_d\sqrt{2g}\tan\frac{\theta}{2}H^{5/2} \qquad [3.40]$$

Equation 3.40 reduces to that for a suppressed rectangular weir (weir with no end contractions) if the reduction in discharge due to the presence of end contractions is compensated by the increase provided by the presence of two half V-notches.

Therefore we can write

$$\frac{2}{3}C_d\sqrt{2g} \times 0.2H \times H^{3/2} = \frac{8}{15}C_d\sqrt{2g}\tan\frac{\theta}{2}H^{5/2}$$

Assuming C_d is constant throughout, we get

$$\tan\frac{\theta}{2} = \frac{1}{4} \Rightarrow \frac{\theta}{2} = 14°2'$$

3.17.4 Proportional or Sutro weir

In general, the discharge through any type of weir may be expressed as $Q \propto H^n$. A weir with $n = 1$ (i.e. the discharge is proportional to the head over the weir's crest), is called a **proportional weir** (Figure 3.17).

Sutro's analytical approach resulted in the relationship

$$x \propto y^{-1/2}$$

for the proportional weir profile, and to overcome the practical limitation (as $y \to 0, x \to \infty$) he proposed the weir shape in the form of hyperbolic curves of the equation

$$\frac{2x}{L} = \left[1 - \frac{2}{\pi}\tan^{-1}\sqrt{\frac{y}{a}}\right] \qquad [3.41]$$

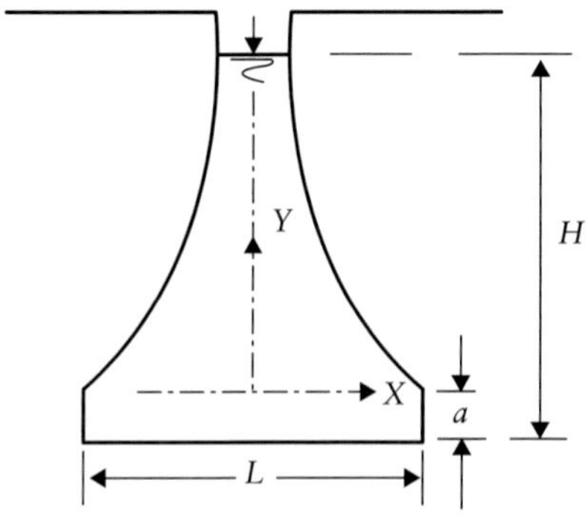

Figure 3.17 Sutro or proportional weir.

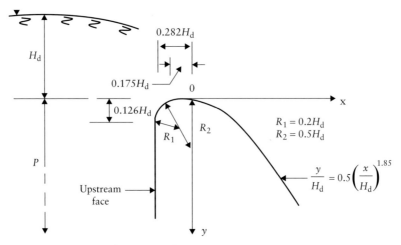

Figure 3.18 Cross section of an ogee spillway.

where a and L are the height and width of the rectangular aperture forming the base of the weir.

$$\text{Discharge, } Q = C_d L (2ga)^{1/2} \left(H - \frac{a}{3} \right) \qquad [3.42]$$

The proportional weir is a very useful device, for example, in chemical dosing and sampling and irrigation outlets.

3.17.5 Ogee spillway

Excess flood flows behind dams are normally discharged by providing spillways. The profile of an ogee spillway conforms to the shape of a sharp-crested weir (see Figure 3.18) at a design head H_d. A discharge equation similar to that of the weir, but with a higher discharge coefficient C_{d0} (since the reference sill level for the spillway is slightly shifted), written in the form

$$Q = \frac{2}{3} C_{d0} B \sqrt{2g} H_{de}^{3/2} \qquad [3.43]$$

is applicable, in which $H_{de} = H_d + V_a^2/2g$, V_a being the velocity of approach. For spillways of $P/H_{de} \geq 3$, the value of $C_{d0} \approx 0.75$. Figure 3.19 shows the variation of C_{d0} with P/H_{de}. For heads other than the design head, the discharge coefficient varies as the underside of the nappe no longer conforms to the spillway profile. Figure 3.20 shows the variation of C_d/C_{d0} with H_e/H_{de}, H_e being any other energy head with a corresponding discharge coefficient C_d. For larger values of P/H, the approach velocity V_a may be negligible, leading to $H_e \approx H$.

3.17.6 Other forms of flow-measuring devices

Open channel flows may also be measured by broad-crested weir and venturi flume (see Chapter 8) and some special structures like Crump weir (see e.g. Novak et al., 2007).

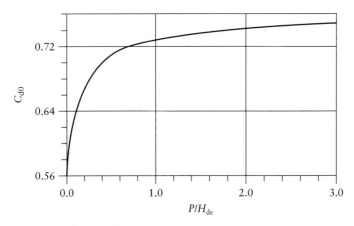

Figure 3.19 Variation of C_{d0} with P/H_{de}.

Other open channel flow-measuring methods range from straightforward velocity–area methods to ultrasonic flow gauging stations (see e.g. Herschy, 2009).

3.17.7 Effect of submergence of flow-measuring structures

If the water level (H_2) downstream of a measuring device is below the sill level, the discharge is said to be modular (free flow, Q_f) and the above equations are valid to compute the free flows. When the downstream water level is above the sill level, the structure is said to be drowned and the discharge (non-modular or drowned flow) is affected (i.e. reduced). The non-modular flow Q_s is given by the equation

$$Q_s = Q_f \left[1 - \left(\frac{H_2}{H_1}\right)^m\right]^{0.385}$$ [3.44]

where m is the exponent of H_1 (upstream water level above sill) in the weir equations: $m = 1.5$ for a rectangular weir and $m = 2.5$ for a triangular weir.

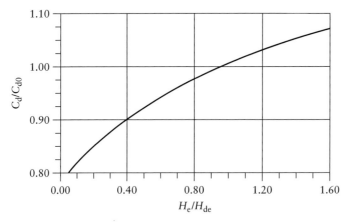

Figure 3.20 Variation in C_d/C_{d0} with H_e/H_{de}.

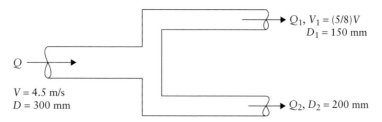

Figure 3.21 Branching pipeline.

Worked examples

Example 3.1

A pipeline of 300 mm diameter carrying water at an average velocity of 4.5 m/s branches into two pipes of 150 mm and 200 mm diameters. If the average velocity in the 150 mm pipe is 5/8 of the velocity in the main pipeline, determine the average velocity of flow in the 200 mm pipe and the total flow rate in the system (in litres per second) (see Figure 3.21).

Solution:

$$\text{Discharge, } Q = AV = Q_1 + Q_2 \quad \text{(by continuity)}$$
$$\Rightarrow AV = A_1 V_1 + A_2 V_2$$
$$\frac{1}{4}\pi(0.3)^2 \times 4.5 = \frac{1}{4}\pi(0.15)^2 \times \frac{5}{8} \times 4.5 + \frac{1}{4}\pi(0.2)^2 \times V_2$$
$$\text{or } V_2 = 8.54 \text{ m/s}$$
$$\text{and total flow rate, } Q = \frac{1}{4}\pi(0.3)^2 \times 4.5$$
$$= 0.318 \text{ m}^3/\text{s}$$
$$= 318 \text{ L/s}$$

Example 3.2

A storage reservoir supplies water to a pressure turbine (Figure 3.22) under a head of 20 m. When the turbine draws 500 L/s of water, the head loss in the 300 mm diameter supply line amounts to 2.5 m. Determine the pressure intensity at the entrance to the turbine. If a negative pressure of 30 kN/m² exists at the 600 mm diameter section of the draft tube 1.5 m below the supply line, estimate the energy absorbed by the turbine (in kilowatts) neglecting all frictional losses between the entrance and exit of the turbine. Hence find the output of the turbine assuming an efficiency of 85%.

Solution:

Referring to Figure 3.19, by Bernoulli's equation (between points 1 and 2)

$$z_1 + \frac{p_1}{\rho g} + \frac{V_1^2}{2g} = z_2 + \frac{p_2}{\rho g} + \frac{V_2^2}{2g} + \text{loss} \tag{i}$$

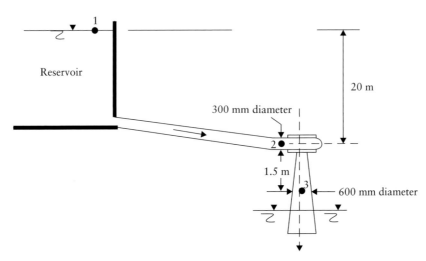

Chapter 3

Figure 3.22   Flow through a hydraulic turbine.

With section 2 as datum, Equation (i) becomes ($p_1 = 0$ and $V_1 = 0$)

$$20 = \frac{p_2}{\rho g} + \frac{V_2^2}{2g} + 2.5$$

By the continuity equation,

$$Q = A_2 V_2 = A_3 V_3$$

$$\text{Average velocities, } V_2 = \frac{Q}{A_2} = \frac{0.5}{(\pi/4)(0.3)^2} = 7.07 \text{ m/s}$$

$$\text{and } V_3 = \frac{Q}{A_3} = \frac{0.5}{(\pi/4)(0.6)^2} = 1.77 \text{ m/s}$$

$$\Rightarrow \frac{p_2}{\rho g} = 20 - 2.5 - \frac{(7.07)^2}{2g}$$

$$= 14.95 \text{ m of water}$$

$$\text{or } p_2 = \rho g \times 14.95 = 9.81 \times 14.95$$

$$= 146.95 \text{ kN/m}^2$$

Between sections 2 and 3, we can write

$$z_2 + \frac{p_2}{\rho g} + \frac{V_2^2}{2g} = z_3 + \frac{p_3}{\rho g} + \frac{V_3^2}{2g} + E_t + \text{losses} \tag{ii}$$

where E_t it the energy absorbed by the machine/unit weight of water flowing.
Assuming no losses between sections 2 and 3, Equation (ii) reduces to

$$1.5 + 14.95 + \frac{(7.07)^2}{2g} = \frac{-30 \times 10^3}{\rho g} + \frac{(1.77)^2}{2g} + E_t$$

$$\Rightarrow E_t = 1.5 + 14.95 + 2.55 + 3.06 - 0.16$$

$$= 21.9 \text{ N m/N}$$

Weight of water flowing through the turbine per second, $W = \rho g Q$

$$= 10^3 \times 9.81 \times 0.5$$
$$= 4.905 \text{ kN/s}$$

Total energy absorbed by the machine $= E_t \times W$

$$= 21.9 \times 4.905$$
$$= 107.42 \text{ kW}$$

Hence its output $=$ efficiency $\times$ input

$$= 0.85 \times 107.42$$
$$= 91.31 \text{ kW}$$

Example 3.3

A 500 mm diameter vertical water pipeline discharges water through a constriction of 250 mm diameter (Figure 3.23). The pressure difference between the normal and constricted sections of the pipe is measured by an inverted U-tube. Determine (i) the difference in pressure between these two sections when the discharge through the system is 600 L/s, and (ii) the manometer deflection h if the inverted U-tube contains air.

Solution:

Discharge, $Q = 600 \text{ L/s} = 0.6 \text{ m}^3\text{/s}$

$$\Rightarrow V_a = \frac{0.6}{(\pi/4)(0.5)^2} = 3.056 \text{ m/s} \quad \text{(by continuity)}$$

$$\text{and } V_b = \frac{0.6}{(\pi/4)(0.25)^2} = 7.54 \text{ m/s}$$

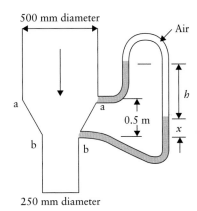

500 mm diameter

Air

h

a

a

0.5 m

x

b

b

250 mm diameter

Figure 3.23   Flow through a vertical constriction.

By Bernoulli's equation between aa and bb (assuming no losses),

$$0.50 + \frac{p_a}{\rho g} + \frac{(3.056)^2}{2g} = 0 + \frac{p_b}{\rho g} + \frac{(7.54)^2}{2g}$$

$$\text{or } \frac{p_a - p_b}{\rho g} = \frac{(7.54)^2 - (3.056)^2}{2g} - 0.50$$

$$= 2.42 - 0.50$$

$$= 1.92 \text{ m of water} \tag{i}$$

$$\Rightarrow p_a - p_b = 10^3 \times 9.81 \times 1.92$$

$$= 18.8 \text{ kN/m}^2$$

Manometer equation:

$$\frac{p_a}{\rho g} - (h + x - 0.50) + x = \frac{p_b}{\rho g}$$

$$\Rightarrow \frac{p_a - p_b}{\rho g} = h - 0.50 = 1.92 \quad \text{(from Equation (i))}$$

$$\text{or } h = 1.92 + 0.50$$

$$= 2.42 \text{ m}$$

Example 3.4

A drainage pump having a tapered suction pipe discharges water out of a sump. The pipe diameters at the inlet and at the upper end are 1 and 0.5 m, respectively. The free water surface in the sump is 2 m above the centre of the inlet, and the pipe is laid at a slope of 1 (vertical) : 4 (along pipeline). The pressure at the top end of the pipe is 0.25 m of mercury below atmosphere and it is known that the loss of head due to friction between the two sections is 1/10 of the velocity head at the top section. Compute the discharge (in litres per second) through the pipe if its length is 20 m (see Figure 3.24).

Solution:

By the continuity equation,

$$Q = a_2 v_2 = a_3 v_3 \tag{i}$$

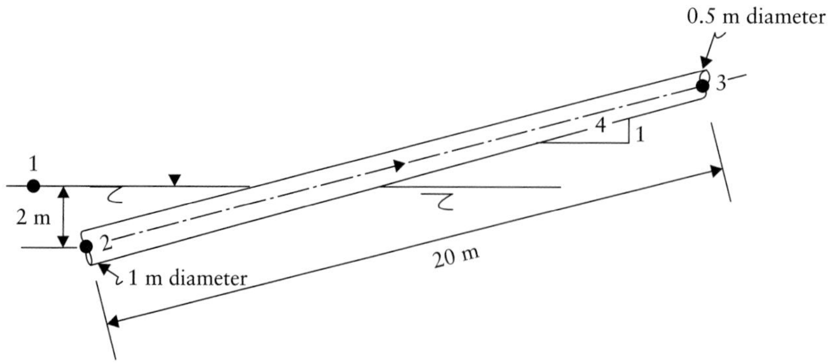

Figure 3.24   Flow through the suction pipe of a pump.

By Bernoulli's equation between 1, 2 and 3,

$$2 + 0 + 0 = 0 + \frac{p_2}{\rho g} + \frac{v_2^2}{2g} = 20 \times \frac{1}{4} + \frac{p_3}{\rho g} + \frac{v_3^2}{2g} + \left(\frac{1}{10}\right)\frac{v_3^2}{2g} \qquad \text{(ii)}$$

assuming the velocity in the sump at 1 as zero and a datum through 2.

Pressure at the top end, $\dfrac{p_3}{\rho g} = 0.25$ m of mercury below atmosphere

$$= -0.25 \times 13.6$$

$$= -3.4 \text{ m of water}$$

$$\Rightarrow 1.1 \times \frac{v_3^2}{2g} = 2 - 5 + 3.4 = 0.4$$

$$\text{or} \quad \frac{v_3^2}{2g} = \frac{0.4}{1.1} = 0.364 \text{ m}$$

$$v_3 = 2.67 \text{ ms}$$

Hence

$$\text{discharge}, Q = a_3 \times v_3 = \frac{1}{4}\pi(0.5)^2 \times 2.67$$

$$= 0.524 \text{ m}^3/\text{s}$$

$$= 524 \text{ L/s}$$

Example 3.5

A jet of water issues out from a fire hydrant nozzle fitted at a height of 3 m from the ground at an angle of 45° with the horizontal. If the jet under a particular flow condition strikes the ground at a horizontal distance of 15 m from the nozzle, find (i) the jet velocity and (ii) the maximum height the jet can reach and its horizontal distance from the nozzle. Neglect air resistance (see Figure 3.25).

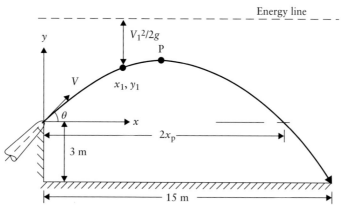

Figure 3.25 Jet dynamics.

Chapter 3

Solution:

In the horizontal direction, acceleration $a = 0$.

$$\Rightarrow V_1 \cos\theta_1 = V\cos\theta = \text{Constant} \tag{i}$$

and in time t, horizontal distance covered

$$x_1 = V\cos\theta \times t \quad\text{or}\quad t = \frac{x_1}{V\cos\theta} \tag{ii}$$

and vertical distance

$$y_1 = V\sin\theta \times t - \frac{1}{2}gt^2 \quad (\text{since } a = -g) \tag{iii}$$

$$\Rightarrow y_1 = V\sin\theta \times \frac{x_1}{V\cos\theta} - \frac{1}{2}g\left(\frac{x_1}{V\cos\theta}\right)^2 \quad (\text{from Equation (ii)})$$

$$= x_1\tan\theta - \frac{1}{2}gx_1^2\frac{\sec^2\theta}{V^2} \tag{iv}$$

Coordinates of the point where the jet strikes the ground are

$$y = -3\text{ m} \quad\text{and}\quad x = 15\text{ m}$$

From Equation (iii), $V = 11.07$ m/s.

Highest point is P: velocity vector is horizontal and is $V\cos\theta$ (from Equation (i)).
In the vertical direction,

$$\text{initial velocity} = V\sin\theta$$
$$\text{final velocity} = 0$$

giving

$$0 - V^2\sin^2\theta = -2gy_{max} \quad (\text{since } a = -g)$$

$$\text{or } y_{max} = V^2\frac{\sin^2\theta}{2g} = 3.12\text{ m}$$

We can also write

$$0 = V\sin\theta - gt_p$$

$$\text{or } t_p = \frac{V\sin\theta}{g} \tag{v}$$

and horizontal distance, $x_p = V\cos\theta \times t_p$

$$= \frac{V^2\sin 2\theta}{2g}$$

$$= 6.24\text{ m} \tag{vi}$$

Note: Total horizontal distance traversed by the jet $= 2x_p$

$$= \frac{V^2\sin 2\theta}{g} \tag{vii}$$

Example 3.6

A 500 mm diameter siphon pipeline discharges water from a large reservoir. Determine (i) the maximum possible elevation of its summit for a discharge of 2.15 m³/s without the pressure becoming less than 20 kN/m² absolute and (ii) the corresponding elevation of its discharge end. Take atmospheric pressure as 1 bar and neglect all losses.

Solution:

Consider the three points A, B and C along the siphon system, as shown in Figure 3.26.

$$\text{Discharge, } Q = av = 2.15 \text{ m}^3/\text{s}$$

$$\text{Velocity, } v = \frac{2.15}{(\pi/4)(0.5)^2} = 10.95 \text{ m/s}$$

$$\text{and } \frac{v^2}{2g} = 6.11 \text{ m}$$

$$\text{Atmospheric pressure} = 1 \text{ bar} = 10^5 \text{ N/m}^2$$

$$= \text{pressures at A and C}$$

$$\text{Minimum pressure at B} = 20 \text{ kN/m}^2 \quad \text{absolute (given)}$$

By Bernoulli's equation between A and B (reservoir water surface as datum),

$$0 + \frac{10^5}{\rho g} + 0 = z_B + \frac{20 \times 10^3}{\rho g} + 6.11$$

$$\Rightarrow z_B = \frac{10^5}{\rho g} - \frac{20 \times 10^3}{\rho g} - 6.11 = 2.04 \text{ m}$$

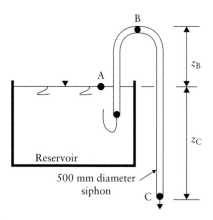

Figure 3.26 Siphon pipeline.

and between A and C (with the exit end as datum)

$$z_C + \frac{p_A}{\rho g} + 0 = 0 + \frac{p_C}{\rho g} + 6.11$$

$$\Rightarrow z_C = 6.11 \text{ m} \,(p_A = p_C = \text{atmospheric pressure})$$

Hence the exit end is to be 6.11 m below the reservoir level.

Example 3.7

A horizontal bend in a pipeline conveying 1 cumec of water gradually reduces from 600 mm to 300 mm in diameter and deflects the flow through an angle of 60°. At the larger end the pressure is 170 kN/m². Determine the magnitude and direction of the force exerted on the bend. Assume $\beta = 1.0$ (see Figure 3.27).

Solution:

Discharge, $Q = 1 \text{ m}^3/\text{s} = A_1 V_1 = A_2 V_2$ (continuity equation)

$$\Rightarrow V_1 = \frac{1}{(\pi/4)(0.6)^2} = 3.54 \text{ m/s}$$

$$\text{and } V_2 = \frac{1}{(\pi/4)(0.3)^2} = 14.15 \text{ m/s}$$

Energy equation neglecting friction losses:

$$\frac{p_1}{\rho g} + \frac{V_1^2}{2g} = \frac{p_2}{\rho g} + \frac{V_2^2}{2g}$$

Pressure at 1, $p_1 = 170 \times 10^3 \text{ N/m}^2$

$$\Rightarrow \frac{p_2}{\rho g} = \frac{170 \times 10^3}{10^3 \times 9.81} + \frac{(3.54)^2}{19.62} - \frac{(14.15)^2}{19.62}$$

$$\text{or } p_2 = 7.62 \times 10^4 \text{ N/m}^2$$

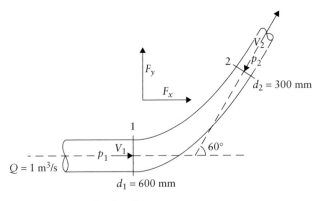

Figure 3.27   Forces on a converging bend.

Momentum equation: Gravity forces are zero along the horizontal plane and the only forces acting on the fluid mass are pressure and momentum forces.

Let F_x and F_y be the two components of the total force F exerted by the bent boundary surface on the fluid mass; these are considered positive if F_x is left to right and F_y upwards.

In the x-direction,

$$p_1 A_1 + F_x - p_2 A_2 \cos \theta = \rho Q \, (V_2 \cos \theta - V_1)$$

and in the y-direction,

$$0 + F_y - p_2 A_2 \sin \theta = \rho Q \, (V_2 \sin \theta - 0)$$

$$\Rightarrow F_x = 10^3 \times 1(14.15 \cos 60° - 3.54) + 7.62 \times 10^4 \times \frac{1}{4}\pi(0.3)^2 \cos 60°$$

$$-17 \times 10^4 \times \frac{1}{4}\pi(0.6)^2$$

$$= -4.2 \times 10^4 \text{ N} \quad \text{(negative sign indicates } F_x \text{ is right to left)}$$

and $F_y = 10^3 \times 1(14.15 \sin 60°) + 7.62 \times 10^4 \times \frac{1}{4}\pi(0.3)^2 \sin 60°$

$$= 1.7 \times 10^4 \text{ N} \quad \text{(upwards)}$$

According to Newton's third law of motion, the forces R_x and R_y exerted by the fluid on the bend will be equal and opposite to F_x and F_y.

$$\Rightarrow R_x = -F_x = 4.2 \times 10^4 \text{ N} \quad \text{(left to right)}$$
$$\text{and } R_y = -F_y = -1.7 \times 10^4 \text{ N} \quad \text{(downwards)}$$

Resultant force on the bend,

$$R = \sqrt{R_x^2 + R_y^2}$$
$$= 4.53 \times 10^4 \text{ N or } 45.3 \text{ kN}$$

$$\text{Acting at an angle, } \theta = \tan^{-1}\left(\frac{R_y}{R_x}\right)$$
$$= 22° \text{ to the } x\text{-direction}$$

Example 3.8

Derive an expression for the normal force on a plate inclined at $\theta°$ to the jet.

A 150 mm $\times$ 150 mm square metal plate, 10 mm thick, is hinged about a horizontal edge. If a 10 mm diameter horizontal jet of water impinging 50 mm below the hinge keeps the plate inclined at 30° to the vertical, find the velocity of the jet. Take the specific weight of the metal as 75 kN/m³.

Solution:

Referring to Figure 3.28a, force in the normal direction to the plate,

$$F = (\text{mass} \times \text{change in velocity normal to the plate}) \text{ of jet}$$
$$F = \rho a V \, [V \cos(90° - \theta) - 0]$$
$$= \rho a V^2 \sin \theta \text{ N}$$

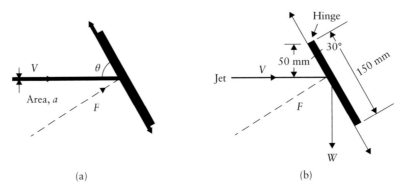

Figure 3.28 Forces on flat plates: (a) inclined plate and (b) hinged plate.

Now, referring to Figure 3.28b,

$$F = \rho a V^2 \sin 60°$$
$$= 10^3 \times \frac{\pi}{4}(0.01)^2 \times \sin 60° \times V^2$$
$$= 6.8 \times 10^{-2} V^2 \text{ N}$$

$$\text{Weight of the plate, } W = 0.150 \times 0.150 \times 0.010 \times 75\,000$$
$$= 16.87 \text{ N}$$

Taking moments about the hinge,

$$F \times 50 \sec 30° = W \times 75 \sin 30°$$
$$\text{or } 6.8 \times 10^{-2}V^2 \times 50 \sec 30° = 16.87 \times 75 \sin 30°$$
$$\Rightarrow V = 12.7 \text{ m/s}$$

Example 3.9

Estimate the energy (head) loss along a short length of pipe suddenly enlarging from a diameter of 350 mm to 700 mm and conveying 300 L/s of water. If the pressure at the entrance of the flow is 10^5 N/m², find the pressure at the exit of the pipe. What would be the energy loss if the flow were to be reversed with a contraction coefficient of 0.62?

Solution:

Case of sudden expansion:

$$Q = 0.3 \text{ m}^3/\text{s} = a_1v_1 = a_2v_2$$
$$\Rightarrow v_1 = 3.12 \text{ m/s} \quad \text{and} \quad v_2 = 0.78 \text{ m/s}$$

and hence

$$h_L = \frac{(3.12 - 0.78)^2}{2g}$$
$$= 0.28 \text{ m of water}$$
$$\text{Pressure, } p_1 = 10^5 \text{ N/m}^2$$

By energy equation,

$$\frac{p_1}{\rho g} + \frac{v_1^2}{2g} = \frac{p_2}{\rho g} + \frac{v_2^2}{2g} + \frac{(v_1 - v_2)^2}{2g}$$

$$10.2 + 0.5 = \frac{p_2}{\rho g} + 0.03 + 0.28$$

$$\Rightarrow \frac{p_2}{\rho g} = 10.39 \text{ m} \quad \text{or} \quad p_2 = 1.02 \times 10^5 \text{ N/m}^2$$

Case of sudden contraction:

$$h_L = \frac{[(1/C_c) - 1]^2 v^2}{2g} \quad \text{(where } v \text{ is the velocity in the smaller pipe)}$$

$$= \frac{[(1/0.62) - 1]^2 (3.12)^2}{2g}$$

$$= 0.186 \text{ m of water}$$

Example 3.10

A venturi meter is introduced in a 300 mm diameter horizontal pipeline carrying water under a pressure of 150 kN/m². The throat diameter of the meter is 100 mm and the pressure at the throat is 400 mm of mercury below atmosphere. If 3% of the differential pressure is lost between inlet and throat, determine the flow rate in the pipeline.

Solution:

Bernoulli's equation between inlet and throat:

$$\frac{p_1}{\rho g} + \frac{v_1^2}{2g} = \frac{p_2}{\rho g} + \frac{v_2^2}{2g} + 0.03 \left(\frac{p_1}{\rho g} - \frac{p_2}{\rho g} \right)$$

$$\Rightarrow \frac{0.97(p_1 - p_2)}{\rho g} = \frac{(v_2^2 - v_1^2)}{2g}$$

$$p_1 = 150 \times 10^3 \text{ N/m}^2 = 15.29 \text{ m of water}$$

$$p_2 = -400 \text{ mm of mercury} = -0.4 \times 13.6 \text{ m of water}$$

$$= -5.44 \text{ m of water}$$

$$\Rightarrow \frac{p_1 - p_2}{\rho g} = 15.29 - (-5.44)$$

$$= 20.73 \text{ m}$$

and hence

$$\frac{v_2^2 - v_1^2}{2g} = 0.97 \times 20.73$$

$$= 20.11 \text{ m} \tag{i}$$

From the continuity equation,

$$a_1 v_1 = a_2 v_2$$

$$\Rightarrow v_1 = \left(\frac{a_2}{a_1} \right) v_2 = \left(\frac{d_2}{d_1} \right)^2 v_2$$

$$= \left(\frac{10}{30} \right)^2 v_2$$

$$= \left(\frac{1}{9} \right) v_2 \tag{ii}$$

From Equations (i) and (ii),

$$\frac{v_2^2\,(1 - 1/81)}{2g} = 20.11$$

or

$$v_2 = \sqrt{\frac{2g \times 20.11}{1 - 1/81}} = 19.89 \text{ m/s}$$

$$\text{Flow rate, } Q = a_2 v_2 = \frac{1}{4}\pi(0.1)^2 \times 19.98$$

$$= 0.157 \text{ m}^3/\text{s or } 157 \text{ L/s}$$

Example 3.11

A 50 mm × 25 mm venturi meter with a coefficient of discharge of 0.98 is to be replaced by an orifice meter having a coefficient of discharge of 0.6. If both meters are to give the same differential mercury manometer reading for a discharge of 10 L/s, determine the diameter of the orifice.

Solution:

Discharge through venturi meter = discharge through orifice meter

$k = a_1/a_2 = (50/25)^2 = 4$ for the venturi meter and k_o for the orifice meter $= (50/d_o)^2$, where d_o is the diameter of orifice.

$$Q = 0.01 = 0.98 \times \frac{1}{4}\pi(0.05)^2 \sqrt{\frac{2gh}{4^2 - 1}}$$

$$= 0.6 \times \frac{1}{4}\pi(0.05)^2 \sqrt{\frac{2gh}{k_o^2 - 1}}$$

or

$$\sqrt{k_o^2 - 1} = \frac{0.6\sqrt{15}}{0.98}$$

$$\Rightarrow k_o = \left(\frac{50}{d_o}\right)^2 = 2.57 \quad \text{or} \quad d_o = \frac{50}{\sqrt{2.57}} = 31.2 \text{ mm}$$

Example 3.12

A Pitot tube was used to measure the quantity of water flowing in a pipe of 300 mm diameter. The stagnation pressure at the centre line of the pipe is 250 mm of water more than the static pressure. If the mean velocity is 0.78 times the centre line velocity and the coefficient of the Pitot tube is 0.98, find the rate of flow in litres per second.

Solution:

The centre line velocity in the pipe, $v = K\sqrt{2gh}$

$$= 0.98 \sqrt{2 \times 9.81 \times 0.25}$$

$$= 2.17 \text{ m/s}$$

Mean velocity of flow $= 0.78 \times 2.17$

$$= 1.693 \text{ m/s}$$

Hence

discharge, $Q = av$

$$= \frac{1}{4}\pi(0.3)^2 \times 1.693$$

$$= 0.12 \text{ m}^3/\text{s or } 120 \text{ L/s}$$

Example 3.13

A large rectangular orifice 0.40 m wide and 0.60 m deep, placed with the upper edge in a horizontal position 0.90 m vertically below the water surface in a vertical side wall of a large tank, is discharging to atmosphere. Calculate the rate of flow through the orifice if its discharge coefficient is 0.65.

Solution:

The discharge rate when $b = 0.4$ m, $H_1 = 0.90 + 0.60 = 1.5$ m, $H_2 = 0.90$ m and $C_d = 0.65$ from Equation 3.27,

$$Q = \frac{2}{3} \times 0.65 \times \sqrt{2g} \times 0.40 \times \left[(1.5)^{3/2} - (0.9)^{3/2}\right]$$

$$= 0.755 \text{ m}^3/\text{s}$$

Example 3.14

A vertical circular tank 1.25 m diameter is fitted with a sharp-edged circular orifice 50 mm in diameter in its base. When the flow of water into the tank was shut off, the time taken to lower the head from 2 to 0.75 m was 253 s. Determine the rate of flow (in litres per second) through the orifice under a steady head of 1.5 m.

Solution:

$T = 253$ s, $H_1 = 2$ m, $H_2 = 0.75$ m, $a = \frac{1}{4}\pi(0.05)^2 = 1.96 \times 10^{-3}$ m^2 and $A = \frac{1}{4}\pi(1.25)^2 = 1.228$ m^2.

From Equation 3.32, $C_d = 0.61$.

Hence steady discharge under a head of 1.5 m

$$= C_d a \sqrt{2gH}$$
$$= 0.61 \times 1.96 \times 10^{-3} \times \sqrt{2g} \times (1.5)^{1/2}$$
$$= 0.0065 \text{ m}^3/\text{s or } 6.5 \text{ L/s}$$

Example 3.15

Determine the discharge over a sharp-crested weir 4.5 m long with no lateral contractions, the measured head over the crest being 0.45 m. The width of the approach channel is 4.5 m and the sill height of the weir is 1 m.

Solution:

Equation 3.35 is rewritten as

$$Q = 1.84b \left[\left(H + \frac{V_a^2}{2g} \right)^{3/2} - \left(\frac{V_a^2}{2g} \right)^{3/2} \right] \tag{i}$$

for a weir with no lateral contractions (suppressed weir) and $\alpha = 1$.
 Equation (i) reduces to

$$Q = 1.84b(H)^{3/2} \tag{ii}$$

neglecting velocity of approach as a first approximation.
 From Equation (ii),

$$Q = 1.84 \times 4.5 \times (0.45)^{3/2}$$
$$= 2.5 \text{ m}^3/\text{s}$$

$$\text{Now velocity of approach, } V_a = \frac{2.5}{4.5 \, (1 + 0.45)}$$
$$= 0.383 \text{ m/s}$$
$$\text{and } \frac{V_a^2}{2g} = 7.48 \times 10^{-3} \text{ m}$$

$$\Rightarrow Q = 1.84 \times 4.5 \left[(0.45 + 0.00748)^{3/2} - (0.00748)^{3/2} \right]$$
$$= 2.556 \text{ m}^3/\text{s}$$

Example 3.16

The discharge over a triangular notch can be written as

$$Q = \left(\frac{8}{15} \right) C_d \sqrt{2g} \, \tan \frac{\theta}{2} H^{5/2}$$

 If an error of 1% in measuring H is introduced, determine the corresponding error in the computed discharge.

A right-angled triangular notch is used for gauging the flow of a laboratory flume. If the coefficient of discharge of the notch is 0.593 and an error of 2 mm is suspected in observing the head, find the percentage error in computing an estimated discharge of 20 L/s.

Solution:

We can write

$$Q = KH^{5/2}$$

$$\Rightarrow dQ = \left(\frac{5}{2}\right) KH^{3/2}\, dH$$

and

$$\frac{dQ}{Q} = \frac{(5/2)\, KH^{3/2}\, dH}{KH^{5/2}}$$

$$= \left(\frac{5}{2}\right) \frac{dH}{H} \qquad (i)$$

If dH/H is 1%, the error in the discharge $dQ/Q = 2.5\%$ (from Equation (i)).

$$Q = 0.02 = \frac{8}{15} \times 0.593 \times \sqrt{2g} \times 1 \times H^{5/2}$$

or

$$H^{5/2} = 1.4275 \times 10^{-2}$$

or

$$H = 0.183 \text{ m or } 183 \text{ mm}$$

and

$$\frac{dQ}{Q} = 2.5 \times \frac{dH}{H}$$

$$= \frac{2.5 \times 2}{183}$$

$$= 2.73\%$$

Example 3.17

If the velocity distribution of a turbulent flow in an open channel is given by a power law

$$\frac{v}{v_{\max}} = \left(\frac{y}{y_0}\right)^{1/7}$$

where v is the velocity at a distance y from the bed and $v_{\max}$ is the maximum velocity in the channel with a flow depth of y_0, determine the average velocity and the energy (α) and momentum (β) correction factors. Assume the flow to be two-dimensional.

Solution:

If the mean velocity of flow is V, the discharge per unit width of the channel is

$$y_0 V = q = \int_0^{y_0} v\, dy$$

which gives $q = (7/8)v_{max}y_0$.
Therefore

$$V = \frac{q}{y_0} = \left(\frac{7}{8}\right) v_{max}$$

The kinetic energy correction factor α, given by Equation 3.11, can be written as

$$\alpha = \frac{1}{A} \int_A \left(\frac{v}{V}\right)^3 dA = \frac{1}{y_0 V^3} \int_0^{y_0} \left[v_{max}\left(\frac{y}{y_0}\right)^{1/7}\right]^3 dy$$

Replacing v_{max} [$=(8/7)V$] and integrating, we obtain $\alpha = 1.045$. The momentum correction factor β given by Equation 3.14 can be written as

$$\beta = \frac{1}{A} \int_A \left(\frac{v}{V}\right)^2 dA = \frac{1}{y_0 V^2} \int_0^{y_0} \left[v_{max}\left(\frac{y}{y_0}\right)^{1/7}\right]^2 dy$$

Again replacing v_{max} and integrating, we obtain $\beta = 1.016$.
    Note: The energy and momentum correction factors α and β for open channels may be computed by the equations

$$\alpha = 1 + 3\varepsilon^2 - 2\varepsilon^3 \quad \text{and} \quad \beta = 1 + \varepsilon^2 \quad \text{where } \varepsilon = \frac{v_{max}}{V} - 1$$

If the velocity distributions are not described by any equation and if the measured data are available, α and β values may be computed by graphical methods; plots of $\int v\, dy$, $\int v^3\, dy$ and $\int v^2\, dy$ will help to give V, α and β, respectively.

Example 3.18

An ogee spillway of large height is to be designed to evacuate a flood discharge of 200 m³/s under a head of 2 m. The spillway is spanned by piers to support a bridge deck above. The clear span between piers is limited to 6 m. Determine the number of spans required in order to pass the flood discharge with the head not exceeding 2 m. Assume the pier contraction coefficient $k_p = 0.01$ and the abutment contraction coefficient $k_a = 0.10$.

Solution:

The flow between the piers and abutments is contracted, thus reducing the spillway width for the flow to B_e. Each pier has two end contractions and one abutment, and hence the effective width is given by

$$B_e = B - 2(nk_p + k_a)H_e$$

n being the number of piers.

The pier contraction coefficient depends on the shape of its nose ($k_p = 0$ for a pointed nose and $k_p = 0.02$ for a square nose), whereas the abutment contraction coefficient may be as high as 0.2 for a square abutment, reducing to zero for a rounded abutment. If the velocity of approach V_a is not negligible, a trial-and-error procedure is to be used for the discharge computations; for large heights (P), $V_a \approx 0$ and hence $H_e \approx H$. Here, assuming $V_a \approx 0$, we can write Equation 3.43 as

$$Q = 200 = \frac{2}{3}C_{d0} \sqrt{(2g)} \, [6(n + 1) - 2(n \times 0.01 + 0.10)2.0]2^{3/2}$$

which gives $n = 4.36$ with $C_{d0} = 0.75$ ($P/H > 3$). Therefore round up to $n = 5$, and so provide five piers. Thus the clear span of the spillway (for flow) = 36 m. From the discharge equation we can now compute the corresponding head for this flow. In fact the spillway is capable of discharging a larger flood flow at the specified design head of 2 m. A stage (head)–discharge relationship can be established by using appropriate discharge coefficients (refer Figure 3.20).

References and recommended reading

Baker, R. C. (2000) *Flow Measurement Handbook: Industrial Design, Operating Principles, Performance and Applications*, Cambridge University Press, Cambridge, UK.

Baker, R. C., Moore, P. I. and Thomas, A. (2006) Electronic verification of flowmeters in the water industry. *Proceedings of the Institution of Civil Engineers Water Management*, 159, WM3 December, pp. 245–251.

British Standards Institution (1981) *Methods of Measurement of Liquid Flow in Open Channels – Thin Plate Weirs*, BS 3680, Part 4A, British Standards Institution, London.

Herschy, R. W. (2009) *Streamflow Measurement*, 3rd edn, Taylor & Francis, Abingdon, UK.

Herschy, R. W. (ed) (1999) *Hydrometry – Principles and Practices*, 2nd edn, Wiley, Chichester.

International Organization for Standardization (2003) *Measurement of Fluid Flow by Means of Pressure Differential Devices Inserted in Circular Cross-section Conduits Running Full*, ISO 5167, International Organization for Standardization, Geneva.

Novak, P., Moffat, A. I. B., Nalluri, C. and Narayanan, R. (2007) *Hydraulic Structures*, 4th edn, Taylor & Francis, Abingdon, UK.

Problems

1. A tapered nozzle is so shaped that the velocity of flow along its axis changes from 1.5 to 15 m/s in a length of 1.35 m. Determine the magnitude of the convective acceleration at the beginning and end of this length.

2. The spillway section of a dam ends in a curved shape (known as the bucket) deflecting water away from the dam. The radius of this bucket is 5 m and when the spillway is discharging 5 cumecs of water per metre length of crest, the average thickness of the sheet of water over the bucket is 0.5 m. Compare the resulting normal or centripetal acceleration with the acceleration due to gravity.

3. The velocity distribution of a real fluid flow in a pipe is given by the equation $v = V_{max}(1 - r^2/R^2)$, where V_{max} is the velocity at the centre of the pipe, R is the pipe radius, and v is the velocity at radius r from the centre of the pipe. Show that the kinetic energy correction factor for this flow is 2.

Chapter 3

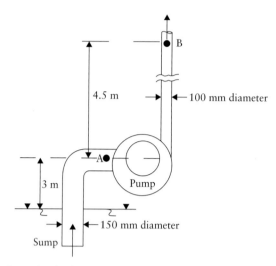

Figure 3.29   Flow through a hydraulic pump.

4. A pipe carrying oil of relative density 0.8 changes in diameter from 150 mm to 450 mm, the pressures at these sections being 90 kN/m² and 60 kN/m², respectively. The smaller section is 4 m below the other, and if the discharge is 145 L/s, determine the energy loss and the direction of flow.

5. Water is pumped from a sump (see Figure 3.29) to a higher elevation by installing a hydraulic pump with the data:
    Discharge of water = 6.9 m³/min
    Diameter of suction pipe = 150 mm
    Diameter of delivery pipe = 100 mm
    Energy supplied by the pump = 25 kW
    (i) Determine the pressure (in kilonewtons per square metre) at points A and B neglecting all losses.
    (ii) If the actual pressure at B is 25 kN/m², determine the total energy loss (in kilowatts) between the sump and the point B.

6. A fire-brigade man intends to reach a window 10 m above the ground with a fire stream from a nozzle of 40 mm diameter held at a height of 1.5 m above the ground. If the jet is discharging 1000 L/min, determine the maximum distance from the building at which the fireman can stand to hit the target. Hence find the angle of inclination with which the jet issues from the nozzle.

7. A 600 mm diameter pipeline conveying oil of relative density 0.85 at the rate of 2 cumecs has a 90° bend in a horizontal plane. The pressure at the inlet to the bend is 2 m of oil. Find the magnitude and direction of the force exerted by the oil on the bend. If the ends of the bend are anchored by tie-rods at right angles to the pipeline, determine tension in each tie-rod.

8. The diameter of pipe bend is 300 mm at inlet and 150 mm at outlet, and the flow is turned through 120° in a vertical plane. The axis at inlet is horizontal and the centre of the outlet section is 1.5 m below the centre of the inlet section. The total volume of

fluid contained in the bend is 8.5×10^{-2} m^3. Neglecting friction, calculate the magnitude and direction of the force exerted on the bend by water flowing through it at a rate of 0.225 m^3/s when the inlet pressure is 140 kN/m^2.

9. A sluice gate is used to control the flow of water in a horizontal rectangular channel, 6 m wide. The gate is lowered so that the stream flowing under it has a depth of 800 mm and a velocity of 12 m/s. The depth upstream of the sluice gate is 7 m. Determine the force exerted by the water on the sluice gate, assuming uniform velocity distribution in the channel and neglecting frictional losses.

10. A 50 mm diameter jet of water strikes a curved vane at rest with a velocity of 30 m/s and is deflected through 45° from its original direction. Neglecting friction, compute the resultant force on the vane in magnitude and direction.

11. A horizontal rectangular outlet downstream of a dam, 2.5 m high and 1.5 m wide, discharges 70 m^3/s of water on to a concave concrete floor of 12 m radius and 6 m length, deflecting the water away from the outlet to dissipate energy. Calculate the resultant thrust the fluid exerts on the floor.

12. A venturi meter is to be fitted to a pipe of 250 mm diameter where the pressure head is 6 m of water and the maximum flow is 9 m^3/min. Find the smallest diameter of the throat to ensure that the pressure head does not become negative.

13. (a) Determine the diameter of throat of a venturi meter to be introduced in a horizontal section of a 100 mm diameter main so that deflection of a differential mercury manometer connected between the inlet and throat is 600 mm when the discharge is 20 L/s of water. The discharge coefficient of the meter is 0.95.
    (b) What difference will it make to the manometer reading if the meter is introduced in a vertical length of the pipeline, with water flowing upwards, and the distance from inlet to throat of the meter is 200 mm?

14. A Pitot tube placed in front of a submarine moving horizontally in sea water 16 m below the water surface is connected to the two limbs of a U-tube mercury manometer. Find the speed of the submarine for a manometer deflection of 200 mm. Relative densities of mercury and sea water are 13.6 and 1.026, respectively.

15. In an experiment to determine the hydraulic coefficients of a 25 mm diameter sharp-edged orifice, it was found that the jet issuing horizontally under a head of 1 m travelled a horizontal distance of 1.5 m from the vena contracta in the course of a vertical drop of 612 mm from the same point. Furthermore, the impact force of the jet on a flat plate held normal to it at the vena contracta was measured as 5.5 N. Determine the three coefficients of the orifice, assuming an impact coefficient of unity.

16. A swimming pool with vertical sides is 25 m long and 10 m wide. Water at the deep end is 2.5 m and shallow end 1 m. If there are two outlets each 500 mm diameter, one at each of the deep and shallow ends, find the time taken to empty the pool. Assume the discharge coefficients for both the outlets as 0.8.

17. A convergent–divergent nozzle is fitted to the vertical side of a tank containing water to a height of 2 m above the centre line of the nozzle. Find the throat and exit diameters

of the nozzle if it discharges 7 L/s of water into the atmosphere, assuming that (i) the pressure head in the throat is 2.5 m of water absolute, (ii) atmospheric pressure is 10 m of water, (iii) there is no hydraulic loss in the convergent part of the nozzle, and (iv) the head loss in the divergent part is 1/5 of exit velocity head.

18.   When water flows through a right-angled V-notch, show that the discharge is given by $Q = KH^{5/2}$, in which k is a dimensional constant and H is the height of water surface above the bottom of the notch. (i) What are the dimensions of K if H is in metres and Q in metres cubed per second? (ii) Determine the head causing flow when the discharge through this notch is 1.42 L/s. Take $C_d = 0.62$. (iii) Find the accuracy with which the head in (ii) must be measured if the error in the estimation of discharge is not to exceed 1.5%.

19.   (a) What is meant by a 'suppressed' weir? Explain the precautions that you would take in using such a weir as a discharge-measuring structure.
       (b) A suppressed weir with two ventilating pipes is installed in a laboratory flume with the following data:
       Width of flume = 1000 mm
       Height of weir, P = 300 mm
       Diameter of ventilating pipes = 30 mm
       Pressure difference between the two sides of the nappe = 1 N/m²
       Head over sill, h = 150 mm
       Density of air = 1.25 kg/m³
       Coefficient of discharge, $C_d = 0.611 + 0.075(h/P)$
       Assuming a smooth entrance to the ventilating pipes and neglecting the velocity of approach, find the air demand in terms of percentage of water discharge.

20.   State the advantages of a triangular weir over a rectangular one, for measuring discharges.
       The following observations of head and the corresponding discharge were made in a laboratory to calibrate a 90° V-notch.

Head (mm)	50	75	100	125	150
Discharge (L/s)	0.81	2.24	4.76	8.03	12.66

Determine K and n in the discharge equation, $Q = KH^n$ (H in metres and Q in metres cubed per second) and hence find the value of the coefficient of discharge.

21.   A reservoir has an area of 8.5 ha and is provided with a weir 4.5 m long ($C_d = 0.6$). Find how long it will take for the water level above the sill to fall from 0.60 to 0.30 m.

Chapter 4
Flow of Incompressible Fluids in Pipelines

4.1 Resistance in circular pipelines flowing full

A fluid moving through a pipeline is subjected to energy losses from various sources. A continuous resistance is exerted by the pipe walls due to the formation of a boundary layer in which the velocity decreases from the centre of the pipe to zero at the boundary. In steady flow in a uniform pipeline the boundary shear stress τ_0 is constant along the pipe, since the boundary layer is of constant thickness, and this resistance results in a uniform rate of total energy or head degradation along the pipeline. The total head loss along a specified length of pipeline is commonly referred to as the 'head loss due to friction' and denoted by h_f. The rate of energy loss or energy gradient $S_f = h_f/L$.

The hydraulic grade line shows the elevation of the pressure head along the pipe. In a uniform pipe the velocity head $\alpha V^2/2g$ is constant and the energy grade line is parallel to the hydraulic grade line (Figure 4.1). Applying Bernoulli's equation to sections 1 and 2,

$$z_1 + \frac{p_1}{\rho g} + \frac{\alpha V_1^2}{2g} = z_2 + \frac{p_2}{\rho g} + \frac{\alpha V_2^2}{2g} + h_f$$

and since $V_1 = V_2$,

$$z_1 + \frac{p_1}{\rho g} = z_2 + \frac{p_2}{\rho g} + h_f \qquad [4.1]$$

In steady uniform flow the motivating and drag forces are exactly balanced. Equating between sections 1 and 2,

$$(p_1 - p_2)A + \rho g A L \sin \theta = \tau_0 P L \qquad [4.2]$$

where A is the area of cross section, P the wetted perimeter and τ_0 the boundary shear stress.

Nalluri & Featherstone's Civil Engineering Hydraulics: Essential Theory with Worked Examples,
Sixth Edition. Martin Marriott.
© 2016 John Wiley & Sons, Ltd. Published 2016 by John Wiley & Sons, Ltd.
Companion Website: www.wiley.com/go/Marriott

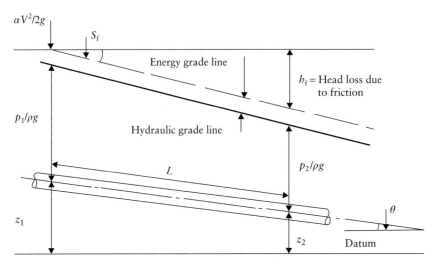

Figure 4.1　Pressure head and energy gradients in full, uniform pipe flow.

Rearranging Equation 4.2 and noting that $L \sin \theta = z_1 - z_2$,

$$\frac{p_1 - p_2}{\rho g} + z_1 - z_2 = \frac{\tau_0 PL}{\rho g A}$$

and from Equation 4.1,

$$h_f = \frac{p_1 - p_2}{\rho g} + z_1 - z_2$$

$$\text{whence } h_f = \frac{\tau_0 PL}{\rho g A}$$

$$\text{or } \tau_0 = \rho g R \frac{h_f}{L} = \rho g R S_f \qquad [4.3]$$

where R (hydraulic radius) $= A/P$ ($= D/4$ for a circular pipe of diameter D).

The head loss due to friction in steady uniform flow is given by the Darcy–Weisbach equation:

$$h_f = \frac{\lambda L V^2}{2gD} \qquad [4.4]$$

where λ is a non-dimensional coefficient which, for turbulent flow, can be shown to be a function of k/D, the relative roughness, and the Reynolds number, $Re = VD/\nu$. The effective roughness size of the pipe wall is denoted by k. For laminar flow ($Re \leq 2000$), h_f can be obtained theoretically in the form of the Hagen–Poiseuille equation:

$$h_f = \frac{32 \mu L V}{\rho g D^2} \qquad [4.5]$$

Thus in Equation 4.4, $\lambda = 64/Re$ for laminar flow.

In the case of turbulent flow, experimental work on smooth pipes by Blasius yielded the relationship

$$\lambda = \frac{0.3164}{Re^{1/4}} \qquad [4.6]$$

Later work by Prandtl and Nikuradse on smooth and artificially roughened pipes revealed three zones of turbulent flow:

(i) a smooth turbulent zone in which the friction factor λ is a function of the Reynolds number only and expressed by

$$\frac{1}{\sqrt{\lambda}} = 2\log\frac{Re\sqrt{\lambda}}{2.51} \qquad [4.7]$$

(ii) a transitional turbulent zone in which λ is a function of both k/D and Re
(iii) a rough turbulent zone in which λ is a function of k/D only and expressed by

$$\frac{1}{\sqrt{\lambda}} = 2\log\frac{3.7D}{k} \qquad [4.8]$$

Equations 4.7 and 4.8 are known as the Kármán–Prandtl equations. Colebrook (1939) and White found that the function resulting from the addition of the rough and smooth Equations 4.7 and 4.8 in the form

$$\frac{1}{\sqrt{\lambda}} = -2\log\left(\frac{k}{3.7D} + \frac{2.51}{Re\sqrt{\lambda}}\right) \qquad [4.9]$$

fitted the observed data on commercial pipes over the three zones of turbulent flow. Further background notes on the development of the form of the Kármán–Prandtl equations are given in Chapter 7. The Colebrook–White equation (Equation 4.9) was first plotted in the form of a λ–Re diagram by Moody (1944) (Figure 4.2) and hence is generally referred to as the 'Moody diagram'. This was presented originally with a logarithmic scale of λ. Figure 4.2 has been drawn, from computation of Equation 4.9, with an arithmetic scale of λ for more accurate interpolation.

Combining the Darcy–Weisbach and Colebrook–White equations, Equation 4.4 and Equation 4.9 yield an explicit expression for V:

$$V = -2\sqrt{2gDS_f}\log\left(\frac{k}{3.7D} + \frac{2.51\nu}{D\sqrt{2gDS_f}}\right) \qquad [4.10]$$

This equation forms the basis of the design charts produced by HR Wallingford (1990). A typical chart is reproduced as Figure 4.3.

Due to the implicit form of the Colebrook–White equation, a number of approximations in explicit form in λ have been proposed.

Moody produced the following formulation:

$$\lambda = 0.0055\left[1 + \left(20\,000\frac{k}{D} + \frac{10^6}{Re}\right)^{1/3}\right] \qquad [4.11]$$

This is claimed to give values of λ within $\pm5\%$ for Reynolds numbers between 4×10^3 and 1×10^7 and for k/D up to 0.01.

Barr (1975) proposed the following form based partly on an approximation to the logarithmic smooth turbulent element in the Colebrook–White function by White:

$$\frac{1}{\sqrt{\lambda}} = -2\log\left(\frac{k}{3.7D} + \frac{5.1286}{Re^{0.89}}\right) \qquad [4.12]$$

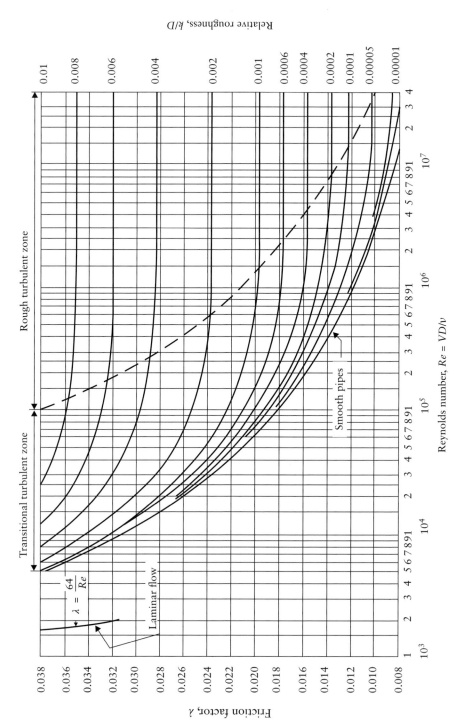

Figure 4.2   The Moody diagram.

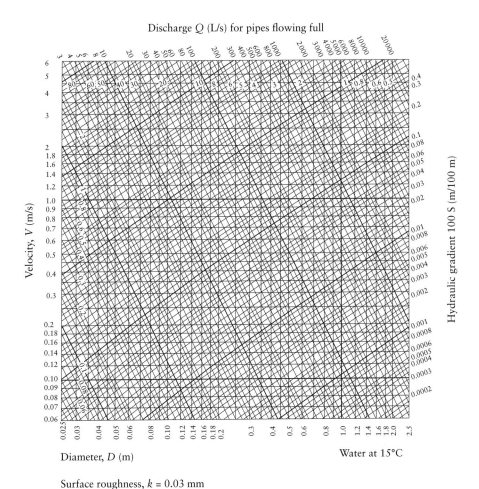

Figure 4.3 Extract from *Charts for the Hydraulic Design of Channels and Pipes* (1990). Reproduced by permission of HR Wallingford.

Further development by Barr (1981) led to an even closer approximation which was expressed as

$$\frac{1}{\sqrt{\lambda}} = -2 \log \left[\frac{k}{3.7D} + \frac{5.02 \log \left(Re/4.518 \log \left(Re/7 \right) \right)}{Re(1 + Re^{0.52}/29 \, (D/k)^{0.7})} \right] \qquad [4.13]$$

Typical percentage errors in λ given by Equation 4.13 compared with the solution of the Colebrook–White function are as follows:

k/D	$Re = 3 \times 10^4$	$Re = 3 \times 10^5$	$Re = 3 \times 10^6$
10^{-3}	−0.12	0.00	−0.07
10^{-4}	−0.16	−0.07	+0.03

The values given by Equation 4.13 should be sufficiently accurate for most purposes, but substitution of these values once into the right-hand side of the Colebrook–White function produces λ values with a maximum discrepancy of +0.04%.

4.2 Resistance to flow in non-circular sections

In order to use the same form of resistance equations such as the Darcy (Equation 4.4) and Colebrook–White (Equation 4.9) equations, it is convenient to treat the non-circular section as an equivalent hypothetical circular section yielding the same hydraulic gradient at the same discharge.

The 'transformation' is achieved by expressing the diameter D in terms of the hydraulic radius $R = A/P$, and since for circular pipes $R = D/4$, Equations 4.4 and 4.9 become

$$h_f = \frac{\lambda L V^2}{8gR} \qquad\qquad [4.14]$$

$$\text{and } \frac{1}{\sqrt{\lambda}} = -2 \log\left(\frac{k}{14.8R} + \frac{2.51v}{4VR\sqrt{\lambda}}\right) \qquad\qquad [4.15]$$

Because in the actual non-circular section the boundary shear stress is not constant around the wetted perimeter whereas it is in the equivalent circular section, the 'transformation' is not exact but experiments have shown that the error is small.

It is important to note that the equivalent circular pipe does not have the same area as the actual conduit; their hydraulic radii are equal. Tables by HR Wallingford and Barr (2006) cover a number of non-circular cross sections, and the hydraulic performance of ovoid pipes as used in sewerage is discussed by Marriott (1996).

4.3 Local losses

In addition to the spatially continuous head loss due to friction, local head losses occur at changes of cross section, at valves and at bends. These local losses are sometimes referred to as 'minor' losses since in long pipelines their effect may be small in relation to the friction loss. However, the head loss at a control valve has a primary effect in regulating the discharge in a pipeline.

4.3.1 Typical values for circular pipelines

$$\text{Head loss at abrupt contraction} = K_c \frac{V_2^2}{2g}$$

where V_2 is the mean velocity in a downstream section of diameter D_2, and D_1 is the upstream diameter.

D_2/D_1	0	0.2	0.4	0.6	0.8	1.0
K_c	0.5	0.45	0.38	0.28	0.14	0

Note that the value of $K_c = 0.5$ relates to the abrupt entry from a tank into a circular pipeline.

$$\text{Head loss at abrupt enlargement} = \frac{V_2^2}{2g}\left(\frac{A_2}{A_1} - 1\right)^2$$

$$\text{Head loss at } 90° \text{ elbow} = 1.0\frac{V^2}{2g}$$

$$\text{Head loss at } 90° \text{ smooth bend} = \frac{V^2}{2g}$$

$$\text{Head loss at a valve} = K_v\frac{V^2}{2g}$$

where K_v depends on the type of valve and percentage of closure.

The following examples demonstrate the application of the above theory and equations to the analysis and design of pipelines.

Worked examples

Example 4.1

Crude oil of density 925 kg/m³ and absolute viscosity 0.065 N s/m² at 20°C is pumped through a horizontal pipeline 100 mm in diameter, at a rate of 10 L/s. Determine the head loss in each kilometre of pipeline and the shear stress at the pipe wall. What power is supplied by the pumps per kilometre length?

Solution:

Determine if the flow is laminar.

$$\text{Area of pipe} = 0.00786 \text{ m}^2$$
$$\text{Mean velocity of oil} = 1.27 \text{ m/s}$$

$$\text{Reynolds number, } Re = \frac{VD}{v} = \frac{1.27 \times 0.1 \times 925}{0.065}$$
$$= 1807$$

Thus the flow may therefore be assumed to be laminar.
    Hence

$$\lambda = \frac{64}{Re} = 0.0354$$

$$\text{Friction head loss per kilometre} = \frac{\lambda L V^2}{2gD}$$
$$= \frac{0.0354 \times 1000 \times 1.27^2}{19.62 \times 0.1}$$
$$= 29.2$$

$$\text{Boundary shear stress, } \tau_0 = \rho g \, R \, S_f$$
$$= 925 \times 9.81 \times \frac{0.1}{4} \times \frac{29.2}{1000}$$
$$= 6.62 \text{ N/m}^2$$

Chapter 4

$$\text{Power consumed} = \rho g Q h_f$$
$$= 925 \times 9.81 \times 0.01 \times 29.2 \text{ W}$$
$$= 2.65 \text{ kW/km}$$

Note that if the outlet end of the pipeline were elevated above the head of oil at inlet, the pumps would have to deliver more power to overcome the static lift. This is dealt with more fully in Chapter 6, which covers pumps.

Example 4.2

A uniform pipeline, 5000 m long, 200 mm in diameter and roughness size 0.03 mm, conveys water at 15°C between two reservoirs, the difference in water level between which is maintained constant at 50 m. In addition to the entry loss of $0.5V^2/2g$, a valve produces a head loss of $10V^2/2g$. Take $\alpha = 1.0$. Determine the steady discharge between the reservoirs using

(a) the Colebrook–White equation
(b) the Moody diagram
(c) the HR Wallingford charts
(d) an explicit function for λ.

Solution:

Apply Bernoulli's equation to A and B in Figure 4.4:

$$\underset{\substack{\text{Gross} \\ \text{head}}}{H} = \underset{\substack{\text{Entry} \\ \text{loss}}}{\frac{0.5V^2}{2g}} + \underset{\substack{\text{Velocity} \\ \text{head}}}{\frac{V^2}{2g}} + \underset{\substack{\text{Valve} \\ \text{head} \\ \text{loss}}}{\frac{10V^2}{2g}} + \underset{\substack{\text{Friction} \\ \text{head} \\ \text{loss}}}{\frac{\lambda L V^2}{2gD}} \qquad (i)$$

(a) Using the Colebrook–White equation:

$$\frac{1}{\sqrt{\lambda}} = -2\log\left(\frac{k}{3.7D} + \frac{2.51}{Re\sqrt{\lambda}}\right) \qquad (ii)$$

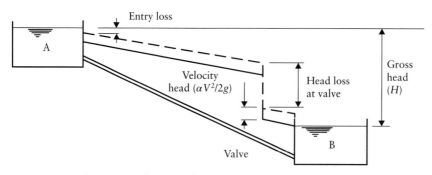

Figure 4.4   Energy losses in uniform pipeline.

The solution to the problem is obtained by solving Equations (i) and (ii) simultaneously. However, direct substitution of λ from Equations (i) into (ii) yields a complex implicit function in V which can only be evaluated by trial or graphical interpolation.

A simpler computational procedure is obtained if terms other than the friction head loss in Equation (i) are initially ignored; in other words, the gross head is assumed to be totally absorbed in overcoming friction. Then Equation 4.10 can be used to obtain an approximate value of V. That is,

$$V = -2 \sqrt{2gD\frac{h_f}{L}} \log \left(\frac{k}{3.7D} + \frac{2.51\nu}{D\sqrt{2gD\frac{h_f}{L}}} \right) \tag{iii}$$

Writing $h_f = H = 50\,\text{m}$, $h_f/L = 0.01$.

$$\text{Thus } V = -2\sqrt{19.62 \times 0.2 \times 0.01}$$

$$\times \log \left(\frac{0.03 \times 10^{-3}}{3.7 \times 0.2} + \frac{2.51 \times 1.13 \times 10^{-6}}{0.2\sqrt{19.62 \times 0.2 \times 0.01}} \right)$$

$$V = 1.564 \text{ m/s}$$

The terms other than friction loss in Equation (i) can now be evaluated:

$$h_m = 11.5 \frac{V^2}{2g} = 1.435 \text{ m}$$

where h_m denotes the sum of the minor head loss.

A better estimate of h_f is thus $h_f = 50 - 1.435 = 48.565$ m.
Thus from Equation (iii), $V = 1.544$ m/s.
Repeating until successive values of V are sufficiently close yields

$$V = 1.541 \text{ m/s} \quad \text{and} \quad Q = 48.41 \text{ L/s}$$
$$\text{with } h_f = 48.61\,\text{m} \quad \text{and} \quad h_m = 1.39\,\text{m}$$

Convergence is usually rapid since the friction loss usually predominates.

(b) The use of the Moody chart (Figure 4.2) involves the determination of the Darcy friction factor. In this case the minor losses need not be neglected initially. However, the solution is still iterative and an estimate of the mean velocity is needed.

$$\text{Estimate } V = 2.0 \text{ m/s}; \quad Re = \frac{2 \times 0.2}{1.13 \times 10^{-6}} = 3.54 \times 10^5$$

$$\text{Relative roughness, } \frac{k}{D} = 0.00015$$

From the Moody chart, $\lambda = 0.015$

Rearranging Equation (i) yields

$$V = \sqrt{\frac{2gH}{11.5 + (\lambda L/D)}} \tag{iv}$$

Chapter 4

And a better estimate of mean velocity is given by

$$V = \sqrt{\frac{19.62 \times 50}{11.5 + (0.015 \times 5000/0.2)}} = 1.593 \text{ m/s}$$

$$\text{Revised } Re = \frac{1.593 \times 0.2}{1.13 \times 10^{-6}} = 2.82 \times 10^5$$

Whence $\lambda = 0.016$, Equation (iv) yields

$$V = 1.54 \text{ m/s}$$

Further change in λ due to the small change in V will be undetectable in the Moody diagram.

Thus accept $V = 1.54$ m/s and $Q = 48.41$ L/s.

(c) The solution by the use of the Wallingford charts is basically the same as method (a) except that values of V are obtained directly from the chart instead of from Equation 4.10.

Making the initial assumption that $h_f = H$, the hydraulic gradient (m/100 m) 100 $S = 1.0$.

Entering Figure 4.3 with $D = 0.2$ m and 100 $S = 1.0$ yields $V = 1.55$ m/s.

The minor loss term $11.5 V^2/2g = 1.41$ m, and a better estimate of h_f is therefore 0.972 m/100 m.

Whence from Figure 4.3 $V = 1.5$ m/s and $Q = 47$ L/s. Note the loss of fine accuracy due to the graphical interpolation in Figure 4.3.

(d) Using Equation 4.12,

$$\frac{1}{\sqrt{\lambda}} = -2 \log \left(\frac{k}{3.7D} + \frac{5.1286}{Re^{0.89}} \right) \tag{v}$$

Assuming $V = 2.0$ m/s, $\lambda = 0.0156$ (from Equation (v)).
Using Equation (iv),

$$V = \sqrt{\frac{2gH}{11.5 + \lambda L/D}}$$

$$V = 1.563 \text{ m/s}; \quad \lambda = 0.0161 \quad \text{(from Equation (v))}$$

whence $V = 1.54$ m/s.

Thus accept $V = 1.54$ m/s, which is essentially identical with that obtained using the other methods.

Example 4.3 (Pipes in series)

Reservoir A delivers to Reservoir B through two uniform pipelines AJ and JB of diameters 300 mm and 200 mm, respectively. Just upstream of the change in section, which is assumed gradual, a controlled discharge of 30 L/s is taken off.

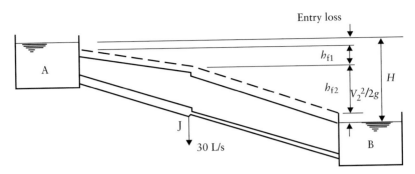

Figure 4.5 Energy losses in pipes in series with outflow.

Length of AJ = 3000 m; length of JB = 4000 m; effective roughness size of both pipes = 0.015 mm; gross head = 25.0 m. Determine the discharge to B, neglecting the loss at J (see Figure 4.5).

Solution:

Apply the energy equation between A and B.

$$H = \frac{0.5\,V_1^2}{2g} + h_{f1} + h_{f2} + \frac{V_2^2}{2g}$$

i.e. $$H = \frac{0.5\,V_1^2}{2g} + \frac{\lambda_1 L_1 V_1^2}{2gD_1} + \frac{\lambda_2 L_2 V_2^2}{2gD_2} + \frac{V_2^2}{2g} \qquad (i)$$

Since λ_1 and λ_2 are initially unknown, the simplest method of solution is to input a series of trial values of Q_1.

Since $Q_2 = Q_1 - 30$ (L/s), the corresponding values of Reynolds number can be calculated and hence λ_1 and λ_2 can be obtained from the Moody diagram (Figure 4.2). The total head loss H corresponding with each trial value of Q_1 is then evaluated directly from Equation (i). From a graph of H versus Q_1 the value of Q_1 corresponding with $H = 25$ m can be read off.

$$\frac{k_1}{D_2} = 0.00005; \quad \frac{k_2}{D_2} = 0.000075$$

Q_1 (L/s)	50	60	80
V_1 (m/s)	0.707	0.849	1.132
V_2 (m/s)	0.637	0.955	1.591
$Re_1\ (\times 10^5)$	1.88	2.25	3.00
$Re_2\ (\times 10^5)$	1.13	1.69	2.81
λ_1	0.0164	0.016	0.0156
λ_2	0.0184	0.018	0.016
H (m)	11.82	22.67	51.66

From Figure 4.6, $Q_1 = 62.5$ L/s, whence $Q_2 = 32.5$ L/s.

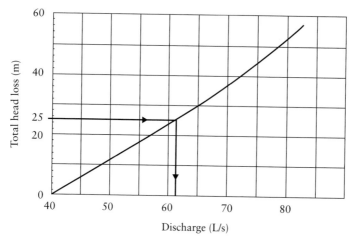

Figure 4.6   Head loss versus discharge.

Note: This problem could also be solved by the 'quantity balance' method of pipe network analysis (see Chapter 5).

Example 4.4 (Head loss in pipe with uniform lateral outflow)

Determine the total head loss due to friction over a 100 m length of a 200 mm diameter pipeline of roughness size 0.03 mm which receives an inflow of 150 L/s and releases a uniform lateral outflow of 1.0 L/(s m) (see Figure 4.7).

Solution:

Note that the pressure head $(p/\rho g)_X$ at any section is not simply $h_1 - h_{f,X}$ since momentum effects occur along the pipe due to the continual withdrawal of water. In addition, the

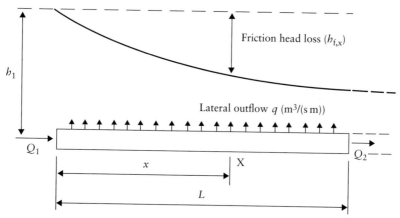

Figure 4.7   Head loss in pipe with uniform lateral outflow.

velocity head decreases along the pipe. Thus applying the energy equation to 1 and section X

$$\frac{p_1}{\rho g} + \frac{V_1^2}{2g} = \frac{p_X}{\rho g} + PD_X + h_{f,X} + \frac{V_X^2}{2g}$$

where PD_X is the increase in pressure head due to the change in momentum between 1 and X. However the present example will deal only with the evaluation of the friction head loss $h_{f,X}$.

The flow rate in the pipe at section X is

$$Q_X = Q_1 - qx$$

The hydraulic gradient at X is

$$\frac{dh_f}{dX} = \frac{\lambda_X Q_X^2}{2gDA^2} = B\lambda_X Q_X^2 \quad \text{where } B = \frac{1}{2gDA^2}$$

$$\frac{dh_f}{dx} = B\lambda_X (Q_1 - qx)^2$$

and the total head loss due to friction between the inlet and outlet is

$$h_f = B \int_0^L \lambda_X (Q_1 - qx)^2 \, dx \tag{i}$$

Now λ_X is given by

$$\frac{1}{\sqrt{\lambda_X}} = -2 \log \left(\frac{k}{3.7D} + \frac{2.51v}{V_X D \sqrt{\lambda_X}} \right)$$

Thus an exact analytical solution to (i) is not possible, but it could be evaluated approximately by summation over finite intervals δx.

However, if we take a constant value of λ_X, based on the average of the inlet and outlet values, an approximate, explicit solution is obtained. Thus the solution to (i) is

$$h_f = B\bar{\lambda} \left(Q_1^2 x - qQ_1 x^2 + \frac{q^2 x^3}{3} \right)_0^L$$

$$= B\bar{\lambda}L \left(Q_1 - qLQ_1 + \frac{q^2 L^2}{3} \right) \tag{ii}$$

$$\frac{k}{D} = \frac{0.03}{200} = 0.00015 \tag{iii}$$

$$Q_1 = 150 \text{ L/s}; \qquad Q_2 = 50 \text{ L/s}$$
$$V_1 = 4.775 \text{ m/s}; \qquad V_2 = 1.59 \text{ m/s}$$
$$Re_1 = 8.45 \times 10^5; \quad Re_1 = 2.82 \times 10^5$$
$$\text{whence } \lambda_1 = 0.014; \qquad \lambda_2 = 0.016$$
$$\text{(from the Moody diagram).}$$

Taking $\bar{\lambda}_1 = 0.015$ and substituting into (ii),

$$h_f = 4.195 \text{ m}$$

Chapter 4

Alternatively, by calculating the head loss in each 10 m interval and summating, the variation of λ along the pipe can be included, and a more accurate result should be obtained.

Then, using Equation (ii) with subscripts 1 and 2 indicating the upstream and downstream ends of each section and with $L = 10$ m, the below table shows the head loss in each section.

x (m)	λ_1	λ_2	Δh_f (m)
10	0.0140	0.0140	0.760
20	0.0140	0.0140	0.659
30	0.0140	0.0144	0.573
40	0.0144	0.0148	0.499
50	0.0148	0.0152	0.427
60	0.0152	0.0152	0.355
70	0.0152	0.0154	0.287
80	0.0154	0.0156	0.225
90	0.0156	0.0160	0.173
100	0.0160	0.0164	0.127

$$h_f = \sum \Delta h_f = 4.086 \text{ m}$$

Example 4.5 (Flow between tanks where the level in the lower tank is dependent upon discharge)

A constant-head tank delivers water through a uniform pipeline to a tank, at a lower level, from which the water discharges over a rectangular weir. The length of the pipeline is 20.0 m, diameter 100 mm and roughness size 0.2 mm. The length of the weir crest is 0.25 m, discharge coefficient 0.6 and crest level 2.5 m below water level in the header tank. Calculate the steady discharge and the head of water over the weir crest (see Figure 4.8).

Solution:

$$\text{For pipeline,} \, H = \frac{1.5V^2}{2g} + \frac{\lambda L V^2}{2gD} = (2.5 - h) \tag{i}$$

$$\text{or } H = \frac{Q^2}{2gA^2}\left(1.5 + \frac{\lambda L}{D}\right) = (2.5 - h) \tag{ii}$$

$$\text{Discharge over weir,} \, Q = \frac{2}{3}C_d\sqrt{2g}Bh^{3/2} \tag{iii}$$

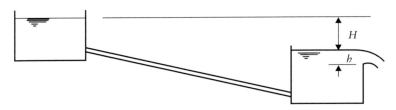

Figure 4.8 Flow through pipeline between two reservoirs, with outflow from receiving reservoir.

$$\Rightarrow Q = \frac{2}{3} \times 0.6 \times \sqrt{19.62} \times 0.25 \times h^{3/2}$$

$$= 0.443 h^{3/2}$$

$$h = \left(\frac{Q}{0.443}\right)^{2/3} \tag{iv}$$

Then in Equation (ii)

$$\frac{Q^2}{2gA^2}\left(1.5 + \frac{\lambda L}{D}\right) = 2.5 - \left(\frac{Q}{0.443}\right)^{2/3}$$

$$\Rightarrow \frac{Q^2}{2gA^2}\left(1.5 + \frac{\lambda L}{D}\right) + \left(\frac{Q}{0.443}\right)^{2/3} = 2.5 \tag{v}$$

Since λ is unknown this equation can be solved by trial or interpolation, that is, inputting a number of trial Q values and evaluating the left-hand side of Equation (v):

$$H_1 = \frac{Q^2}{2gA^2}\left(1.5 + \frac{\lambda L}{D}\right) + \left(\frac{Q}{0.443}\right)^{2/3}$$

For the same values of Q, the corresponding values of h are evaluated from Equation (iv).
For each trial value of Q, the Reynolds number is calculated and the friction factor obtained from the Moody diagram, for $k/D = 0.0002$. See table below.
Hence $Q = 0.0213$ m^3/s (21.3 L/s) when $H_1 = 2.5$ m and $h = 0.132$ m.

Q (m^3/s)	Re	λ	H_1 (m)	h (m)
0.010	1.13×10^5	0.0250	0.617	0.080
0.015	1.69×10^5	0.0243	1.287	0.105
0.018	2.03×10^5	0.0241	1.810	0.118
0.020	2.25×10^5	0.0241	2.215	0.126
0.022	2.48×10^5	0.0240	2.655	0.135

Example 4.6 (Pipes in parallel)

A 200 mm diameter pipeline, 5000 m long and of effective roughness 0.03 mm, delivers water between reservoirs. The minimum difference in water level between reservoirs is 40 m.

(a) Taking only friction, entry and velocity head losses into account, determine the steady discharge between the reservoirs.
(b) If the discharge is to be increased to 50 L/s without increase in gross head, determine the length of a 200 mm diameter pipeline of effective roughness 0.015 mm to be fitted in parallel. Consider only friction losses.

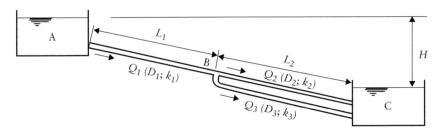

Figure 4.9   Pipes in parallel.

Solution:

(a) Using the technique of Example 4.2,

$$40 = \frac{\lambda L V^2}{2gD} + \frac{1.5 V^2}{2g}$$

yields $Q = 43.52$ L/s.

(b) See Figure 4.9.

In the general case where local losses (h_m) occur in each branch

$$H = \frac{0.5 V_1^2}{2g} + h_{m,1} + h_{f,1} + h_{m,B} + h_{f,2} + h_{m,2} + \frac{V_2^2}{2g} \tag{i}$$

$$\text{also } H = \frac{0.5 V_1^2}{2g} + h_{m,1} + h_{f,1} + h_{m,B} + h_{f,3} + h_{m,3} + \frac{V_3^2}{2g} \tag{ii}$$

where $h_{m,B}$ is the head loss at junction.

Note that the head loss along branch 2 is equal to that along branch 3.

The local losses can be expressed in terms of the velocity heads. Thus Equation (i) or (ii) can be solved simultaneously with the continuity equation at B,

that is, $Q_1 = Q_2 + Q_3$ (iii)

$$\text{and } h_{L,2} = h_{L,3} \tag{iv}$$

and using the Colebrook–White equation (or the Moody chart) for λ.

If friction losses predominate, Equation (i) reduces to

$$H = h_{f,1} + h_{f,2}$$

$$\Rightarrow H = \frac{\lambda_1 L_1 Q_1^2}{2gD_1 A_1^2} + \frac{\lambda_2 L_2 Q_2^2}{2gD_2 A_2^2} \tag{v}$$

Equation (iv) becomes

$$\frac{\lambda_2 L_2 Q_2^2}{2gD_2 A_2^2} = \frac{\lambda_3 L_3 Q_3^2}{2gD_3 A_3^2} \tag{vi}$$

Since $D_2 = D_3$ and $L_2 = L_3$, we have

$$\lambda_2 Q_2^2 = \lambda_3 Q_3^2 \tag{vii}$$

$$\text{also } Q_3 = 0.05 - Q_2$$

Equation (vii) can be solved by trial of Q_2 and using the Moody diagram to obtain the corresponding λ values.

For example

$$\frac{k_2}{D_2} = 0.00015; \quad \frac{k_3}{D_3} = 0.000075$$

$$
\begin{aligned}
\text{Try } Q_2 &= 0.022 \text{ m}^3/\text{s}; & Q_3 &= 0.028 \text{ m}^3/\text{s} \\
Re_2 &= 1.24 \times 10^5; & Re_3 &= 1.58 \times 10^5 \\
\lambda_2 &= 0.0182; & \lambda_3 &= 0.017 \\
\lambda_2 Q_2^2 &= 8.81 \times 10^{-6}; & \lambda_3 Q_3^2 &= 1.33 \times 10^{-5}
\end{aligned}
$$

By adjusting Q_2 and repeating, Equation (vii) is satisfied when $\lambda_2 Q_2^2 \simeq 1.10 \times 10^{-5}$.

Now Equation (v) can be solved:

$$Q_1 = 0.05 \text{ m}^3/\text{s}; \quad Re_1 = 2.816 \times 10^5; \quad \frac{k_1}{D_1} = 0.00015$$

$$\text{whence } \lambda_1 = 0.0161; \quad \frac{\lambda_1 Q_1^2}{2gDA_1^2} = 0.01039$$

Substituting into (v),

$$40 = 0.01039 \times L_1 + \frac{1.10 \times 10^{-5}(5000 - L_1)}{19.62 \times 0.2 \times 0.03142^2}$$

$$\text{whence } L_1 = 3355 \text{ m}$$
$$\text{and } L_2 = 1645 \text{ m}$$
$$\text{or duplicated length} = 1645 \text{ m}$$

Example 4.7 (Design of a uniform pipeline)

A uniform pipeline of length 20 km is to be designed to convey water at a minimum rate of 250 L/s from an impounding reservoir to a service reservoir, the minimum difference in water level between which is 160 m. Local losses including entry loss and velocity head total $10V^2/2g$.

(a) Determine the diameter of a standard commercially available lined spun iron pipeline which will provide the required flow when in new condition ($k = 0.03$ mm).

(b) Calculate also the additional head loss to be provided by a control valve such that with the selected pipe size installed the discharge will be regulated exactly to 250 L/s.

(c) An existing pipeline in a neighbouring scheme, conveying water of the same quality, has been found to lose 5% of its discharge capacity, annually, due to wall deposits (which are removed annually).

(i) Check the capacity of the proposed pipeline after 1 year of use assuming the same percentage reduction.

(ii) Determine the corresponding effective roughness size.

Solution:

(a)

$$160 = \frac{\lambda L V^2}{2gD} + \frac{10V^2}{2g} \tag{i}$$

Neglecting minor losses in the first instance,

$$h_f = H = \frac{\lambda L V^2}{2gD} \tag{ii}$$

$$\Rightarrow \frac{1}{\sqrt{\lambda}} = -2 \log \left(\frac{k}{3.7D} + \frac{2.51}{Re\sqrt{\lambda}} \right) \tag{iii}$$

Combining (ii) and (iii),

$$V = -2\sqrt{2gD\frac{h_f}{L}} \log \left(\frac{k}{3.7D} + \frac{2.51v}{D\sqrt{2gDh_f/L}} \right) \tag{iv}$$

Substituting $h_f = 160$ in Equation (iv), and calculating the corresponding discharge capacity for a series of standard pipe diameters (and noting that there is no need to correct for the reduction due to minor losses each time since there is a considerable percentage increase in capacity between adjacent pipe sizes), the following table is produced:

D (mm)	150	200	250	300	350	400
Q (L/s)	20.3	43.6	78.6	127.3	191.1	271.5

Thus a 400 mm diameter pipeline is required.
  Now check the effect of minor losses:

$$Q = 271.5 \text{ L/s}; \quad V = 2.16 \text{ m/s}; \quad h_m = \frac{10V^2}{2g} = 2.38 \text{ m}$$

$$h_f = 157.6; \quad \text{revised } Q = 269.4 \text{ L/s}$$

The 400 mm diameter is satisfactory.
(b) To calculate the head loss at a valve to control the flow to 250 L/s, calculate the hydraulic gradient corresponding with this discharge:

$$V = 1.99 \text{ m/s}$$

h_f may be obtained by trial in Equation (iv) until the right-hand side = 1.99

$$\text{Thus } h_f = 137 \text{ m}$$

$$\text{Minor head loss, } h_m = \frac{10V^2}{2g} = 2.0 \text{ m}$$

$$\text{Thus valve loss} = 160 - 137 - 2.0 = 21.0 \text{ m}$$

Alternatively using the Moody chart,

$$\frac{k}{D} = \frac{0.03}{400} = 0.000075$$

$$Re = \frac{1.99 \times 0.4}{1.13 \times 10^{-6}} = 7.04 \times 10^5$$

$$\lambda = 0.0136; \quad h_f = \frac{0.0136 \times 20\,000 \times 1.99^2}{19.62 \times 0.4} = 137.25 \text{ m}$$

Adopting $h_f = 137$ m; $10V^2/2g = 2.0$ m

Additional loss by valve $= 160 - (137 + 2.0) = 21$ m.

(c) (i) 5% annual reduction in capacity.

Capacity at the end of 1 year $= 0.95 \times 269.4 = 255.9$ L/s

Pipe will be satisfactory if cleaned each year.

(ii) To calculate the effective roughness size after 1 year's operation, use Equation (iv)

$$\text{i.e. } \frac{k}{3.7D} = \text{antilog}\left(-\frac{V}{\sqrt{2gDh_f/L}}\right) - \frac{2.5\,v}{D\sqrt{2gDh_f/L}} \tag{v}$$

$$Q = 255.9 \text{ L/s}; \quad V = 2.036 \text{ m/s}$$

$$h_f = 160 - \frac{10V^2}{2g} = 157.89 \text{ m}$$

whence from Equation (v), $k = 0.0795$ mm.

Example 4.8 (Effect of booster pump in pipeline)

In the gravity supply system illustrated in Example 4.6, as an alternative to the duplicated pipeline, calculate the head to be provided by a pump to be installed on the pipeline and the power delivered by the pump (see Figure 4.10).

Solution:

$$L = 5000 \text{ m}; \quad D = 200 \text{ mm}; \quad k = 0.03 \text{ mm};$$

$$H = 40 \text{ m}; \quad Q = 50 \text{ L/s}$$

H_m = manometric head to be delivered by the pump

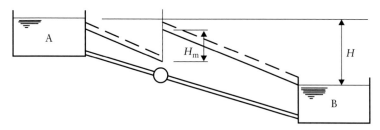

Figure 4.10   Pipeline with booster pump.

Chapter 4

$$\text{Total head} = H_m + H$$

$$H + H_m = \frac{1.5V^2}{2g} + \frac{\lambda LV^2}{2gD}$$

$$V = 1.59 \text{ m/s}; \quad Re = 2.83 \times 10^5; \quad \frac{k}{D} = 0.00015$$

whence $\lambda = 0.0162$ from the Moody chart.

$$\text{Hence } H + H_m = 52.48 \text{ m}$$
$$\text{and } H_m = 12.48 \text{ m}$$

$$\text{Hydraulic power delivered} = \frac{\rho g Q H_m}{1000} \text{ kW}$$
$$P = 9.81 \times 0.05 \times 12.48$$
$$P = 6.12 \text{ kW}$$

Note that the power consumed P_c will be greater than this.

$$P_c = \frac{P}{\eta}$$

where η is the overall efficiency of the pump and motor unit.
(Pump–pipeline combinations are dealt with in more detail in Chapter 6.)

Example 4.9 (Resistance in a non-circular conduit)

A rectangular culvert to be constructed in reinforced concrete is being designed to convey a stream through a highway embankment. For short distances upstream and downstream of the culvert, the existing stream channel will be improved to become rectangular and 6 m wide. The proposed culvert having a bed slope of 1:500 and length 100 m is 4 m wide and 2 m deep and is assumed to have an effective roughness of 0.6 mm. The design discharge is 40 m³/s at which flow the depth in the stream is 3.0 m. Water temperature is 4°C. Entry and exit from the culvert will be taken to be abrupt (although in the final design, transitions at entry and exit would probably be adopted). Determine the depth at the entrance to the culvert at a flow of 40 m³/s (see Figure 4.11).

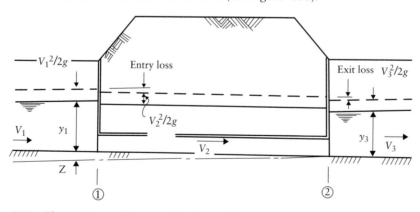

Figure 4.11   Flow resistance in a non-circular conduit.

Solution:

Referring to Figure 4.11, apply the energy equation to 1 and 2

$$Z + y_1 + \frac{V_1^2}{2g} = y_3 + \frac{V_3^2}{2g} + \text{entry loss} + \text{friction loss in culvert} + \text{exit loss} \qquad \text{(i)}$$

The entry loss coefficient K_c may be less than 0.5, which is commonly adopted for entry from a reservoir. Assuming that the loss at the contraction is similar to that for concentric pipes, K_c will depend on the ratio of upstream and downstream areas of flow, derived from the table in Section 4.3 as follows:

A_2/A_1	0.00	0.04	0.16	0.36	1.00
K_c	0.50	0.45	0.38	0.28	0.00

Assume $y_1 = 4.0$ m, then $A_2/A_1 = 8/24 = 0.3$;
whence $K_c = 0.3$.

$$\text{Discharge} = 40.0 \, \text{m}^3/\text{s}; \quad V_2 = \frac{40}{8} = 5.0 \, \text{m/s}$$

$$\text{Entry loss} = 0.3 \frac{V_2^2}{2g} = 0.38 \, \text{m}$$

The exit loss (expansion) is expressed as $(V_2 - V_3)^2/2g$.

$$V_3 = \frac{40}{80} = 2.22 \, \text{m/s} \quad \text{whence exit loss} = 0.39 \, \text{m}$$

Friction head loss in culvert: referring to Section 4.2, the resistance in the duct can be calculated by 'transforming' the cross section into an equivalent circular section by equating the hydraulic radii. For the culvert, $R = A/P = 8/12$ m and the equivalent diameter D_e is therefore 2.67 m $(=4R)$.

Note that either the Colebrook–White equation 4.9 or its graphical form (Figure 4.2) can now be used with $D = 2.67$ m.

$$\text{Kinematic viscosity of water at } 4°C = 1.568 \times 10^{-6} \, \text{m}^2/\text{s}$$

$$V_2 = 5 \, \text{m/s}; \quad Re = \frac{5 \times 2.67}{1.568 \times 10^{-6}} = 8.5 \times 10^6$$

$$\frac{k}{D} = \frac{0.6 \times 10^{-3}}{2.67} = 0.000225$$

From the Moody chart, $\lambda = 0.014$.

$$h_f = \frac{0.014 \times 100 \times 5^2}{19.62 \times 2.67} = 0.668 \, \text{m} \quad \text{(say 0.67 m)}$$

$$\frac{V_3^2}{2g} = \frac{2.22^2}{19.62} = 0.25 \, \text{m}$$

Chapter 4

In Equation (i),

$$0.2 + y_1 + \frac{V_1^2}{2g} = 3.0 + 0.25 + 0.38 + 0.67 + 0.39$$

$$\text{i.e. } 0.2 + y_1 + \frac{Q^2}{2g(6y_1)^2} = 4.49 \text{ m}$$

$$\text{whence by trial } y_1 = 4.37 \text{ m}$$

Since this is close to the assumed value of y_1, the entry loss will not be significantly altered.

Example 4.10 (Pumped-storage power scheme — pipeline design)

The four pump–turbine units of a pumped storage hydroelectric scheme are each to be supplied by a high-pressure pipeline of length 2000 m. The minimum gross head (difference in level between upper and lower reservoirs) is 310 m and the maximum head 340 m.

The upper reservoir has a usable volume of 3.25×10^6 m^3 which could be released to the turbines in a minimum period of 4 h.

$$\text{Maximum power output required/turbine} = 110 \text{ MW}$$
$$\text{Turbogenerator efficiency} = 80\%$$
$$\text{Effective roughness of pipeline} = 0.6 \text{ mm}$$

Taking minor losses in the pipeline, power station and draft tube to be 3.0 m,

(a) determine the minimum diameter of pipeline to enable the maximum specified power to be developed.
(b) determine the pressure head to be developed by the pump–turbine units when reversed to act in the pumping mode to return a total volume of 3.25×10^6 m^3 to the upper reservoir uniformly during 6 h in the off-peak period (see Figure 4.12).

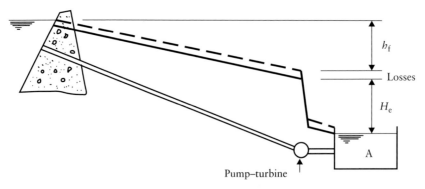

Figure 4.12 Pumped-storage power scheme in generating mode.

Solution:

(a) Pipe capacity must be adequate to convey the required flow under minimum head conditions:

$$\frac{Q_{max}}{unit} = \frac{3.25 \times 10^6}{4 \times 4 \times 3600} = 56.42 \text{ m}^3/s$$

$$\text{Power generated, } P = \frac{\text{efficiency} \times \rho g Q H_e}{10^6} \text{ MW} = 110 \text{ MW}$$

where H_e is the effective head at the turbines.

$$\Rightarrow 110 = \frac{0.8 \times 1000 \times 9.81 \times 56.42 \times H_e}{10^6}$$

whence $H_e = 248 \text{ m}$

Total head loss $= 310 - 248 = 62 \text{ m}$

Head loss due to friction $= 62 - \text{minor losses} = 62 - 3.0 = 59 \text{ m}$

$$h_f = \frac{\lambda L V^2}{2gD} \quad \text{and} \quad \frac{1}{\sqrt{\lambda}} = -2 \log\left(\frac{k}{3.7D} + \frac{2.51}{Re\sqrt{\lambda}}\right)$$

Since the hydraulic gradient is known but D is unknown, it is preferable to use Equation 4.10 in this case rather than use the Moody chart. That is,

$$V = -2\sqrt{2gD\frac{h_f}{L}} \log\left(\frac{k}{3.7D} + \frac{251v}{D\sqrt{2gDh_f/L}}\right)$$

$$\text{and } Q = \frac{\pi D^2 V}{4}$$

Substituting values of D yields the corresponding discharge under the available hydraulic gradient.

D (m)	1.0	2.0	2.5	2.6	2.65
Q (m³/s)	4.47	27.32	48.87	54.123	56.875

Hence required diameter $= 2.65 \text{ m}$.

(b) In pumping mode, static lift $= 340 \text{ m}$

$$Q = \frac{3.25 \times 10^6}{6 \times 4 \times 3600} = 37.616 \text{ m}^3/s$$

Since the diameter of the pipeline is known, it is more straighforward to use the Moody chart in this case.

$$V = 6.82 \text{ m/s}$$

$$Re = \frac{VD}{\nu} = 6.82 \times 2.65 \times 10^6$$

$$= 1.8 \times 10^7$$

$$\frac{k}{D} = \frac{0.6 \times 10^{-3}}{2.65} = 0.000226$$

$$\lambda = 0.0138$$

$$h_f = \frac{\lambda LV^2}{2gD} = \frac{0.0138 \times 2000 \times 6.82^2}{19.62 \times 2.65} = 24.69 \text{ m}$$

$$\text{Total head on pumps} = 340 + 24.69 + 3.0$$

$$= 367.69 \text{ m}$$

Example 4.11

A high-head hydroelectric scheme consists of an impounding reservoir from which the water is delivered to four Pelton wheel turbines through a low-pressure tunnel, 10 000 m long, 4.0 m in diameter, lined with concrete, which splits into four steel pipelines (penstocks) 600 m long, 2.0 m in diameter, each terminating in a single nozzle the area of which is varied by a spear valve. The maximum diameter of each nozzle is 0.8 m and the coefficient of velocity (C_v) is 0.98. The difference in level between reservoir and jets is 550 m. Roughness sizes of the tunnel and pipelines are 0.1 mm and 0.3 mm, respectively.

(a) Determine the effective area of the jets for maximum power and the corresponding total power generated.
(b) A surge chamber is constructed at the downstream end of the tunnel. What is the difference in level between the water in the chamber and that in the reservoir under the condition of maximum power? (See Figure 4.13.)

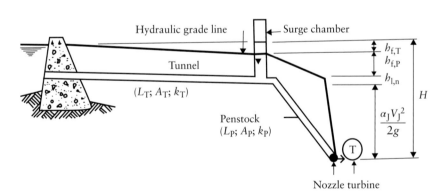

Figure 4.13   Tunnel and penstock.

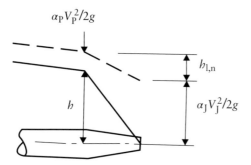

Figure 4.14 Pressure and velocity conditions at nozzle.

Solution:

Let subscript T relate to tunnel and P to pipeline.

$$H = \frac{0.5Q_T^2}{2gA_T^2} + \frac{\lambda_T L_T}{2gD_T}\frac{Q_T^2}{A_T^2} + \frac{\lambda_P L_P Q_P^2}{2gD_P A_P^2} + \frac{\alpha_J V_J^2}{2g} + h_{l,n} \qquad (i)$$

where $h_{l,n}$ is head loss in nozzle and V_J velocity of jet issuing from nozzle. $h_{l,n}$ can be related to the coefficient of velocity C_v (see Figure 4.14).

$$h + \frac{\alpha_P V_P^2}{2g} = \frac{\alpha_J V_J^2}{2g} + h_{l,n} \qquad (ii)$$

$$\text{and } V_J = C_v\sqrt{\frac{2g\left(h + (\alpha_P V_P^2/2g)\right)}{\alpha_J}} \qquad (iii)$$

$$\text{whence } h + \frac{\alpha_P V_P^2}{2g} = \frac{\alpha_J V_J^2}{C_v^2 2g}$$

and substituting in (ii),

$$h_{l,n} = \frac{\alpha_j V_J^2}{2g}\left(\frac{1}{C_v^2} - 1\right) \qquad (iv)$$

Let a be the area of each jet and N be the number of jets (nozzles) per turbine. Since $Q_P = Q_T/4$ and $V_J = Q_P/Na = Q_T/4Na$, Equation (i) becomes

$$2gH = Q_T^2\left[\frac{0.5 + (\lambda_T L_T/D_T)}{A_T^2}\right] + \frac{Q_T^2}{16}\left[\frac{\lambda_P L_P}{D_P A_P^2}\right] + \frac{\alpha_J Q_T^2}{N^2 \times 16 \times C_v^2 \times a^2} \qquad (v)$$

Write

$$E = \frac{0.5 + (\lambda_T L_T/D)}{A_T^2}; \quad F = \frac{\lambda_P L_P}{16D_P A_P^2}; \quad G = \frac{\alpha_J}{16N^2 C_v^2}$$

Chapter 4

Equation (v) becomes

$$2gH = Q_T^2 \left(E + F + \frac{G}{a^2} \right) = Q_T^2 \left(C + \frac{G}{a^2} \right)$$

where $C = E + F$

and $Q_T = \sqrt{\dfrac{2gH}{C + (G/a^2)}}$ \hfill (vi)

$$\text{Power of each jet, } P = \rho g Q_J \frac{\alpha_J V_J^2}{2g}$$

and since $Q_J = Q_P/N = Q_T/4N$ and $V_J = Q_T/4Na$

$$P = \frac{\alpha_J \rho Q_T^3}{128N^3 a^2}$$

Substituting for Q_T from Equation (vi),

$$P = \frac{\rho}{128N^3 a^2} \left[\frac{2gH}{C + (G/a^2)} \right]^{3/2}$$

$$\text{Thus } P \propto \frac{1}{a^2} \left(\frac{a^2}{Ca^2 + G} \right)^{3/2} \propto \frac{a^{2/3}}{Ca^2 + G}$$

For max $dP/da = 0$, that is,

$$-a^{2/3}(Ca^2 + G)^{-2} \times 2Ca + (Ca^2 + G)^{-1} \frac{2}{3} a^{-1/3} = 0$$

$$\text{whence } \frac{-3a^2 C}{(Ca^2 + G)} + 1 = 0$$

$$\text{or } a = \sqrt{\frac{G}{2C}} \hfill \text{(vii)}$$

$$\text{and } D_J = \sqrt{\frac{4a}{\pi}} \hfill \text{(viii)}$$

To evaluate λ_T and λ_P, assume $V_T = V_P = 5$ m/s.

$$Re_T = 17.6 \times 10^6; \qquad (k/D)_T = \frac{0.0003}{4} = 0.000025$$

$$Re_P = 8.8 \times 10^6; \qquad (k/D)_P = \frac{0.0003}{2} = 0.00015$$

$$\lambda_T = 0.0095; \qquad \lambda_P = 0.013$$

Noting that in this example $N = 1$ and taking $\alpha_J = 1.0$,

$$E = 0.1536; \quad F = 0.0247; \quad C = 0.1783; \quad G = 0.065$$

whence from Equations (vii) and (viii),

$$D_J = 0.737 \text{ m}$$

From Equation (vi),

$$Q_T = 142.0 \text{ m}^3/\text{s}; \quad V_T = 11.3 \text{ m/s}; \quad V_P = 11.3 \text{ m/s}$$

Using revised estimates of V_T and $V_P = 11.3$ m/s,

$$Re_T = 40 \times 10^6; \quad Re_P = 20 \times 10^6$$
$$\lambda_T = 0.0092; \quad \lambda_P = 0.013$$

whence $D_J = 0.742$ m

$$Q_T = 143.98 \text{ m}^3/\text{s}; \quad V_T = 11.46 \text{ m/s} = V_P$$

$$\text{Power} = 497.4 \text{ MW}$$

Head loss in tunnel = 157.24 m

= difference in elevation between water in reservoir

and surge shaft

References and recommended reading

Barr, D. I. H. (1975) Two additional methods of direct solution of the Colebrook–White function. *Proceedings of the Institution of Civil Engineers Part 2*, 59, 827–835.

Barr, D. I. H. (1981) Solutions of the Colebrook–White function for the resistance to uniform turbulent flow. *Proceedings of the Institution of Civil Engineers Part 2*, 71, 529–535.

Colebrook, C. F. (1939) Turbulent flow in pipes with particular reference to the transition between the smooth and rough pipe laws, *Journal of the Institution of Civil Engineers*, 11, 133–156.

HR Wallingford (1990) *Charts for the Hydraulic Design of Channels and Pipes*, 6th edn, Thomas Telford, London.

HR Wallingford and D. I. H. Barr (2006) *Tables for the Hydraulic Design of Pipes, Sewers and Channels*, 8th edn, Thomas Telford, London.

Marriott M. J. (1996) The hydraulic characteristics of ovoid sewers, *Journal of the Chartered Institution of Water and Environmental Management*, 10, 5, 365–368; and corrigendum 11, 1/1997, 77.

Moody, L. F. (1944) Friction factors for pipe flows, *Transactions of the American Society of Mechanical Engineers*, 66, 671–684.

Problems

(*Note:* Unless otherwise stated assume the kinetic viscosity of water $\nu = 1.13 \times 10^{-6} \text{ m}^2/\text{s}$).

1. A pipeline 20 km long delivers water from an impounding reservoir to a service reservoir, the minimum difference in level between which is 100 m. The pipe of uncoated cast iron

Chapter 4

($k = 0.3$ mm) is 400 mm in diameter. Local losses, including entry loss, and velocity head amount to $10V^2/2g$.

(a) Calculate the minimum uncontrolled discharge to the service reservoir.

(b) What additional head loss would need to be created by a valve to regulate the discharge to 160 L/s?

2. A long, straight horizontal pipeline of diameter 350 mm and effective roughness size 0.03 mm is to be constructed to convey crude oil of density 860 kg/m³ and absolute viscosity 0.0064 N s/m² from the oilfield to a port at a steady rate of 7000 m³/day. Booster pumps, each providing a total head of 20 m with an overall efficiency of 60%, are to be installed at regular intervals. Determine the required spacing of the pumps and the power consumption of each.

3. A service reservoir, A, delivers water through a trunk pipeline ABC to the distribution network having inlets at B and C.

Pipe AB: length = 1000 m; diameter = 400 mm; $k = 0.06$ mm
Pipe BC: length = 600 m; diameter = 300 mm; $k = 0.06$ mm

The water surface elevation in the reservoir is 110 m o.d. Determine the maximum permissible outflow at B such that the pressure head elevation at C does not fall below 90.0 m o.d. Neglect losses other than friction, entry and velocity head. Outflow at C = 160 L/s.

4. (a) Determine the diameter of commercially available spun iron pipe ($k = 0.03$ mm) for a pipeline 10 km long to convey a steady flow of at least 200 L/s of water at 15°C from an impounding reservoir to a service reservoir under a gross head of 100 m. Allow for entry loss and velocity head. What is the unregulated discharge in the pipeline?

   (b) Calculate the head loss to be provided by a valve to regulate the flow to 200 L/s.

5. Booster pumps are installed at 2 km intervals on a horizontal sewage pipeline of diameter 200 mm and effective roughness size, when new, of 0.06 mm. Each pump was found to deliver a head of 30 m when the pipeline was new. At the end of 1 year, the discharge was found to have decreased by 10% due to pipe wall deposits while the head at the pumps increased to 32 m. Considering only friction losses, determine the discharge when the pipeline was in new condition and the effective roughness size after 1 year.

6. An existing spun iron trunk pipeline of length 15 km, diameter 400 mm and effective roughness size 0.10 mm delivers water from an impounding reservoir to a service reservoir under a minimum gross head of 90 m. Losses in bends and valves are estimated to total $12V^2/2g$ in addition to the entry loss and velocity head.

   (a) Determine the minimum discharge to the service reservoir.

   (b) The impounding reservoir can provide a safe yield of 300 L/s. Determine the minimum length of 400 mm diameter unplasticised polyvinyl chloride (PVCu) pipeline ($k = 0.03$ mm) to be laid in parallel with the existing line so that a discharge equal to the safe yield could be delivered under the available head. Neglect local losses in the new pipe and assume local losses of $12V^2/2g$ in the duplicated length of the original pipeline.

7. A proposed small-scale hydropower installation will utilise a single Pelton Wheel supplied with water by a 500 m long, 300 mm diameter pipeline of effective roughness 0.03 mm.

The pipeline terminates in a nozzle ($C_v = 0.98$) which is 15 m below the level in the reservoir. Determine the nozzle diameter such that the jet will have the maximum possible power using the available head, and determine the jet power.

8. Oil of absolute viscosity 0.07 N s/m^2 and density 925 kg/m^3 is to be pumped by a roto-dynamic pump along a uniform pipeline 500 m long to discharge to atmosphere at an elevation of +80 m o.d. The pressure head elevation at the pump delivery is 95 m o.d. Neglecting minor losses, compare the discharges attained when the pipe of roughness 0.06 mm is (a) 100 mm and (b) 150 mm diameter, and state in each whether the flow is laminar or turbulent.

9. A pipeline 10 km long is to be designed to deliver water from a river through a pumping station to the inlet tank of a treatment works. Elevation of delivery pressure head at the pumping station is 50 m o.d.; elevation of water in tank is 30 m o.d. Neglecting minor losses, compare the discharges obtainable using
   (a) a 300 mm diameter plastic pipeline which may be considered to be smooth
      (i) using the Colebrook–White equation
      (ii) using the Blasius equation
   (b) a 300 mm diameter pipeline with an effective roughness of 0.6 mm
      (i) using the Kármán–Prandtl rough law
      (ii) using the Colebrook–White equation.

10. Determine the hydraulic gradient in a rectangular concrete culvert 1 m wide and 0.6 m high of roughness size 0.06 mm when running full and conveying water at a rate of 2.5 m^3/s.

Chapter 4

Chapter 5
Pipe Network Analysis

5.1 Introduction

Water distribution network analysis provides the basis for the design of new systems and the extension of existing systems. Design criteria are that specified minimum flow rates and pressure heads must be attained at the outflow points of the network. The flow and pressure distributions across a network are affected by the arrangement and sizes of the pipes and the distribution of the outflows. Since a change of diameter in one pipe length will affect the flow and pressure distribution everywhere, network design is not an explicit process. Optimal design methods almost invariably incorporate the hydraulic analysis of the system in which the pipe diameters are systematically altered (see e.g. Featherstone and El Jumailly, 1983).

Pipe network analysis involves the determination of the pipe flow rates and pressure heads which satisfy the continuity and energy conservation equations. These may be stated as follows:

(i) *Continuity*: The algebraic sum of the flow rates in the pipes meeting at a junction, together with any external flows, is zero:

$$\sum_{I=1}^{I=NP(J)} Q_{IJ} - F_J = 0, \quad J = 1, NJ \tag{5.1}$$

in which Q_{IJ} is the flow rate in pipe IJ at junction J, $NP(J)$ the number of pipes meeting at junction J, F_J the external flow rate (outflow) at J and NJ the total number of junctions in the network.

(ii) *Energy conservation*: The algebraic sum of the head losses in the pipes, together with any heads generated by inline booster pumps, around any closed loop formed by pipes is zero.

$$\sum_{J=1}^{J=NP(I)} h_{L,IJ} - H_{m,IJ} = 0, \quad I = 1, NL \tag{5.2}$$

Nalluri & Featherstone's Civil Engineering Hydraulics: Essential Theory with Worked Examples, Sixth Edition. Martin Marriott.
© 2016 John Wiley & Sons, Ltd. Published 2016 by John Wiley & Sons, Ltd.
Companion Website: www.wiley.com/go/Marriott

in which $h_{L,IJ}$ is the head loss in pipe J of loop I and $H_{m,IJ}$ is the manometric head generated by a pump in line IJ.

When the equation relating energy losses to pipe flow rate is introduced into Equations 5.1 or 5.2, systems of non-linear equations are produced. No method exists for the direct solution of such sets of equations and all methods of pipe network analysis are iterative. Pipe network analysis is therefore ideally suited for computer application but simple networks can be analysed with the aid of a calculator.

The earliest systematic method of network analysis, due to Professor Hardy-Cross, known as the head balance or 'loop' method is applicable to systems in which the pipes form closed loops. Assumed pipe flow rates, complying with the continuity requirement, Equation 5.1, are successively adjusted, loop by loop, until in every loop Equation 5.2 is satisfied within a specified small tolerance. In a similar later method, due to Cornish, assumed junction head elevations are systematically adjusted until Equation 5.1 is satisfied at every junction within a small tolerance; it is applicable to both open- and closed-loop networks. These methods are amenable to desk calculation but can also be programmed for computer analysis. However convergence is slow since the hydraulic parameter is adjusted at one element (either loop or junction) at a time. In later methods systems of simultaneous linear equations, derived from Equations 5.1 and 5.2 and the head loss–flow rate relationship, are formed, enabling corrections to the hydraulic parameters (flows or heads) to be made over the whole network simultaneously. Convergence is much more rapid but since a number of simultaneous linear equations, depending on the size of the network, have to be solved, these methods are only realistically applicable to computer evaluation.

The majority of the worked examples in this chapter illustrate the use of Equations 5.1 and 5.2 in systems which can be analysed by desk calculation using either the head balance or quantity balance methods. In addition to friction losses, the effect of local losses and booster pumps is shown. The networks illustrated have been analysed by computer but the intermediate steps in the computations have been reproduced, enabling the reader to follow the process as though it were by desk calculation; the numbers have been rounded to an appropriate number of decimal places. An example showing the gradient method is also given.

5.2 The head balance method ('loop' method)

This method is applicable to closed-loop pipe networks. It is probably more widely applied to this type of network than is the quantity balance method. The head balance method was originally devised by Professor Hardy-Cross and is often referred to as the Hardy-Cross method. Figure 5.1 represents the main pipes in a water distribution network.

The outflows from the system are generally assumed to occur at the nodes (junctions); this assumption results in uniform flows in the pipelines, which simplifies the analysis.

For a given pipe system with known junction outflows, the head balance method is an iterative procedure based on initially estimated flows in the pipes. At each junction these flows must satisfy the continuity criterion.

The head balance criterion is that the algebraic sum of the head losses around any closed loop is zero; the sign convention that clockwise flows (and the associated head losses) are positive is adopted.

The head loss along a single pipe is

$$h = KQ^2$$

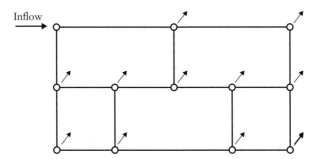

Figure 5.1   Closed-loop pipe network.

If the flow is estimated with an error ΔQ,

$$h = K(Q + \Delta Q)^2 = K[Q^2 + 2Q\Delta Q + \Delta Q^2]$$

Neglecting ΔQ^2 and assuming ΔQ to be small,

$$h = K(Q^2 + 2Q\Delta Q)$$

Now round a closed loop $\sum h = 0$ and ΔQ is the same for each pipe to maintain continuity.

$$\sum h = \sum KQ^2 + 2\Delta Q \sum KQ = 0$$
$$\Rightarrow \Delta Q = -\frac{\sum KQ^2}{2\sum KQ} = -\frac{\sum KQ^2}{2\sum KQ^2/Q}$$

which may be written as $\Delta Q = -\frac{\sum h}{2\sum h/Q}$, where h is the head loss in a pipe based on the estimated flow Q.

5.3 The quantity balance method ('nodal' method)

Figure 5.2 shows a branched-type pipe system delivering water from the impounding reservoir A to the service reservoirs B, C and D. F is a known direct outflow from the node J.

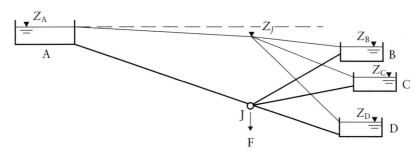

Figure 5.2   Branched-type pipe network.

If Z_J is the true elevation of the pressure head at J, the head loss along each pipe can be expressed in terms of the difference between Z_J and the pressure head elevation at the other end.

For example: $h_{L,AJ} = Z_A - Z_J$.

Expressing the head loss in the form $h = KQ^2$, N such equations can be written as (where N is the number of pipes)

$$\begin{bmatrix} Z_A - Z_J \\ Z_B - Z_J \\ \vdots \\ Z_I - Z_J \end{bmatrix} = \begin{bmatrix} (\text{SIGN})\ K_{AJ}(|Q_{AJ}|)^2 \\ (\text{SIGN})\ K_{BJ}(|Q_{BJ}|)^2 \\ \vdots \\ (\text{SIGN})\ K_{IJ}(|Q_{IJ}|)^2 \end{bmatrix} \qquad [5.3]$$

and in general, (SIGN) is $+$ or $-$ according to the sign of $(Z_I - Z_J)$. Thus flows towards the junction are positive and flows away from the junction are negative.

K_{IJ} is composed of the friction loss and minor loss coefficients.

The continuity equation for flow rates at J is

$$\sum Q_{IJ} - F = Q_{AJ} + Q_{BJ} + Q_{CJ} + Q_{DJ} - F = 0 \qquad [5.4]$$

Examination of Equations 5.3 and 5.4 shows that the correct value of Z_J will result in values of Q_{IJ}, calculated from Equation 5.3, which will satisfy Equation 5.4.

Rearranging Equation 5.3 we have

$$[Q_{IJ}] = \left[(\text{SIGN})\ \left(\frac{|Z_I - Z_J|}{K_{IJ}} \right)^{1/2} \right] \qquad [5.5]$$

The value of Z_J can be found using an iterative method by making an initial estimate of Z_J, calculating the pipe discharges from Equation 5.5 and testing the continuity condition in Equation 5.4.

If $(\sum Q_{IJ} - E) \neq 0$ (with acceptable limits), a correction ΔZ_J is made to Z_J and the procedure repeated until Equation 5.4 is reasonably satisfied. A systematic correction for ΔZ_J can be developed: expressing the head loss along a pipe as $h = KQ^2$, for a small error in the estimate Z_J, the correction ΔZ_J can be derived as

$$\Delta Z_J = \frac{2(\sum Q_{IJ} - F)}{\sum Q_{IJ}/h_{IJ}}$$

Example 5.7 shows the procedure for networks with multiple unknown junction head elevations.

Evaluation of K_{IJ}:

$$K_{IJ} = \frac{\lambda L}{2gDA^2} + \frac{C_m}{2gA^2} \quad (= K_f + K_m)$$

where C_m is the sum of the minor loss coefficients. λ can be obtained from the Moody chart using an initially assumed value of velocity in the pipe (say 1 m/s). A closer approximation to the velocity is obtained when the discharge is calculated. For automatic computer analysis Equation 5.5 should be replaced by the Darcy–Colebrook–White combination:

$$Q = -2A\sqrt{2gD\frac{h_f}{L}} \log\left(\frac{k}{3.7D} + \frac{2.51v}{D\sqrt{2gDh_f/L}}\right)$$ [5.6]

For each pipe, $h_{f,IJ}$ (friction head loss) is initialised to $Z_I - Z_J$, Q_{IJ} calculated from Equation 5.6 and $h_{f,IJ}$ re-evaluated from $h_{f,IJ} = (Z_I - Z_J) - K_m Q_{IJ}^2$. This subroutine follows the procedure of Example 4.2.

5.4 The gradient method

In addition to Equations 5.1–5.6, the gradient method needs the following vector and matrix definitions:

NT = number of pipelines in the network

NN = number of unknown piezometric head nodes

[A12] = 'connectivity matrix' associated with each one of the nodes. Its dimension is NT × NN with only two non-zero elements in the ith row:

-1 in the column corresponding to the initial node of pipe i

1 in the column corresponding to the final node of pipe i

NS = number of fixed head nodes

[A10] = topologic matrix: pipe to node for the NS fixed head nodes. Its dimension is NT × NS with a −1 value in rows corresponding to pipelines connected to fixed head nodes

Thus, the head loss in each pipe between two nodes is

$$[A11][Q] + [A12][H] = -[A10][H_0]$$ [5.7]

where

[A11] = diagonal matrix of NT × NT dimension, defined as

$$[A11] = \begin{bmatrix} \alpha_1 Q_1^{(n_1-1)} + \beta_1 + \frac{\gamma_1}{Q_1} & 0 & \cdots & 0 \\ 0 & \alpha_2 Q_2^{(n_2-1)} + \beta_2 + \frac{\gamma_2}{Q_2} & \cdots & 0 \\ \vdots & \vdots & \vdots\vdots\vdots & \vdots \\ 0 & 0 & \cdots & \alpha_{NT} Q_{NT}^{(n_{NT}-1)} + \beta_{NT} + \frac{\gamma_{NT}}{Q_{NT}} \end{bmatrix}$$ [5.8]

[Q] = discharge vector with NT × 1 dimension

[H] = unknown piezometric head vector with NN × 1 dimension

[H_0] = fixed piezometric head vector with NS × 1 dimension

Equation 5.7 is an energy conservation equation. The continuity equation for all nodes in the network is

$$[A21][Q] = [q]$$ [5.9]

Chapter 5

where [A21] is the transpose matrix of [A12] and [q] is the water consumption and water supply vector in each node with $NN \times 1$ dimension.

In matrix form, Equations 5.7 and 5.9 are

$$\begin{bmatrix} [A11] & [A12] \\ [A21] & [0] \end{bmatrix} \begin{bmatrix} [Q] \\ [H] \end{bmatrix} = \begin{bmatrix} -[A10][H_0] \\ [q] \end{bmatrix} \qquad [5.10]$$

The upper part is *nonlinear*, which implies that Equation 5.10 must use some *iterative* algorithm for its solution. The gradient method consists of a truncated Taylor expansion. Operating simultaneously on the ([Q], [H]) field and applying the gradient operator, we can write

$$\begin{bmatrix} [N][A11]' & [A12] \\ [A21] & [0] \end{bmatrix} \begin{bmatrix} [dQ] \\ [dH] \end{bmatrix} = \begin{bmatrix} [dE] \\ [dq] \end{bmatrix} \qquad [5.11]$$

where [N] is the diagonal matrix $(n_1, n_2, \ldots, n_{NT})$ with $NT \times NT$ dimension and $[A11]' = NT \times NT$ matrix defined as

$$[A11]' = \begin{bmatrix} \alpha_1 Q_1^{(n_1-1)} & 0 & 0 & \cdots & 0 \\ 0 & \alpha_2 Q_2^{(n_2-1)} & 0 & \cdots & 0 \\ 0 & 0 & \alpha_3 Q_3^{(n_3-1)} & \cdots & 0 \\ \vdots & \vdots & \vdots & \vdots\vdots\vdots & \vdots \\ 0 & 0 & 0 & \cdots & \alpha_{NT} Q_{NT}^{(n_{NT}-1)} \end{bmatrix} \qquad [5.12]$$

In any iteration **i**, [dE] is the energy imbalance in each pipe and [dq] is the discharge imbalance in each node. These are given by

$$[dE] = [A11][Q_i] + [A12][H_i] + [A10][H_0] \qquad [5.13]$$

and

$$[dq] = [A21][Q_i] - [q] \qquad [5.14]$$

The objective of the gradient method is to solve the system described by Equation 5.11, taking into account that in each iteration

$$[dQ] = [Q_{i+1}] - [Q_i] \qquad [5.15]$$

and

$$[dH] - [H_{i+1}] - [H_i] \qquad [5.16]$$

Using matrix algebra, it is possible to show that the solution to the system represented by Equation 5.11 is

$$[H_{i+1}] = -\{[A21]([N][A11]')^{-1}[A12]\}^{-1}\{[A21]([N][A11]')^{-1}$$
$$([A11][Q_i]) + [A10][H_0] - ([A21][Q_i]) - [q]\} \qquad [5.17]$$
$$[Q_{i+1}] = \{[I] - ([N][A11]') - [A11]\}[Q_i] - \{((N][A11]')^{-1}([A12]$$
$$[H_{i+1}] + [A10][H_0])\} \qquad [5.18]$$

The method has the advantage of fast convergence and does not need continuity balancing in each node to begin the process. The method is not suited for hand calculation. Example 5.8 illustrates the methodology.

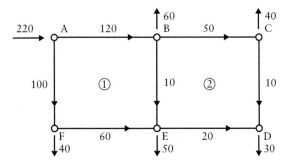

Figure 5.3   Two-loop network.

Worked examples

Example 5.1

Neglecting minor losses in the pipes, determine the flows in the pipes and the pressure heads at the nodes (see Figure 5.3).

				Data				
Pipe	AB	BC	CD	DE	EF	AF	BE
Length (m)	600	600	200	600	600	200	200
Diameter (mm)	250	150	100	150	150	200	100

Roughness size of all pipes = 0.06 mm
Pressure head elevation at A = 70 m.

			Elevation of pipe nodes			
Node	A	B	C	D	E	F
Elevation (m)	30	25	20	20	22	25

Procedure:

1. Identify loops. When using hand calculation the simplest way is to employ adjacent loops, for example Loop 1: ABEFA; Loop 2: BCDEB.
2. Allocate estimated flows in the pipes. Only one estimated flow in each loop is required; the remaining flows follow automatically from the continuity condition at the nodes; for example, since the total required inflow is 220 L/s, if Q_{AB} is estimated at 120 L/s, then $Q_{AF} = 100$ L/s. The initial flows are shown in Figure 5.3.
3. The head loss coefficient $K = \lambda L/2gDA^2$ is evaluated for each pipe, λ being obtained from the λ versus Re diagram (Figure 4.2) corresponding to the flow in the pipe. Alternatively, Barr's equation (Equation 4.12) may be used.

If the Reynolds numbers are fairly high ($\nless 10^5$), it may be possible to proceed with the iterations using the initial λ values, making better estimates as the solution nears convergence.

The calculations proceed in tabular form. Note that Q is written in litres per second simply for convenience; all computations are based on Q in cubic metres per second. However, h/Q could have been expressed in m/(L/s) yielding ΔQ directly in litres per second.

	Pipe	k/D	Q (L/s)	Re ($\times 10^5$)	λ	K	h (m)	$h/Q\left(\frac{m}{m^3/s}\right)$
	AB	0.00024	120.00	5.41	0.0157	797.0	11.48	95.64
Loop 1	BE	0.00060	10.00	1.31	0.0205	33 877.0	3.39	338.77
	EF	0.00040	−60.00	4.51	0.0172	11 229.1	−40.42	673.75
	FA	0.00030	−100.00	5.63	0.0162	336.6	−8.36	83.66
						Σ	−33.91	1191.82

$$\Rightarrow \Delta Q = \frac{-\sum h}{2\sum h/Q} = \frac{-(-33.91)}{2 \times 1191.82} = 0.01423 = 14.23 \text{ L/s.}$$

	Pipe	Q (L/s)	Re ($\times 10^5$)	λ	K	h (m)	$h/Q\left(\frac{m}{m^3/s}\right)$
	BC	50.0	3.76	0.0174	11 359.7	28.40	567.98
Loop 2	CD	10.0	1.13	0.0205	33 877.0	3.39	338.77
	DE	−20.0	1.50	0.0189	12 338.9	−4.94	246.78
	EB	−24.23	2.73	0.0189	31 232.9	−18.34	756.77
					Σ	−8.51	1910.30

$$\Rightarrow \Delta Q = -2.23 \text{ L/s.}$$

(Note that the previously corrected value of flow in the 'common' pipe EB has been used in Loop 2.)

	Pipe	Q (L/s)	Re ($\times 10^5$)	λ	K	h (m)	$h/Q\left(\frac{m}{m^3/s}\right)$
	AB	134.23	6.05	0.0156	791.9	14.27	106.30
Loop 1	BE	26.46	2.98	0.0188	31 067.7	21.75	822.05
	EF	−45.77	3.44	0.0175	11 424.9	−23.93	522.92
	FA	−85.77	4.83	0.0164	846.9	−6.23	72.64
					Σ	5.86	1523.91

$$\Rightarrow \Delta Q = -1.92 \text{ L/s.}$$

Proceed to Loop 2 again, and continuing in this way the solution is obtained within the required specified limit on $\sum h$ in any loop after several further iterations. The solution given is obtained for $\sum h < 0.01$ m but an acceptable result may be achieved with a larger tolerance.

	Final values			Pressure heads	
Pipe	Q (L/s)		h (m)	Node	Pressure head (m)
AB	131.55		13.70	A	40.00
BE	25.02		19.55	B	31.29
FE	48.45		26.67	C	11.57
AF	88.45		6.59	D	10.05
BC	46.53		24.74	E	14.74
CD	6.55		1.52	F	38.41
ED	23.47		6.69		

Note: Flows in direction of pipe identifier (e.g. A → B).

Example 5.2

In the network shown in Figure 5.4 a valve in BC is partially closed to produce a local head loss of $10V_{BC}^2/2g$. Analyse the flows in the network.

Pipe	AB	BC	CD	DE	BE	EF	AF
Length (m)	500	400	200	400	200	600	300
Diameter (mm)	250	150	100	150	150	200	250

Note: Roughness of all pipes is 0.06 mm.

Solution:

The procedure is identical with that of the previous problem. K_{BC} is now composed of the valve loss coefficient and the friction loss coefficient.

With the initial assumed flows shown in the table below, $Q_{BC} = 50$ L/s; $Re = 3.7 \times 10^5$; $k/D = 0.0004$; and $\lambda = 0.0174$ (from the Moody chart). Hence, $K_f = 7573$, $K_m = 1632$ and $K_{BC} = 9205$.

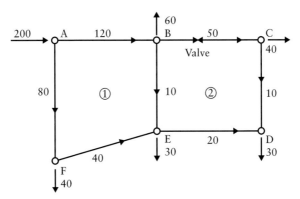

Figure 5.4 Pipe network with valve losses.

Pipe	k/D	Q (L/s)	Re (×10⁵)	λ	K	h (m)	$h/Q\left(\frac{m}{m^3/s}\right)$
AB	0.00024	120.00	5.41	0.0157	664.2	9.56	79.70
Loop 1 BE	0.00040	10.00	0.75	0.0208	4 526.5	0.45	45.26
EF	0.00030	−40.00	2.25	0.0175	2 711.2	−4.34	108.45
FA	0.00024	−80.00	3.61	0.0163	413.7	−2.65	33.10
					Σ	3.03	266.51

$\Rightarrow \Delta Q = -5.69$ L/s.

Pipe	k/D	Q (L/s)	Re (×10⁵)	λ	K	h (m)	$h/Q\left(\frac{m}{m^3/s}\right)$
BC	0.0004	50.00	3.75	0.0174	9 205.2	23.01	460.26
Loop 2 CD	0.0006	10.00	1.13	0.0205	33 877.0	3.39	338.77
DE	0.0004	−20.00	1.50	0.0190	8 226.0	−3.29	164.52
EB	0.0004	−4.31	0.32	0.0242	5 266.4	−0.10	22.70
					Σ	23.01	986.25

$\Rightarrow \Delta Q = -11.67$ L/s.

Proceeding in this way the solution is obtained within a small limit on $\sum h$ in any loop:

	Final values						
Pipe	AB	BE	FE	FA	BC	CD	ED
Q (L/s)	111.52	16.48	48.48	88.48	35.05	4.95	34.95
h_L (m)	8.31	1.15	6.26	3.20	11.57	0.91	9.52

Example 5.3

If in the network shown in Example 5.2 a pump is installed in line BC boosting the flow towards C and the valve removed, analyse the network. Assume that the pump delivers a head of 10 m. (*Note:* In reality, it would not be possible to predict the head generated by the pump since this will depend upon the discharge. The head–discharge relationship for the pump, e.g. $H = AQ^2 + BQ + C$, must therefore be solved for the discharge in the pipe at each iteration. However, for the purpose of illustration of the basic effect of a pump, the head in this case is assumed to be known.) An example of a network analysis in which the pump head–discharge curve is used is given in Chapter 6 (Example 6.8). Consider length BC (see Figure 5.5).

The net loss of head along BC($Z_B - Z_C$) is ($h_f - H_p$), where H_p is the total head delivered by pump. The value of K for BC is now due to friction only; the head loss for BC in the table now becomes $h_{L,BC} = (KQ_{BC}^2 - 10)$ m. Otherwise the iterative procedure is as before.

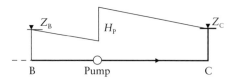

Figure 5.5   Network of Example 5.2 with pump.

Solution:

	Pipe	k/D	Q (L/s)	Re ($\times 10^5$)	λ	K	h (m)	$h/Q\left(\frac{m}{m^3/s}\right)$
	AB	0.00024	120.00	5.41	0.0157	664.2	9.56	79.70
Loop 1	BE	0.00040	10.00	0.75	0.0208	4526.5	0.45	45.26
	EF	0.00030	−40.00	2.25	0.0175	2711.2	−4.34	108.45
	FA	0.00024	−80.00	3.61	0.0163	413.7	−2.65	33.10
						Σ	3.03	266.51

$\Rightarrow \Delta Q = -5.69$ L/s.

	Pipe	k/D	Q (L/s)	Re ($\times 10^5$)	λ	K	h (m)	$h/Q\left(\frac{m}{m^3/s}\right)$
	BC	0.00040	50.00	3.76	0.0174	7573.0	8.93	178.66
Loop 2	CD	0.00060	10.00	1.13	0.0205	33877.0	3.39	333.77
	DE	0.00040	−20.00	1.50	0.0189	8225.96	−3.29	164.52
	EB	0.00040	−4.31	0.32	0.0242	5266.4	−0.10	22.70
						Σ	8.93	704.65

$\Rightarrow \Delta Q = -6.34$ L/s.

	Pipe	Q (L/s)	Re ($\times 10^5$)	λ	K	h (m)	$h/Q\left(\frac{m}{m^3/s}\right)$
	AB	114.31	5.15	0.0158	668.4	8.73	76.41
Loop 1	BE	10.65	0.80	0.0206	4482.9	0.51	47.74
	EF	−45.69	2.57	0.0173	2680.2	−5.59	122.46
	FA	−85.69	3.66	0.0162	411.2	−3.02	35.24
					Σ	0.63	281.85

$\Rightarrow \Delta Q = -1.11$ L/s.

After similar further iterations:

			Final values				
Pipe	AB	BE	FE	FA	BC	CD	ED
Q (L/s)	113.21	8.90	46.79	86.79	44.30	4.30	25.70
h_L (m)	8.57	0.37	5.83	3.10	4.95	0.71	5.29

Chapter 5

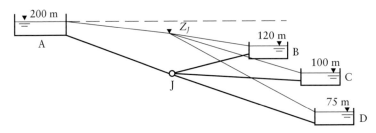

Figure 5.6   Network connecting multi-reservoirs.

Example 5.4

Determine the discharges in the pipes of the network shown in Figure 5.6, neglecting minor losses. Figure 5.6 shows reservoir water surface elevations in metres, relative to a common datum.

Pipe	Length (m)	Diameter (mm)
AJ	10 000	450
BJ	2 000	350
CJ	3 000	300
DJ	3 000	250

Note: Roughness size of all pipes is 0.06 mm.

The friction factor λ may be obtained from the Moody diagram, or using Barr's equation, using an initially estimated velocity in each pipe. Subsequently, λ can be based on the previously calculated discharges. However, unless there is a serious error in the initial velocity estimates, much effort is saved by retaining the initial λ values until perhaps the penultimate or final correction.

Solution:

Estimate Z_J (pressure head elevation at J) = 150.0 m (*Note*: the elevation of the pipe junction itself does not affect the solution.) See tables below.

First correction

Pipe	Velocity (estimate) (m/s)	$Re\ (\times 10^5)$	λ	K	$Z_I - Z_J$	Q (m/s)	$Q/b\ (\times 10^{-3})$	Q/A (m/s)
AJ	2.0	7.96	0.0145	649	+50	0.2775	5.55	1.75
BJ	2.0	6.20	0.0150	472	−30	−0.2521	8.40	2.62
CJ	2.0	5.31	0.0155	1581	−50	−0.1778	3.56	2.50
DJ	2.0	4.42	0.0165	4188	−75	−0.1338	1.78	2.73

$$\Sigma \qquad -0.2862 \qquad 0.0193$$

$$\Rightarrow \text{Correction to } Z_J = \frac{2(-0.2862)}{0.0193} = -29.67;\ Z_J = 120.33\,\text{m}.$$

Second correction

Pipe	Velocity (estimate) (m/s)	Re (×10⁵)	λ	K	$Z_I - Z_J$	Q (m/s)	Q/h (×10⁻³)	Q/A (m/s)
AJ	As	7.96	0.0145	649	79.67	0.3504	4.39	2.20
BJ	initial	6.20	0.0150	472	−0.33	−0.0264	80.12	0.27
CJ	estimate	5.31	0.0155	1581	−20.33	−0.1134	5.58	1.60
DJ		4.42	0.0165	4188	−45.33	−0.1040	2.29	2.20

$$\Sigma \qquad +0.1066 \qquad +0.092$$

$\Rightarrow \Delta Z_J = +2.30 \text{ m}; \ Z_J = 122.63 \text{ m}.$

Comment: The velocity in BJ has changed significantly but it may oscillate; it is therefore estimated at 1.0 m/s for the next correction. Note that λ (BJ) altered accordingly.

Third correction

Pipe	Velocity (estimate) (m/s)	λ	K	$Z_I - Z_J$	Q (m³/s)	Q/h (×10⁻³)	Q/A (m/s)
AJ	2.0	0.0145	649	77.37	0.3452	4.46	2.17
BJ	1.0	0.016	503	−2.63	−0.0723	27.50	0.75
CJ	1.8	0.0155	1581	−22.63	−0.1196	5.29	1.69
DJ	2.3	0.016	4061	47.63	−0.1083	2.27	2.21

$$\Sigma \qquad +0.0450 \qquad 0.0395$$

$\Rightarrow \Delta Z_J = 2.27 \text{ m}; \ Z_J = 124.90 \text{ m}.$

Final values:

$$Q_{AJ} = 0.344 \text{ m}^3/\text{s}; \quad Q_{JB} = 0.105 \text{ m}^3/\text{s}; \quad Q_{JC} = 0.127 \text{ m}^3/\text{s}; \quad Q_{JD} = 0.112 \text{ m}^3/\text{s}$$

Example 5.5

If in the network of Example 5.4 the flow to C is regulated by a valve to 100 L/s, calculate the effect on the flows to the other reservoirs; determine the head loss to be provided by the valve.

The principle of the solution is identical with that of the previous example except that the flow in JC is prescribed and simply treated as an external outflow at J. In this example the flow rates in the pipes have been evaluated directly from Equation 5.6.

$$Q = -2A\sqrt{2gD\frac{h}{L}} \log \left(\frac{k}{3.7D} + \frac{2.51\nu}{D\sqrt{2gDh_f/L}} \right)$$

in which $h = Z_I - Z_J$, since there are no minor losses. This approach is ideal for computer analysis; if minor losses are present, use the iterative procedure described in Example 4.2.

The method is also suitable for desk analysis using an electronic calculator since for each pipe the only variable is h and Equation 5.6 can be written as

$$Q = -C_1\sqrt{h}\,\log\left(C_2 + \frac{C_3}{\sqrt{h}}\right)$$

in which C_1, C_2 and C_3 are constants for a particular pipe.

The corresponding velocities and λ values have been evaluated and tabulated; these data may be useful for those who wish to work through the example using the Moody diagram as shown in Example 5.4.

Note that Q is expressed in litres per second; in evaluating $\sum Q/h$, the flow is also expressed in litres per second so that the units in the correction term $\Delta Z = 2(\sum Q - F)/(\sum Q/h)$ are consistent.

Example 5.5 calculation

Pipe	AJ	BJ	DJ
k/D	0.000133	0.000171	0.000240

Note: Estimate $Z_J = 150.00$ m.

First correction

	Pipe	$Z_1 - Z_J$ $(=h)$ (m)	Q (L/s)	Q/h	V (m/s)	λ
	AJ	50.00	279.32	5.59	1.76	0.0143
Junction J	BJ	−30.00	−255.95	8.53	2.66	0.0146
	DJ	−75.00	−137.90	1.84	2.81	0.0155
	Σ		−114.53	15.96		

$$\text{Correction to } Z_J = \frac{2(\sum Q - F)}{\sum Q/h} = \frac{2(-144.53 - 100)}{15.96} = -26.89 \text{ m}$$

$$Z_J = 123.11 \text{ m}$$

Second correction

	Pipe	$Z_1 - Z_J$	Q (L/s)	Q/h	V (m/s)	λ
	AJ	76.89	349.70	4.55	2.20	0.0140
Junction J	BJ	−3.11	−77.61	24.96	0.81	0.0164
	DJ	−48.11	−109.50	2.28	2.23	0.0158

$\Rightarrow \Delta Z_J = 3.94$ m; $Z_J = 127.05$ m.

	Pipe	$Z_I - Z_J$	Q (L/s)	Q/h	V (m/s)	λ
		Third correction				
	AJ	72.95	340.2	4.66	2.14	0.0141
Junction J	BJ	−7.05	−119.94	17.01	1.25	0.0156
	DJ	−52.05	−114.08	2.19	2.32	0.0158

$$\Rightarrow \Delta Z_J = 0.52 \text{ m}; \quad Z_J = 127.57 \text{ m}.$$

	Final values		
Pipe	AJ	JB	JD
Q (L/s)	338.98	124.36	114.65

Head loss due to friction along JC:

$$\text{Diameter} = 300 \text{ mm}; \quad A = 0.0707 \text{ m}^2; \quad Q = 0.100 \text{ m}^3/\text{s}; \quad V = 1.415 \text{ m/s}$$

$$Re = \frac{1.415 \times 0.3}{1.13 \times 10^{-6}} = 3.76 \times 10^5; \quad \frac{k}{D} = 0.0002$$

$$\text{whence } \lambda = 0.016; \quad h_f = \frac{0.016 \times 3000 \times 1.415^2}{19.62 \times 0.3} = 16.33 \text{ m}$$

(See Figure 5.7.)

$$\Rightarrow \text{Head loss at valve} = Z_J - Z_C - h_f$$
$$= 127.55 - 100.00 - 16.33$$
$$= 11.22 \text{ m}$$

Example 5.6

In the network as before, a pump P is installed on JB to boost the flow to B. With the flows to C and D uncontrolled and the pump delivering 10 m head, determine the flows in the pipes (see Figure 5.8).

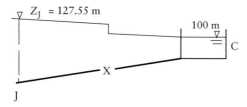

Figure 5.7   Network of Example 5.4 with valve losses.

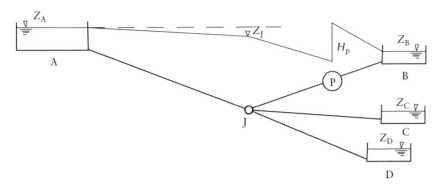

Figure 5.8 Network of with pump.

Note: In the case of rotodynamic pumps the manometric head delivered varies with the discharge (see Chapter 6). Thus it is not strictly possible to specify the head and it is necessary to solve the pump equation $H_p = AQ^2 + BQ + C$ together with the resistance equation for JB. However, to illustrate the effect of a pump in this example, let us assume that the head does not vary with flow.

Solution:

The analysis is straightforward and follows the procedure of Example 5.5.
   The head giving flow along JB is

$$h_{L,JB} = Z_J - Z_B - H_p$$

The final solution is as follows:

Pipe	AJ	JB	JC	JD
Q (L/s)	357.7	141.6	110.8	105.3

Note: $Z_J = 119.66$ m.

Example 5.7

Determine the flows in the network shown in Figure 5.9 neglecting minor losses.

Pipe	AB	BC	BD	BE	EF	EG
Length (m)	10 000	3 000	4 000	6 000	3 000	3 000
Diameter (mm)	450	250	250	350	250	200

Note: Roughness of all pipes is 0.03 mm (=k).

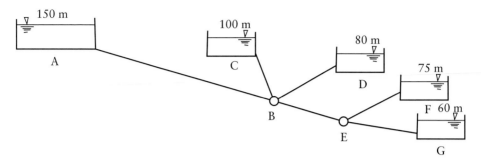

Figure 5.9   Network with multi-reservoirs.

Solution:

In this case there are two unknown pressure head elevations which must therefore be both initially estimated and corrected alternately.

$$\text{Estimate } Z_B = 120.0 \text{ m}; \qquad Z_E = 95.0 \text{ m}$$

First correction

	Pipe	$Z_I - Z_J$ $(=h)$	Q (L/s)	Q/h	V (m/s)	λ
	AB	30.00	219.77	7.33	1.38	0.0139
Junction B	CB	−20.00	−71.38	3.57	1.45	0.0155
	DB	−40.00	−86.75	4.34	1.77	0.0151
	EB	−25.00	−135.00	5.40	1.40	0.0145
	Σ		−73.35	20.63		

$$\Rightarrow \Delta Z_B = +7.11 \text{ m}; \ Z_B = 112.89 \text{ m}.$$

Proceed to Junction E noting that the amended value of Z_B is now used:

	Pipe	$Z_I - Z_J$	Q (L/s)	Q/h	V (m/s)	λ
	BE	17.89	112.81	6.31	1.17	0.0149
Junction E	FE	−20.00	−71.38	3.57	1.45	0.0155
	GE	−35.00	−53.38	1.53	1.70	0.0159
	Σ		11.95	11.40		

$$\Rightarrow \Delta Z_E = -2.1 \text{ m}; \ Z_E = 92.9 \text{ m}.$$

Second correction

	Pipe	$Z_I - Z_J$	Q (L/s)	Q/h	V (m/s)	λ
	AB	37.11	246.21	6.63	1.55	0.0137
Junction B	CB	−12.89	−56.38	4.37	1.15	0.0160
	DB	−32.89	−78.16	6.06	1.59	0.0153
	EB	−19.99	−119.75	5.99	1.25	0.0148
	Σ		−8.07	23.06		

$\Rightarrow \Delta Z_B = -0.7$ m; $Z_B = 112.19$ m.

	Pipe	$Z_I - Z_J$	Q (L/s)	Q/h	V (m/s)	λ
	BE	92.9	117.48	6.09	1.22	0.0148
Junction E	FE	−17.9	−67.26	3.76	1.37	0.0156
	GE	−32.9	−51.64	1.57	1.64	0.0159
	Σ		−1.43	11.42		

$\Rightarrow \Delta Z_E = -0.25$ m; $Z_E = 92.65$ m.

Example 5.8

In the network shown in Figure 5.10, a valve in pipe 2-3 is partially closed, producing a local head loss of $10V^2_{2\text{-}3}/2g$. The head at node 1 is 100 m of water. The roughness of all pipes is 0.06 mm. The pipe lengths are in metres and the demand discharges are in litres per second.

The pipe diameters are pipes 1-2 and 1-6, 250 mm; pipe 6-5, 200 mm; pipes 2-3 and 4-5, 150 mm; and pipes 2-5 and 3-4, 100 mm. Analyse the network using the gradient method.

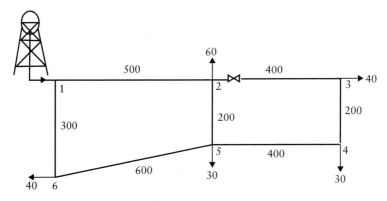

Figure 5.10 Pipe network with valve loss.

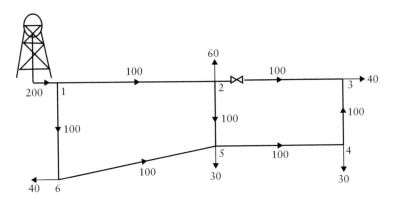

Figure 5.11   Network solution.

The iterative process can be summarised in the following steps:

1. Assume initial discharges in each of the network pipes. (They can be unbalanced at each node.)
2. Solve the system represented by Equation 5.17 using a standard method for the solution of simultaneous linear equations.
3. With the calculated $[H_{i+1}]$ (Step 2), $[Q_{i+1}]$ is solved by Equation 5.18.
4. With the new $[Q_{i+1}]$, Equation 5.17 is solved (Step 2) to find a new $[H_{i+1}]$.
5. Process continues until

$$[H_{i+1}] \approx [H_i]$$

For all pipes initial discharges of 100 L/s have been assumed with the directions as shown in Figure 5.11.

Solution:

All the matrices and vectors needed for the gradient method are as follows:

$$NT = 7$$
$$NN = 5$$
$$NS = 1$$

[A12] = connectivity matrix; dimension (7 × 5)

$$
\begin{vmatrix}
1 & 0 & 0 & 0 & 0 \\
-1 & 1 & 0 & 0 & 0 \\
0 & 1 & -1 & 0 & 0 \\
0 & 0 & 1 & -1 & 0 \\
-1 & 0 & 0 & 1 & 0 \\
0 & 0 & 0 & 1 & -1 \\
0 & 0 & 0 & 0 & 1
\end{vmatrix}
$$

[A21] = transposed matrix of [A12]

$$
\begin{vmatrix}
1 & -1 & 0 & 0 & -1 & 0 & 0 \\
0 & 1 & 1 & 0 & 0 & 0 & 0 \\
0 & 0 & -1 & 1 & 0 & 0 & 0 \\
0 & 0 & 0 & -1 & 1 & 1 & 0 \\
0 & 0 & 0 & 0 & 0 & -1 & 1
\end{vmatrix}
$$

[A10] = topologic matrix node to node; dimension (7×1)

[Q] = discharges vector; dimension (7×1)

[H] = unknown piezometric head vector; dimension (5×1)

[H_0] = fixed piezometric head vector; dimension (1×1)

[q] = water demand vector; dimension (5×1)

[A10]	[Q] (m^3/s)	[H]	[H_0] (m)	[q] (m^3/s)
-1	0.10	H_2	100	0.06
0	0.10	H_3		0.04
0	0.10	H_4		0.03
0	0.10	H_5		0.03
0	0.10	H_6		0.04
0	0.10			
-1	0.10			

[N] = diagonal matrix; dimension (7×7); having 2 in the diagonal (from the Darcy–Weisbach head loss equation)

$$
\begin{vmatrix}
2 & 0 & 0 & 0 & 0 & 0 & 0 \\
0 & 2 & 0 & 0 & 0 & 0 & 0 \\
0 & 0 & 2 & 0 & 0 & 0 & 0 \\
0 & 0 & 0 & 2 & 0 & 0 & 0 \\
0 & 0 & 0 & 0 & 2 & 0 & 0 \\
0 & 0 & 0 & 0 & 0 & 2 & 0 \\
0 & 0 & 0 & 0 & 0 & 0 & 2
\end{vmatrix}
$$

[I] = identity matrix; dimension (7×7)

$$
\begin{vmatrix}
1 & 0 & 0 & 0 & 0 & 0 & 0 \\
0 & 1 & 0 & 0 & 0 & 0 & 0 \\
0 & 0 & 1 & 0 & 0 & 0 & 0 \\
0 & 0 & 0 & 1 & 0 & 0 & 0 \\
0 & 0 & 0 & 0 & 1 & 0 & 0 \\
0 & 0 & 0 & 0 & 0 & 1 & 0 \\
0 & 0 & 0 & 0 & 0 & 0 & 1
\end{vmatrix}
$$

First iteration:

The previous matrices and vectors are valid for all the iterations. The following matrices change in each iteration:

[A11] = diagonal matrix; dimension (7×7); having the value of $\alpha_i Q_i^{(n_i-1)}$ on the diagonal, with coefficients β and γ zero as no pumps exist in the network

The following table shows the calculated values for α:

Pipe	Q (m³/s)	λ	V (m/s)	h_f (m)	$h_f + h_m$ (m)	α
1-2	0.10	0.0159	1.974	6.22	6.22	622.28
2-3	0.10	0.0166	5.482	66.89	82.21	8 220.77
3-4	0.10	0.0178	12.335	271.02	271.02	27 101.65
5-4	0.10	0.0166	5.482	66.89	66.89	6 688.98
2-5	0.10	0.0178	12.335	270.99	270.09	27 098.90
6-5	0.10	0.0161	3.084	23.09	23.09	2 308.78
6-1	0.10	0.0159	1.974	3.73	3.73	373.42

Matrix [A11]:

$$
\begin{vmatrix}
62.23 & 0 & 0 & 0 & 0 & 0 & 0 \\
0 & 822.08 & 0 & 0 & 0 & 0 & 0 \\
0 & 0 & 2710.16 & 0 & 0 & 0 & 0 \\
0 & 0 & 0 & 668.90 & 0 & 0 & 0 \\
0 & 0 & 0 & 0 & 2709.89 & 0 & 0 \\
0 & 0 & 0 & 0 & 0 & 230.88 & 0 \\
0 & 0 & 0 & 0 & 0 & 0 & 37.34
\end{vmatrix}
$$

[A11]′ = diagonal matrix; dimension (7×7); having the value of $\alpha_i Q_i^{(n_i-1)}$ on the diagonal.

For this network, [A11′] = [A11].

$$
\begin{vmatrix}
62.23 & 0 & 0 & 0 & 0 & 0 & 0 \\
0 & 822.08 & 0 & 0 & 0 & 0 & 0 \\
0 & 0 & 2710.16 & 0 & 0 & 0 & 0 \\
0 & 0 & 0 & 668.90 & 0 & 0 & 0 \\
0 & 0 & 0 & 0 & 2709.89 & 0 & 0 \\
0 & 0 & 0 & 0 & 0 & 230.88 & 0 \\
0 & 0 & 0 & 0 & 0 & 0 & 37.34
\end{vmatrix}
$$

To find H_{i+1} by Equation 5.17 following a step-by-step analysis, the following matrices can be found:

$$[N][A11]′$$

$$
\begin{vmatrix}
124.46 & 0 & 0 & 0 & 0 & 0 & 0 \\
0 & 1644.15 & 0 & 0 & 0 & 0 & 0 \\
0 & 0 & 5420.33 & 0 & 0 & 0 & 0 \\
0 & 0 & 0 & 1337.80 & 0 & 0 & 0 \\
0 & 0 & 0 & 0 & 5419.78 & 0 & 0 \\
0 & 0 & 0 & 0 & 0 & 461.76 & 0 \\
0 & 0 & 0 & 0 & 0 & 0 & 74.68
\end{vmatrix}
$$

Chapter 5

$$([N][A11]')^{-1}$$

$$
\begin{vmatrix}
0.00804 & 0 & 0 & 0 & 0 & 0 & 0 \\
0 & 0.00061 & 0 & 0 & 0 & 0 & 0 \\
0 & 0 & 0.00018 & 0 & 0 & 0 & 0 \\
0 & 0 & 0 & 0.00075 & 0 & 0 & 0 \\
0 & 0 & 0 & 0 & 0.00018 & 0 & 0 \\
0 & 0 & 0 & 0 & 0 & 0.00219 & 0 \\
0 & 0 & 0 & 0 & 0 & 0 & 0.01339
\end{vmatrix}
$$

$$[A21]([N][A11]')^{-1}$$

$$
\begin{vmatrix}
0.00804 & -0.00061 & 0 & 0 & -0.00018 & 0 & 0 \\
0 & 0.00061 & 0.00018 & 0 & 0 & 0 & 0 \\
0 & 0 & -0.00018 & 0.00075 & 0 & 0 & 0 \\
0 & 0 & 0 & -0.00075 & 0.00018 & 0.00219 & 0 \\
0 & 0 & 0 & 0 & 0 & -0.00219 & 0.01339
\end{vmatrix}
$$

$$[A21]([N][A11]')^{-1}[A12]$$

$$
\begin{vmatrix}
0.00804 & -0.00061 & 0 & -0.00018 & 0 \\
-0.00061 & 0.00079 & -0.00018 & 0 & 0 \\
0 & -0.00018 & 0.00093 & -0.00075 & 0 \\
-0.00018 & 0 & -0.00075 & 0.00310 & -0.00217 \\
0 & 0 & 0 & -0.00217 & 0.01556
\end{vmatrix}
$$

$$-([A21]([N][A11]')^{-1}[A12])^{-1}$$

$$
\begin{vmatrix}
-120.541 & -100.256 & -33.383 & -16.879 & -2.350 \\
-100.256 & -1423.476 & -365.444 & -104.310 & -14.522 \\
-33.383 & -365.444 & -1460.157 & -392.548 & -54.651 \\
-16.879 & -104.310 & -392.548 & -463.689 & -64.555 \\
-2.350 & -14.522 & -54.651 & -64.555 & -73.273
\end{vmatrix}
$$

$[A11][Q]$	$[A10][H_0]$	$([A11][Q]) + ([A10][H_0])$
6.223	-100	-93.777
82.208	0	82.208
271.016	0	271.016
66.890	0	66.890
270.989	0	270.989
23.088	0	23.088
3.734	-100	-96.266

$([A21]([N][A11]')^{-1}$ $([A11][Q][A10][H_0]))$	$[A21][Q]$	$([A21]([N][A11]')^{-1}([A11]$ $[Q] + [A10][H_0]) - ([A21]$ $[Q] - [q]))$
-0.853	-0.1	-0.6935
0.1	0.2	-0.06
0	0	0.03
0.05	0.1	-0.02
-1.339	0	-1.299

Thus

$$H_{i+1} = -([A21]([N][A11]')^{-1}([A12])^{-1}([A21]([N][A11]')^{-1}([A11][Q]$$
$$+[A10][H_0] - ([A21][Q] - [q]))$$

Node		(m)
2		92.000
3		164.922
4	=	80.115
5		99.317
6		97.335

To find Q_{i+1} by Equation 5.18 following a step-by-step analysis, the following matrices can be found:

$[A12][H_{i+1}]$	$[A12][H_{i+1}]+$ $[A10][H_0]$	$([N][A11']^{-1})(([A12][H_{i+1}])$ $+([A10][H_0]))$
92.00	−8.00	−0.0643
72.92	72.92	0.0444
84.81	84.81	0.0157
−19.20	−19.20	−0.0144
7.32	7.32	0.0014
1.98	1.98	0.0043
97.33	−2.67	−0.0357

$$([N][A11]')^{-1}[A11]$$

0.5	0	0	0	0	0	0
0	0.5	0	0	0	0	0
0	0	0.5	0	0	0	0
0	0	0	0.5	0	0	0
0	0	0	0	0.5	0	0
0	0	0	0	0	0.5	0
0	0	0	0	0	0	0.5

$$([I] - ([N])[A11]')^{-1}[A11])$$

0.5	0	0	0	0	0	0
0	0.5	0	0	0	0	0
0	0	0.5	0	0	0	0
0	0	0	0.5	0	0	0
0	0	0	0	0.5	0	0
0	0	0	0	0	0.5	0
0	0	0	0	0	0	0.5

$$([I] - ([N][A11]')^{-1} \times [A11])[Q]$$

0.05
0.05
0.05
0.05
0.05
0.05
0.05

Thus

$$Q_{i+1} = ([I] - [N][A11'])^{-1}[A11])[Q] - (([N][A11'])^{-1}([A12][H_{i+1}] + [A10][H_0]))$$

Pipe			(m^3/s)
1-2			0.114
2-3			0.006
3-4			0.034
5-4	=		0.064
2-5			0.049
6-5			0.046
6-1			0.086

After only five iterations the following are the results.

Head at each node:

Node			(m)
2			92.960
3			81.358
4	=		81.780
5			89.812
6			96.727

Pipe discharges:

Pipe			(m^3/s)
1-2			0.10667
2-3			0.03658
3-4			0.00342
5-4	=		0.03342
2-5			0.01009
6-5			0.05333
6-1			0.09333

References and recommended reading

Featherstone, R. E. and El Jumailly, K. K. (1983) Optimal diameter selection for pipe networks. *Journal of the Hydraulics Division, American Society of Civil Engineers*, 109, 221–234.

Ratnayaka, D. D., Brandt, M. J. and Johnson, K. M. (2009) *Twort's Water Supply*, 6th edn, Butterworth-Heinemann, Oxford.

Saldarriaga, J. (1998) *Hidraulica de Tuberias*, McGraw-Hill Interamericana, Santafe de Bogota, Colombia.

Walski, T. M., Chase, D. V. and Savic, D. A. (2001) *Water Distribution Modeling*, Haestad Press, Waterbury, CT.

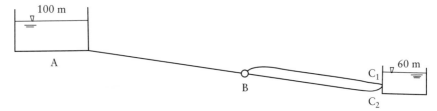

Figure 5.12   Pipes in parallel.

Problems

1. Calculate the flows in the pipes of the pipe system illustrated in Figure 5.12. Minor losses are given by $C_m V^2/2g$.

Pipe	Length (m)	Diameter (mm)	Roughness (mm)	Minor loss coefficients (C_m)
AB	5000	400	0.15	10.0
BC_1	7000	250	0.15	15.0
BC_2	7000	250	0.06	10.0

(*Note:* While this problem could be solved by the method of Example 4.1, the method of quantity balance facilitates a convenient method of solution. Note that the pressure head elevations at the ends of C_1 and C_2 are identical.)

2. In the system shown in Problem 1, an axial flow pump producing a total head of 5.0 m is installed in pipe BC_1 to boost the flow in this branch. Determine the flows in the pipes. (*Note:* Although it is not strictly possible to predict the head generated by a rotodynamic pump since this varies with the discharge (see Chapter 6), axial flow pumps often produce a fairly flat head–discharge curve in the mid-discharge range.)

3. Determine the flows in the network illustrated in Figure 5.13. Minor losses are given by $C_m V^2/2g$.

Pipe	Length (m)	Diameter (mm)	k (mm)	C_m
AB	20 000	500	0.3	20
BC	5 000	350	0.3	10
BD_1	6 000	300	0.3	10
BD_2	6 000	250	0.06	10

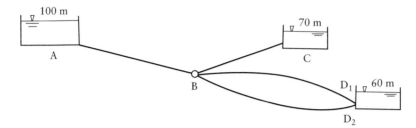

Figure 5.13   Network with reservoirs.

4.   In the system illustrated in Figure 5.14, a pump is installed in pipe BC to provide a flow of 40 L/s to Reservoir C. Neglecting minor losses, calculate the total head to be generated by the pump and the power consumption assuming an overall efficiency of 60%. Determine also the flow rates in the other pipes.

Pipe	Length (m)	Diameter (mm)	Roughness (mm)
AB	10 000	400	0.06
BC	4 000	250	0.06
BD	5 000	250	0.06

5.   Determine the pressure head elevations at B and D and the discharges in the branches in the system illustrated in Figure 5.15. Neglect minor losses.

Pipe	Length (m)	Diameter (mm)	Roughness (mm)
AB	20 000	600	0.06
BC	2 000	250	0.06
BD	2 000	450	0.06
DE	2 000	300	0.06
DF	2 000	250	0.06

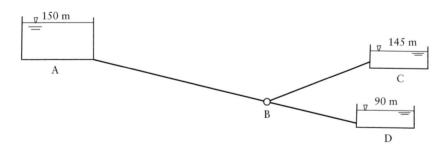

Figure 5.14   Network with reservoirs.

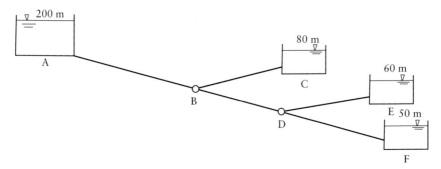

Figure 5.15 Network with reservoirs.

6. Determine the flows in a pipe system similar in configuration to that in Problem 5. A valve is installed in BC producing a minor loss of $20V^2/2g$; otherwise, consider only friction losses.

Pipe	Length (m)	Diameter (mm)	Roughness (mm)
AB	20 000	450	0.06
BC	2 000	300	0.06
BD	10 000	400	0.06
DE	3 000	250	0.06
DF	4 000	300	0.06

7. Determine the flow in the pipes and the pressure head elevations at the junctions of the closed-loop pipe network illustrated, neglecting minor losses. All pipes have the same roughness size of 0.03 mm. The outflows at the junctions are shown in litres per second (see Figure 5.16).

Pipe	AB	BC	CD	DE	EA	BE
Length (m)	500	600	200	600	600	200
Diameter (mm)	200	150	100	150	200	100

Pressure head elevation at A = 60 m.

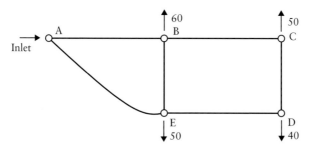

Figure 5.16 Two-loop network.

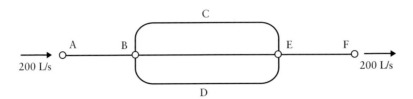

Figure 5.17 Pipes in parallel.

(*Note:* A more rapid solution is obtained by using the head balance method. However, the network can be analysed by the quantity balance method but in this case four unknown pressure heads, at B, C, D and E, are to be corrected. If the quantity balance method is used, set a fixed arbitrary pressure head elevation to A, say 100 m.)

8. Determine the flow distribution in the pipe system illustrated in Figure 5.17 and the total head loss between A and F. Neglect minor losses. A total discharge of 200 L/s passes through the system.

Pipe	AB	BCE	BE	BDE	EF
Length (m)	1000	3000	2000	3000	1000
Diameter (mm)	450	300	250	350	450
Roughness (mm)	0.15	0.06	0.15	0.06	0.15

9. In the system shown in Problem 7 (Figure 5.16) a pump is installed in BC to boost the flow to C. Neglecting minor, losses determine the flow distribution and head elevations at the junctions if the pump delivers a head of 15.0 m.

10. Determine the flows in the pipes and the pressure head elevations at the junctions in the network shown in Figure 5.18. Neglect minor losses and take the pressure head elevation at A to be 100 m. The outflows are in litres per second. All pipes have a roughness of 0.06 mm.

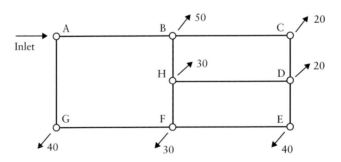

Figure 5.18 Three-loop network.

Pipe	AB	BH	HF	FG	GA
Length (m)	400	150	150	400	300
Diameter (mm)	200	200	150	150	200

Pipe	BC	CD	DH	DE	EF
Length (m)	300	150	300	150	300
Diameter (mm)	150	150	150	150	150

11. Analyse the flows and pressure heads in the pipe system shown in Figure 5.19. Neglect minor losses.

Pipe	AB	BC	CD	DE	EF	EF
Length (m)	1000	400	300	400	800	300
Diameter (mm)	250	200	150	150	250	200
Roughness (mm)	0.06	0.15	0.15	0.15	0.06	0.15

12. Solve the network in Problem 10 using the gradient method.

13. Analyse the network of Example 5.1 by the gradient method.

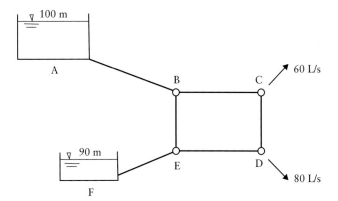

Figure 5.19   Network with reservoirs.

Chapter 5

Chapter 6
Pump–Pipeline System Analysis and Design

6.1 Introduction

This chapter deals with the analysis and design of pipe systems which incorporate rotodynamic pumps. The civil engineer is mostly concerned with pump selection in the design of pumping stations and therefore the design of the shape of pump impellers will not be dealt with here. Rotodynamic pumps can be subclassified into three main categories according to the shape of the impellers:

(i) Centrifugal (radial flow)
(ii) Mixed flow
(iii) Propeller (axial flow)

For the same power input and efficiency the centrifugal type would generate a relatively large pressure head with a low discharge, while the propeller type a relatively large discharge at a low head with the mixed flow having characteristics somewhere between the other two.

Pump types may be more explicitly defined by the parameter called specific speed (N_s), expressed by

$$N_s = \frac{N\sqrt{Q}}{H^{3/4}}$$

where Q is the discharge, H the total head and N the rotational speed (rev/min). This expression is derived from dynamical similarity considerations and may be interpreted as the speed in revolutions per minute at which a geometrically scaled model would have to operate to deliver unit discharge (e.g. 1 L/s) when generating unit head (e.g. 1 m).

Pump type	N_s range (Q, L/s; H, m)
Centrifugal	Up to 2400
Mixed flow	2400 to 5000
Axial flow	Above 3400

Nalluri & Featherstone's Civil Engineering Hydraulics: Essential Theory with Worked Examples, Sixth Edition. Martin Marriott.
© 2016 John Wiley & Sons, Ltd. Published 2016 by John Wiley & Sons, Ltd.
Companion Website: www.wiley.com/go/Marriott

Be aware of the units in this specific speed term, which is not dimensionless, due to the exclusion of the gravity term. See Chadwick *et al.* (2013) for further discussion and details.

The total head generated by a pump is also called the **manometric head** (H_m) since it is the difference in pressure head recorded by pressure gauges connected to the delivery and inlet pipes on either side of the pump, provided that the pipes are of the same diameter.

6.2 Hydraulic gradient in pump–pipeline systems

Figure 6.1 shows a pump delivering a liquid from a lower tank to a higher tank through a static lift H_{ST} at a discharge Q. It is clear that the pump must generate a total head equal to H_{ST} plus the pipeline head losses.

V_s = velocity in suction pipe

V_d = velocity in delivery pipe

h_{ld} = head loss in delivery pipe (friction, valves etc.)

h_{ls} = head loss in suction pipe

h_m = local losses

Manometric head is defined as the rise in total head across the pump:

$$H_m = \frac{p_d}{\rho g} + \frac{V_d^2}{2g} - \left(\frac{p_s}{\rho g} + \frac{V_s^2}{2g}\right) \tag{6.1}$$

$$\text{Now } \frac{p_s}{\rho g} = Z_1 - \frac{V_s^2}{2g} - h_{ls}; \qquad \frac{p_d}{\rho g} = Z_2 + h_{ld} - \frac{V_d^2}{2g}$$

$$\text{Thus } H_m = Z_2 - Z_1 + h_{ld} + h_{ls} \quad \text{or} \quad H_m = H_{ST} + h_{ld} + h_{ls} \tag{6.2}$$

Note that the energy losses within the pump itself are not included; such losses will affect the efficiency of the pump.

Total head–discharge and efficiency–discharge curves (Figure 6.2) for particular pumps are obtained from the manufacturers.

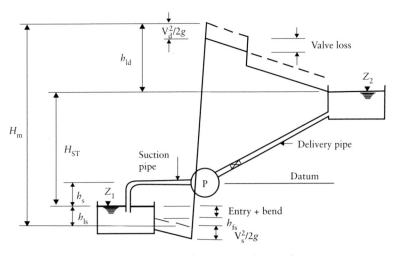

Figure 6.1 Total energy and hydraulic grade lines in pipeline with pump.

Chapter 6

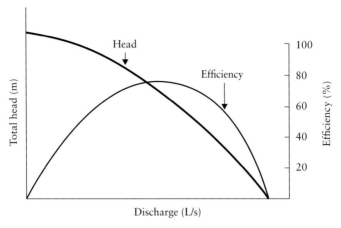

Figure 6.2 Typical performance curve for centrifugal pump.

The total head–discharge curves for a centrifugal pump can generally be expressed in the functional form

$$H_m = AQ^2 + BQ + C \qquad [6.3]$$

The coefficients A, B and C can be evaluated by taking three pairs of H_m and Q from a particular curve and solving Equation 6.3.

The power P (in watts) consumed by a pump when delivering a discharge Q (m³/s) at a head H_m (in metres) with a combined pump–motor efficiency η is

$$P = \frac{\rho g Q H_m}{\eta}$$

6.3 Multiple pump systems

6.3.1 Parallel operation

Pumping stations frequently contain several pumps in a 'parallel' arrangement. In this configuration (Figure 6.3) any number of the pumps can be operated simultaneously, the objective being to deliver a range of discharges. This is a common feature of sewage

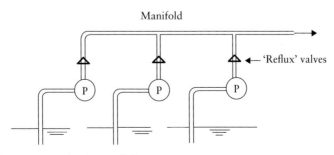

Figure 6.3 Pumps operating in parallel.

Chapter 6

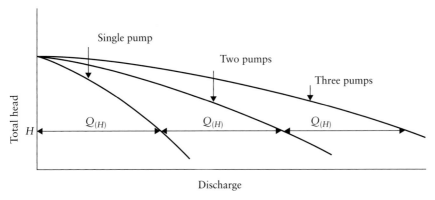

Figure 6.4   Characteristic curves for identical pumps operating in parallel.

pumping stations where the inflow rate varies during the day. By automatic switching according to the level in the suction well, any number of the pumps can be brought into operation.

In predicting the head–discharge curve for parallel operation it is assumed that the head across each pump is the same. Thus at any arbitrary head the individual pump discharges are added as shown in Figure 6.4.

6.3.2 Series operation

This configuration is the basis of multi-stage and borehole pumps; the discharge from the first pump (or stage) is delivered to the inlet of the second pump, and so on. The same discharge passes through each pump, receiving a pressure boost in doing so. Figure 6.5 shows the series configuration together with the resulting head–discharge characteristics which are clearly obtained by adding the individual pump manometric heads at any

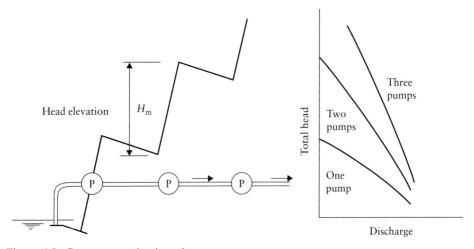

Figure 6.5   Pumps operating in series.

arbitrary discharges. Note that, of course, all pumps in a series system must be operating simultaneously.

6.4 Variable-speed pump operation

By the use of variable-speed motors the discharge of a single pump can be varied to suit the operating requirements of the system.

Using dimensional analysis and dynamic similarity criteria (see Chapter 9) it can be shown that if the pump delivers a discharge Q_1 at manometric head H_1 when running at speed N_1, the corresponding values when the pump is running at speed N_2 are given by the following relationships:

$$Q_2 = Q_1 \left(\frac{N_2}{N_1} \right) \tag{6.4}$$

$$H_2 = H_1 \left(\frac{N_2}{N_1} \right)^2 \tag{6.5}$$

In constructing the characteristic curve for speed N_2, several pairs of values of Q_1, H_1 from the curve for N_1 can be obtained and transformed into homologous points Q_2, H_2 on the N_2 curve (see Figure 6.6).

6.5 Suction lift limitations

Cavitation, the phenomenon which consists of local vaporisation of a liquid and which occurs when the absolute pressure falls to the vapour pressure of the liquid at the operating temperature, can occur at the inlet to a pump and on the impeller blades, particularly if the pump is mounted above the level in the suction well. Cavitation causes physical damage, noise and reduction in discharge, and to avoid it the pressure head at the inlet should not

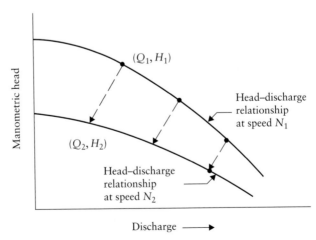

Figure 6.6 Effect of speed change on pump characteristics.

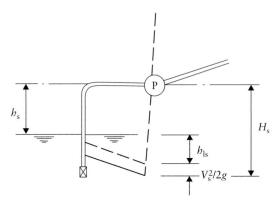

Figure 6.7   Head conditions in suction pipe.

fall below a certain minimum, which is influenced by the further reduction in pressure within the pump impeller (see Figure 6.7).

If p_s represents the pressure at inlet, then $(p_s - p_v)/\rho g$ is the absolute head at the pump inlet above the vapour pressure (p_v) and is known as the net positive suction head (NPSH).

$$\text{Thus, NPSH} = \frac{p_s - p_v}{\rho g} = \frac{p_a}{\rho g} - H_s - \frac{p_v}{\rho g} \qquad [6.6]$$

where p_a is the ambient atmospheric pressure.

$$H_s = \text{manometric suction head} = h_s + h_{ls} + \frac{V_s^2}{2g} \qquad [6.7]$$

where h_s is the suction lift, h_{ls} total head loss in the suction pipe, V_s velocity head in the suction pipe and ρ the density of liquid.

Values of NPSH can be obtained from the pump manufacturer and are derived from full-scale or model tests; these values must not be exceeded if cavitation is to be avoided.

Thoma introduced a cavitation number σ $(=\text{NPSH}/H_m)$ and from physical tests found this to be strongly related to specific speed.

In recent years submersible pumps in the small-to-medium size range have been widely used. This type eliminates the need for suction pipes, and provided that the pump is immersed to the manufacturer's specified depth, the problems of cavitation and cooling are avoided. An example, including many practical details, is given in Water UK/WRc (2012).

Worked examples

Example 6.1

Tests on a physical model pump indicated a cavitation number of 0.12. A homologous (geometrically and dynamically similar) unit is to be installed where the atmospheric pressure is 950 mb and the vapour pressure head 0.2 m. The pump will be situated above the suction well, the suction pipe being 200 mm in diameter, of unplasticised polyvinyl chloride (PVCu), 10 m long; it is vertical with a 90° elbow leading into the pump inlet and is fitted with a foot valve. The foot-valve head loss $(h_v) = 4.5V_s^2/2g$ and bend loss $(h_b) = 1.0V_s^2/2g$. The total head at the operating discharge of 35 L/s is 25 m. Calculate the maximum permissible suction head and suction lift.

Solution:

$$p_a = 950\,\text{mb} = 0.95 \times 10.198 = 9.688 \text{ m of water}$$

$$\Rightarrow \frac{p_a - p_v}{\rho g} = 9.688 - 0.2 = 9.488 \text{ m of water}$$

$$\text{NPSH} = \sigma H_m = 0.12 \times 25 = 3.0 \text{ m}$$

From Equation 6.6, maximum permissible suction head is

$$H_s = 9.488 - 3.0 = 6.488 \text{ m}$$

Now calculate the losses in the suction pipe:

$$V_s = 1.11 \text{ m/s}; \qquad \frac{V_s^2}{2g} = 0.063 \text{ m}; \qquad Re = 1.96 \times 10^5$$

$$k = 0.03 \text{ mm (PVCu)}; \qquad \frac{k}{D} = 0.001, \qquad \text{whence } \lambda = 0.0167$$

$$\Rightarrow h_{fs} = 0.053 \text{ m}; \qquad h_v = 4.5 \times 0.063 = 0.283 \text{ m}; \qquad h_b = 0.063 \text{ m}$$

$$\Rightarrow h_{ls} = 0.4 \text{ m}$$

$$h_s = \text{suction lift} = H_s - h_{ls} - V_s^2/2g = 6.488 - 0.463 = 6.025 \text{ m}$$

(See also Example 6.7.)

Example 6.2

A centrifugal pump has a 100 mm diameter suction pipe and a 75 mm diameter delivery pipe. When discharging 15 L/s of water, the inlet water–mercury manometer with one limb exposed to the atmosphere recorded a vacuum deflection of 198 mm; the mercury level on the suction side was 100 mm below the pipe centre line. The delivery pressure gauge, 200 mm above the pump inlet, recorded a pressure of 0.95 bar. The measured input power was 3.2 kW. Calculate the pump efficiency (see Figure 6.8).

Solution:

$$\text{Manometric head} = \text{rise in total head}$$

$$H_m = \frac{p_2}{\rho g} + \frac{V_2^2}{2g} + z - \left(\frac{p_1}{\rho g} + \frac{V_1^2}{\rho g} \right)$$

$$1 \text{ bar} = 10.198 \text{ m of water}$$

$$\frac{p_2}{\rho g} = 0.95 \times 10.198 = 9.69 \text{ m of water}$$

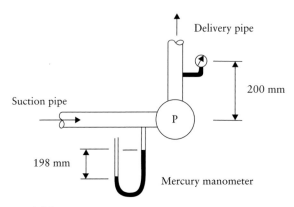

Figure 6.8 Suction and delivery pressures across pump.

$$\frac{p_1}{\rho g} = -0.1 - 0.198 \times 13.6 = -2.793 \text{ m of water}$$

$$V_2 = 3.39 \text{ m/s}; \qquad \frac{V_2^2}{2g} = 0.588 \text{ m}$$

$$V_1 = 1.91 \text{ m/s}; \qquad \frac{V_1^2}{2g} = 0.186 \text{ m}$$

Then, $H_m = 9.69 + 0.588 + 0.2 - (-2.793 + 0.186) = 13.09 \text{ m}$

$$\text{Efficiency}, \eta = \frac{\text{output power}}{\text{input power}} = \frac{\rho g Q H_m \text{ (watts)}}{3200 \text{ (watts)}}$$

$$\eta = \frac{9.81 \times 0.015 \times 13.09}{3.2} = 0.602 \ (60.2\%)$$

Example 6.3

Calculate the steady discharge of water between the tanks in the system shown in Figure 6.1 and the power consumption. Pipe diameter $(D_s = D_d) = 200$ mm; length = 2000 m; $k = 0.03$ mm (PVCu). Losses in valves, bends plus the velocity head, amount to $6.2V^2/2g$. Static lift is 10.0 m.

Pump characteristics

Discharge (L/s)	0	10	20	30	40	50
Total head (m)	25.0	23.2	20.8	16.5	12.4	7.3
Efficiency (%)	–	45	65	71	65	45

The efficiencies given are the overall efficiencies of the pump and motor combined.

Solution:

The solution to such problems is basically to solve simultaneously the head–discharge relationships for the pump and pipeline.

For the pump, head delivered at discharge Q may be expressed by

$$H_m = AQ^2 + BQ + C \tag{i}$$

and for the pipeline, the head required to produce a discharge Q is given by

$$H_m = H_{ST} + \frac{K_m Q^2}{2gA^2} + \frac{\lambda L Q^2}{D2gA^2} \quad \text{(from Equation 6.2)} \tag{ii}$$

where K_m is the minor loss coefficient.

A graphical solution is the simplest method and also gives the engineer a visual interpretation of the 'matching' of the pump and pipeline.

Equation (ii) when plotted (H vs. Q) is called the *system curve*. Values of H corresponding with a range of Q values will be calculated: $k/D = 0.03$; values of λ obtained from the Moody diagram.

Q (L/s)	10	20	30	40	50
Re ($\times 10^5$)	0.56	1.13	1.10	2.25	2.81
λ	0.0210	0.0185	0.0172	0.0165	0.0160
h_f (m)	1.08	3.82	7.99	13.63	20.65
H_m (m)	0.03	0.13	0.29	0.51	0.80
H (m)	11.11	13.95	18.28	24.14	31.45

Alternatively, the combined Darcy–Colebrook–White equation can be used:

$$Q = \frac{-2\pi D^2}{4} \sqrt{2gD \frac{h_f}{L}} \log \left(\frac{k}{3.7D} + \frac{2.51\nu}{D\sqrt{2gDh_f/L}} \right)$$

In evaluating pairs of H and Q, it is now preferable to take discrete values of h_f, calculate Q explicitly from the above equation and add the static lift and minor head loss.

h_f (m)	2.00	4.00	8.00	16.00
Q (L/s)	14.06	20.57	30.00	43.61
H_m (m)	0.06	0.13	0.29	0.61
H (m)	12.06	14.13	18.29	26.61

The computed system curve data and pump characteristic curve data are now plotted in Figure 6.9.

The intersection point gives the operating conditions; in this case, $H_m = 17.5$ m and $Q = 28.0$ L/s. The operating efficiency is 71%. Therefore,

$$\text{power consumption, } P = \frac{1000 \times 9.81 \times 0.028 \times 17.5}{0.71}$$

$$= 6770 \text{ W (6.77 kW)}$$

Chapter 6

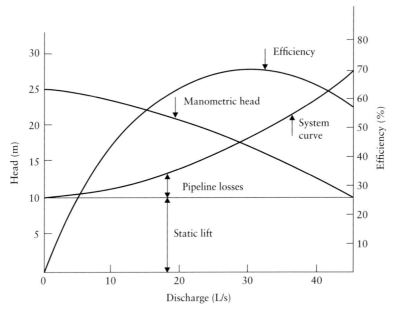

Figure 6.9   Pump and system characteristics.

Example 6.4 (Pipeline selection in pumping system design)

An existing pump having the tabulated characteristics is to be used to pump raw sewage to a treatment plant through a static lift of 20 m. A PVCu pipeline 10 km long is to be used. Allowing for minor losses totalling $10V^2/2g$ and taking an effective roughness of 0.15 mm because of sliming, select a suitable commercially available pipe size to achieve a discharge of 60 L/s. Calculate the power consumption.

Pump characterstics

Discharge (L/s)	0	10	20	30	40	50	60	70
Total head (m)	45.0	44.7	43.7	42.5	40.6	38.0	35.0	31.0
Overall efficiency (%)	–	35	50	57	60	60	53	40

Solution:

At 60 L/s, total head = 35.0 m; therefore, the sum of the static lift and pipeline losses must not exceed 35.0 m.

$$\text{Try 300 mm diameter: } A = 0.0707 \text{ m}^2; \qquad V = 0.85 \text{ m/s}$$
$$Re = 2.25 \times 10^5; \qquad k/D = 0.0005; \qquad \lambda = 0.019$$
$$\text{Friction head loss} = \frac{0.019 \times 10\,000 \times 0.85^2}{0.3 \times 19.62} = 23.32 \text{ m}$$
$$H_s + h_f = 43.32 \; (>35) \quad \text{– pipe diameter too small}$$

Try 350 mm diameter: $A = 0.0962\,\text{m}^2$; $V = 0.624\,\text{m/s}$

$$Re = 1.93 \times 10^5; \qquad k/D = 0.00043; \qquad \lambda = 0.0185$$

$$h_f = 10.48\,\text{m}; \qquad h_m = \frac{10 \times 0.624^2}{19.62} = 0.2\,\text{m}$$

$$H_{ST} + h_f + h_m = 30.68 \ (<35\,\text{m}) \quad - \text{OK}$$

The pump would deliver approximately 70 L/s through the 350 mm pipe, and to regulate the flow to 60 L/s an additional head loss of 4.32 m by valve closure would be required.

$$\text{Power consumption, } P = \frac{1000 \times 9.81 \times 0.06 \times 35}{0.55 \times 1000}$$

$$= 38.85\,\text{kW}$$

Example 6.5 (Pumps in parallel and series)

Two identical pumps having the tabulated characteristics are to be installed in a pumping station to deliver sewage to a settling tank through a 200 mm PVCu pipeline 2.5 km long. The static lift is 15 m. Allowing for minor head losses of $10V^2/2g$ and assuming an effective roughness of 0.15 mm, calculate the discharge and power consumption if the pumps were to be connected (a) in parallel and (b) in series.

Pump characteristics

Discharge (L/s)	0	10	20	30	40
Total head (m)	30.0	27.5	23.5	17.0	7.5
Overall efficiency (%)	–	44	58	50	18

Solution:

The 'system curve' is computed as in the previous examples; this is, of course, independent of the pump characteristics. Calculated system heads (H) are tabulated below for discrete discharges (Q):

$$H = H_{ST} + h_f + H_m$$

Q (L/s)	10	20	30	40
H (m)	16.53	20.80	27.37	36.48

(a) *Parallel operation*: The predicted head–discharge curve for dual-pump operation in parallel mode is obtained as described in Section 6.3.1 (i.e. by doubling the discharge over the range of heads, since the pumps are identical in this case). The system and efficiency curves are added as shown in Figure 6.10. From the intersection of the characteristic and system curves, the following results are obtained:

$$\text{Single-pump operation, } Q = 22.5\,\text{L/s}; \quad H_m = 24\,\text{m}; \quad \eta = 0.58$$

$$\text{Power consumption} = 9.13\,\text{kW}$$

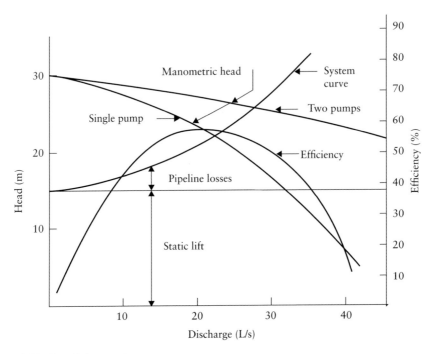

Figure 6.10   Parallel operation.

Parallel operation, $Q = 28.5 \, \text{L/s}$; $H_m = 26 \, \text{m}$; $\eta = 0.51$ (corresponding with 14.25 L/s per pump)

$$\text{Power input} = 14.11 \, \text{kW}$$

(b)   *Series operation*: Using the method described in Section 6.3.2 and plotting the dual-pump characteristic curve, intersection with the system curve yields (see Figure 6.11) the following:

$$Q = 32.5 \, \text{L/s}; \qquad H_m = 25 \, \text{m}; \qquad \eta = 0.41$$
$$\text{Power input} = 21.77 \, \text{kW}$$

Note that for this particular pipe system, comparing the relative power consumptions, the parallel operation is more efficient in producing an increase in discharge than the series operation.

Example 6.6 (Pump operation at different speeds)

A variable-speed pump having the tabulated characteristics, at 1450 rev/min, is installed in a pumping station to handle variable inflows. Static lift = 15 m; diameter of pipeline = 250 mm; length = 2000 m; $k = 0.06$ mm; minor loss = $10V^2/2g$.
Determine the total head of the pump and discharge in the pipeline at pump speeds of 1000 rev/min and 500 rev/min.

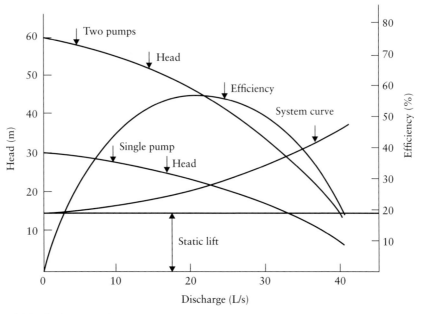

Figure 6.11 Series operation.

Pump characteristics at 1450 rev/min

Discharge (L/s)	0	10	20	30	40	50	60	70
Total head (m)	45.0	44.0	42.5	39.5	35.0	29.0	20.0	6.0

Solution:

The characteristic curve for speed N_2 using that for speed N_1 can be constructed using Equations 6.4 and 6.5 (Section 6.4); in other words,

$$H_2 = H_1 \left(\frac{N_2}{N_1} \right)^2 \tag{i}$$

$$Q_2 = Q_1 \left(\frac{N_2}{N_1} \right) \tag{ii}$$

where Q_1, H_1 are pairs of values taken from the N_1 curve and Q_2, H_2 are the corresponding points on the N_2 curve.

The system curve is computed giving the following data:

Discharge (L/s)	20	40	60	80
System head (m)	16.40	20.08	26.12	34.23

Chapter 6

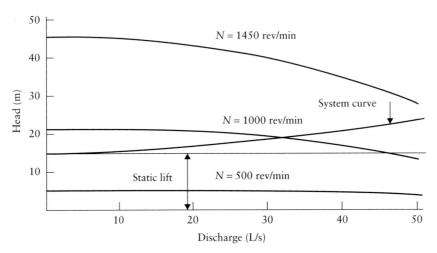

Figure 6.12   Pump operation with variable speed.

Construct the pump head–discharge curves for speeds of 1000 rev/min and 500 rev/min using Equations (i) and (ii). Q_1, H_1 values (at 1450 rev/min) can be taken from the tabulated data (or from the plotted curve in Figure 6.12). For example, taking $Q_1 = 20$ L/s and $H_1 = 42.5$ m at 1450 rev/min, the corresponding values at 1000 rev/min are

$$Q_2 = 20 \times \left(\frac{1000}{1450}\right) = 13.79 \text{ L/s}; \qquad H_2 = 42.5 \left(\frac{1000}{1450}\right)^2 = 20.2 \text{ m}$$

Taking three other pairs of values, the following table can be constructed:

N (rev/min)					
1450	Q_1	0	20	40	60
	H_1	45.0	42.5	35.0	20.0
1000	Q_2	0	13.79	27.59	41.38
	H_2	21.40	20.20	16.65	9.50
500	Q_2	0	6.9	13.90	20.70
	H_2	5.35	5.05	4.16	2.40

The computed values are now plotted together with the system curve (see Figure 6.12). Operating conditions:

At N = 1450 rev/min, $Q = 55$ L/s; $H_m = 25.0$ m

At N = 1000 rev/min, $Q = 33$ L/s; $H_m = 18.5$ m

At N = 500 rev/min,   no discharge produced.

Example 6.7

A laboratory test on a pump revealed that the onset of cavitation occurred at a discharge of 35 L/s when the total head at the inlet was reduced to 2.5 m and the total head across the pump was 32 m. Barometric pressure was 760 mm Hg, and the vapour pressure 17 mm Hg. Calculate the Thoma cavitation number. The pump is to be installed in a situation where the atmospheric pressure is 650 mm Hg and water temperature 10°C (vapour pressure 9.22 mm Hg) to give the same total head and discharge. The losses and velocity head in the suction pipe are estimated to be 0.55 m of water. What is the maximum height of the suction lift?

Solution:

$$\text{NPSH} = \left(\frac{p_a}{\rho g} - H_s\right) - \frac{p_v}{\rho g} \quad \text{(Equation 6.6)} \qquad \text{(i)}$$

where H_s is the manometric suction head.

$$\frac{p_a}{\rho g} = 103 \text{ m of water}; \qquad \frac{p_v}{\rho g} = 0.23 \text{ m of water}$$

$$\frac{p_a}{\rho g} - H_s = 2.5 \text{ m}$$

$$\Rightarrow \text{NPSH} = 2.5 - 0.23 = 2.27 \text{ m}$$

Cavitation number,

$$\sigma = \frac{\text{NPSH}}{H} = \frac{2.27}{32} = 0.071$$

$$\text{Installed conditions:} \quad \frac{p_a}{\rho g} = 8.84 \text{ m}; \qquad \frac{p_v}{\rho g} = 0.1254 \text{ m} \quad \text{(at 10°C)}$$

$$\text{NPSH} = \frac{p_a}{\rho g} - H_s - \frac{p_v}{\rho g} \qquad \text{(from Equation (i))}$$

$$\Rightarrow 2.27 = 8.84 - H_s - 0.1254$$

$$\text{whence } H_s = 6.44 \text{ m}$$

$$H_s = h_s + h_{ls} + \frac{V_s^2}{2g}$$

$$\text{whence } h_s = 6.44 - 0.55 = 5.89 \text{ m}$$

where h_s is the suction lift.

Example 6.8

An impounding reservoir at elevation 200 m delivers water to a service reservoir at elevation 80 m through a 20 km long, 500 mm diameter coated cast iron pipeline ($k = 0.03$ mm). Minor losses amount to $20V^2/2g$. Determine the steady discharge (410.9 L/s). A booster pump having the tabulated characteristics is to be installed on the pipeline. Determine the improved discharge and the power consumption (see Figure 6.13).

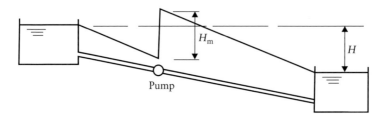

Figure 6.13   Pipe delivery with booster pump.

Q (L/s)	0	100	200	300	400	500	600
H_m (m)	60.0	58.0	54.0	47.0	38.4	26.0	8.0
Overall efficiency (%)	–	33.0	53.0	62.0	62.0	54.0	28.0

The effective gross head is now $H_e = H + H_m$, where H_m is a function of the total discharge passing through the pump.

Thus,

$$H_e = f_1(Q^2) \tag{i}$$

The head H_e is overcome by the pipeline losses

$$H_L = f_2(Q^2) \tag{ii}$$

The discharge in the system is therefore evaluated by equating (i) and (ii); this can be done graphically as in the previous examples.

Compute the head loss–discharge curve ($f_2(Q^2)$) for the pipeline using one of the methods described in Chapter 4. The relationship (Equation (ii)) is as follows:

Total pipeline head loss (m)	120	130	140	150	160
Discharge (L/s)	410.90	428.70	445.75	462.25	478.25

Using a common head datum of 120 m, Equations (i) and (ii) are now plotted (see Figure 6.14):

The point of intersection yields: $Q = 465 \, \text{L/s};$ $H_m = 32 \, \text{m};$ $\eta = 58\%$

$$\text{Power consumption, } P = \frac{9.81 \times 0.465 \times 32.0}{0.58} = 251.67 \, \text{kW}$$

Example 6.9 (Pipe network with pump, using a head–discharge curve)

Neglecting minor losses, determine the discharges in the pipes of the network illustrated in Figure 6.15, (a) with the pump in BC absent and (b) with the pump which boosts the flow to C in operation, and calculate the power consumption.

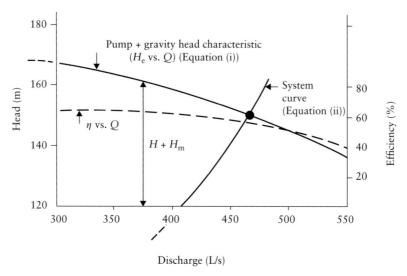

Figure 6.14   Pump and gravity and system characteristics.

Pipe	Length (m)	Diameter (mm)	Roughness (mm)
AB	5000	300	0.06
BC	2000	200	0.06
BD	3000	150	0.06

Pump characteristics

Discharge (L/s)	0	20	40	60	80	100
Total head (m)	40.0	38.8	35.4	29.5	21.0	10.0
Efficiency (%)	–	50.0	70.0	73.0	58.0	22.0

Solution:

Plot the pump head–discharge curve (see Figure 6.16).

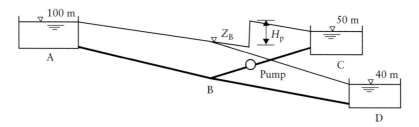

Figure 6.15   Pipe network with pump.

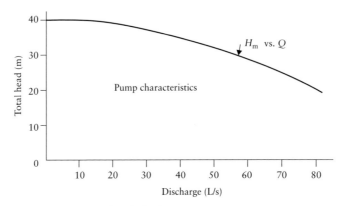

Figure 6.16 Pump characteristic head–discharge curve.

The analysis is carried out using the quantity balance method, noting that with the pump in operation in BC the head producing flow is $H_{BC} = Z_c - Z_B - H_p$, where H_p is the head delivered by the pump at the discharge in the pipeline. Initially, H_p can be obtained using an estimated pipe velocity but subsequently the computed discharges can be used.

(a)

Pipe	AB	BC	BD
k/D	0.0002	0.0003	0.0004

λ values may be obtained from the Moody diagram using initially assumed values of pipe velocity.

With the pump not installed,

$$Z_B = 80.25 \text{ m}$$
$$\text{and } Q_{AB} = 84.87 \text{ L/s}; \quad Q_{BC} = 58.97 \text{ L/s}; \quad Q_{BD} = 25.9 \text{ L/s}$$

(b) Estimate $Z_B = 80$ m.

Units in table: $V = $ m/s; H_p (pump total head) $= $ m; $Q = $ m³/s

'Head' $= $ head loss in pipeline
$A = $ area of pipe (m²)

H_p is initially obtained from an estimated flow in BC of 60 L/s (29 m).

Pipe	V (estimated) (m/s)	Re (×10⁵)	λ	H_p (m)	Head (m)	Q (m³/s)	Q/h (×10⁻³)	Q/A (m/s)
AB	2.0	5.3	0.0155	–	20.0	0.0871	4.36	1.23
BC	2.0	5.3	0.0170	29.0	−59.0	−0.0820	1.39	2.6
BD	2.0	5.3	0.0178	–	−40.0	−0.0262	0.65	1.5
					Σ	−0.0211	6.40	

$\Delta Z_B = -6.6$ m; $Z_B = 73.41$ m

Note that H_P at each step is obtained from the H versus Q curve corresponding with the value of Q_{BC} at the previous step.

Pipe	V (estimated)	Re (×10⁵)	λ	H_P	Head	Q	Q/h (×10⁻³)	Q/A
AB	1.5	3.98	0.016	–	26.59	0.0989	3.72	1.40
BC	2.5	4.42	0.017	22.0	−45.41	−0.0720	1.58	2.30
BD	1.5	1.99	0.018	–	−33.41	−0.0238	0.71	1.35
					Σ		0.0031	6.01

$\Delta Z_B = 1.03$ m; $Z_B = 74.44$ m

Pipe	V (estimated)	Re (×10⁵)	λ	H_P	Head	Q	Q/h (×10⁻³)	Q/A
AB	1.4	3.7	0.016	–	25.56	0.0969	3.79	1.37
BC	2.3	4.07	0.0165	24.8	−49.24	−0.0760	1.54	2.4
BD	1.35	1.79	0.0185	–	−34.44	−0.0239	0.69	1.35
					Σ		−0.003	6.02

$\Delta Z_B = -0.99$ m; $Z_B = 73.45$ m

Pipe	V (estimated)	Re (×10⁵)	λ	H_P	Head	Q	Q/h (×10⁻³)	Q/A
AB	1.4	3.7	0.016	–	26.55	0.0988	3.72	1.4
BC	2.4	4.2	0.0165	23.0	−46.45	−0.0738	1.59	2.35
BD	1.35	1.79	0.0185	–	−33.45	−0.0236	0.70	1.33
					Σ		0.0014	6.01

$\Delta Z_B = 0.47$ m; $Z_B = 73.92$ m

Final values:

$$Z_B = 73.79 \text{ m}$$
$$Q_{AB} = 98.1 \text{ L/s}; \qquad Q_{BC} = 74.5 \text{ L/s}; \qquad Q_{BD} = 23.6 \text{ L/s}$$
$$H_p = 23.5 \text{ m}$$

At the pump discharge of 74.5 L/s, efficiency = 64%.

Whence

$$\text{Power} = \frac{9.81 \times 0.0745 \times 23.5}{0.64}$$
$$= 26.83 \, \text{kW}$$

References and recommended reading

Chadwick, A., Morfett, J. and Borthwick, M. (2013) *Hydraulics in Civil and Environmental Engineering*, 5th edn, Taylor & Francis, Abingdon, UK.

Novak, P., Moffat, A. I. B., Nalluri, C. and Narayanan, R. (2007) *Hydraulic Structures*, 4th edn, Taylor & Francis, Abingdon, UK.

Water UK/WRc (2012) *Sewers for Adoption – A Design and Construction Guide for Developers*, 7th edn, WRc, Swindon.

Problems

1. A rotodynamic pump having the characteristics tabulated below delivers water from a river at elevation 52.0 m to a reservoir with a water level of 85 m through a 350 mm diameter coated cast iron pipeline, 2000 m long ($k = 0.15$ mm). Allowing $10V^2/2g$ for losses in valves, calculate the discharge in the pipeline and the power consumption.

Q (L/s)	0	50	100	150	200
H_m (m)	60	58	52	41	25
η (%)	–	44	65	64	48

2. If, in the system described in Problem 1, the discharge is to be increased to 175 L/s by the installation of a second identical pump,
    (a) Determine the unregulated discharges produced by connecting the pumps (i) in parallel and (ii) in series.
    (b) Calculate the power demand when the discharge is regulated (by valve control) to 175 L/s in the case of (i) parallel operation and (ii) series operation.

3. A pump is required to discharge 250 L/s against a calculated system head of 6.0 m. Assuming that the pump will run at 960 rev/min, what type of pump would be most suitable?

4. The performance characteristics of a variable-speed pump when running at 1450 rev/min are tabulated below, together with the calculated system head losses. The static lift is 8.0 m. Determine the discharge in the pipeline when the pump runs at 1450, 1200 and 1000 rev/min.

Q (L/s)	0	10	20	30	40
H_m (m)	20.0	19.2	17.0	13.7	8.7
System head loss (m)	–	0.7	2.3	4.8	9.0

5. A pump has the characteristics tabulated when operating at 960 rev/min. Calculate the
    specific speed and state what type of pump this is. What discharge will be produced
    when the pump is operating at a speed of 700 rev/min in a pipeline having the system
    characteristics given in the table? Static lift is 2.0 m. What power would be consumed by
    the pump itself?

Q (L/s)	0	50	100	150	200	250	300
H_m (m)	7.0	6.3	5.5	5.0	4.6	4.1	3.5
Pump efficiency (%)	–	20.0	40.0	56.0	71.0	81.0	82.0
System head loss (m)	–	0.10	0.35	0.80	1.40	2.10	3.40

6. (a) Tests on a rotodynamic pump revealed that cavitation started when the manometric
    suction head, H_s, was 5 m, the discharge 60 L/s and the total head 40 m. Barometric
    pressure was 986 mb and the vapour pressure 23.4 mb. Calculate the NPSH and the
    Thoma cavitation number.
    (b) Determine the maximum suction lift if the same pump is to operate at a discharge of
    65 L/s and total head of 35 m under field conditions where the barometric pressure
    is 950 mb and vapour pressure is 12.5 mb. The sum of the suction pipe losses and
    velocity head is estimated to be 0.6 m.

7. The characteristics of a variable-speed rotodynamic pump when operating at 1200
    rev/min are as follows:

Q (L/s)	0	10	20	30	40	50	60
H_m (m)	47.0	46.0	42.5	38.4	34.0	27.2	20.0

    The pump is required to be used to deliver water through a static lift of 10 m in a 300 mm
    diameter pipeline, 5000 m long, and of roughness size 0.15 mm, at a rate of 70 L/s. At
    what speed will the pump have to operate?

8. The steady level below ground level in an abstraction well in a confined aquifer is calcu-
    lated from the equation

$$Z_w = 2.0 + \frac{Q}{2\pi Kb} \log_e \frac{R_o}{r_w} \text{ (m)}$$

    where Q is the abstraction rate (m³/day), K the coefficient of permeability of the aquifer
    (m³/(day m²)), b the aquifer thickness (m), R_o the radius of influence of the well (m) and
    r_w the radius of the well (m).

$$K = 50 \text{ m}^3/(\text{day m}^2); \quad b = 20 \text{ m}; \quad r_w = 0.15 \text{ m}$$

    During a pumping test, the observed Z_w was 5.0 m when an abstraction rate of 30 L/s
    was applied.
    Under operating conditions the submersible borehole pump delivers the groundwater
    to the surface, from where an inline booster pump delivers the water to a reservoir the

level in which is 20 m above the ground level at the well site. The pipeline is of length 500 m, diameter 200 mm and roughness size 0.3 mm. Minor losses total $10V^2/2g$.

Pump characteristics

	(a) Borehole pump					(b) Booster pump			
Discharge (L/s)	0	10	20	30	40	0	10	20	30	40
Total head (m)	10.0	9.6	8.7	7.4	5.6	22.0	21.5	20.4	19.0	17.4

Assuming that the radius of influence of the well is linearly related to the abstraction rate, determine the maximum discharge which the combined pumps would deliver to the reservoir.

9. The discharge in a pipeline delivering water under gravity between two reservoirs at elevations 150 m and 60 m is to be boosted by the installation of a rotodynamic pump, the characteristics of which are shown tabulated.

The pipeline is 15 km in length and 350 mm in diameter, and has a roughness value of 0.3 mm. Determine the discharges (a) under gravity flow conditions and (b) with the pump installed. Assume a minor loss of $20V^2/2g$ in both cases.

Pump characteristics

Discharge (L/s)	0	50	100	150	200	250
Manometric head (m)	50.0	49.0	46.5	42.0	36.0	28.2

10. Reservoir A delivers water to service reservoirs C and D through pipelines AB, BC and BD. A pump is installed in pipeline BD to boost the flow to D.

Elevations of water in reservoirs: $Z_A = 100\,\text{m}$; $Z_C = 60\,\text{m}$; $Z_D = 50\,\text{m}$

Pipe	Length (m)	Diameter (mm)	Roughness (mm)
AB	10 000	350	0.15
BC	4 000	200	0.15
BD	5 000	150	0.15

Pump characteristics

Discharge (L/s)	0	20	40	60	80
Total head (m)	30.0	27.5	23.0	17.0	9.0
Efficiency (%)	–	44.0	68.0	66.0	44.0

Neglecting minor losses, calculate the flows to the service reservoirs:
(a) With the pump not installed
(b) With the pump operating, and determine the power consumption.

Chapter 7
Boundary Layers on Flat Plates and in Ducts

7.1 Introduction

A boundary layer will develop along either side of a flat plate placed edgewise into a fluid stream (Figure 7.1) with free stream velocity U_0 and kinematic viscosity v. Initially the flow in the layer may be laminar with a parabolic velocity distribution. The boundary layer increases in thickness with distance x along the plate, and at a Reynolds number $U_0 x/v \simeq 500\,000$ turbulence develops in the boundary layer. The frictional force due to the turbulent portion of the boundary layer may be considered as that which would be found if the entire length were turbulent *minus* that corresponding to the hypothetical turbulent layer up to the critical point.

For rough plates, Schlichting gave an expression for the maximum height of roughness elements such that the surface may be considered hydraulically smooth:

$$k \le \frac{100v}{U_0} \tag{7.1}$$

The theory of boundary layers on smooth flat plates is to be found in texts such as Massey and Ward-Smith (2012).

7.2 The laminar boundary layer

Blasius developed analytical equations for the flow in a laminar boundary layer formed on a flat plate for the case of zero pressure gradient along the plate. Taking the outer limit of the boundary layer as the position where $v = 0.99U_0$, the boundary layer thickness δ_x at distance x from the leading edge was given as

$$\delta_x = \frac{5x}{Re_x^{1/2}} \quad \text{where } Re_x = \frac{U_0 x}{v} \tag{7.2}$$

$$\text{Local boundary shear stress, } \tau_0 = 0.332\mu \frac{U_0}{x} Re_x^{1/2} \tag{7.3}$$

Nalluri & Featherstone's Civil Engineering Hydraulics: Essential Theory with Worked Examples, Sixth Edition. Martin Marriott.
© 2016 John Wiley & Sons, Ltd. Published 2016 by John Wiley & Sons, Ltd.
Companion Website: www.wiley.com/go/Marriott

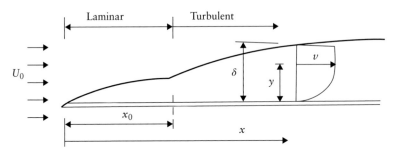

Figure 7.1   Boundary layer formation on a flat plate.

Drag along one side of a plate of length L and width B

$$F_s = \int_0^L \tau_0 B \, dx = 0.664 B \mu U_0 \, Re_L^{1/2} \tag{7.4}$$

$$\text{or } F_s = C_f B L \rho \frac{U_0^2}{2} \tag{7.5}$$

$$\text{where } C_f = \frac{1.33}{Re_L^{1/2}} \tag{7.6}$$

7.3 The turbulent boundary layer

Studies have shown that the velocity profile in the turbulent boundary layer is approximated closely over a wide range of Reynolds numbers by the equation:

$$\frac{v}{U_0} = \left(\frac{y}{\delta}\right)^{1/7} \tag{7.7}$$

If the turbulent boundary layer is assumed to develop at the leading edge of the plate,

$$\delta_x = 0.37x \left(\frac{v}{U_0 x}\right)^{1/5} = \frac{0.37 \, x}{Re_x^{1/5}} \tag{7.8}$$

$$\text{Boundary shear stress, } \tau_0 = 0.0225 \rho U_0^2 \left(\frac{v}{U_0 \delta_x}\right)^{1/4} \tag{7.9}$$

$$\text{Drag on one side of plate, } F_s = C_f B L \rho \frac{U_0^2}{2} \tag{7.10}$$

$$\text{where } C_f = \frac{0.074}{Re_L^{1/5}} \tag{7.11}$$

Chapter 7

7.4 Combined drag due to both laminar and turbulent boundary layers

Where the plate is sufficiently short such that the laminar boundary layer forms over a significant proportion of the length, the total drag may be calculated by considering the turbulent boundary layer to start at the leading edge, deducting the drag in the turbulent boundary layer up to x_0 and adding the drag in the laminar boundary layer.

$$\text{Hence } F_s = \left(\frac{1.33x_0}{Re_{x_0}^{1/2}} + \frac{0.074L}{Re_L^{1/2}} - \frac{0.074x_0}{Re_{x_0}^{1/5}} \right) B\rho \frac{U_0^2}{2} \qquad [7.12]$$

7.5 The displacement thickness

Due to the reduction in velocity in the boundary layer, the discharge past a point on the surface is reduced by an amount

$$\delta q = \int_0^\delta (U_0 - v)\, dy \qquad [7.13]$$

The displacement thickness is the distance δ^*, shown in Figure 7.2, by which the surface would have to be moved in order to reduce the discharge of an ideal fluid at velocity U_0 by the same amount. Then

$$U_0 \delta^* = \int_0^\delta (U_0 - v)\, dy \qquad [7.14]$$

Assuming that the velocity distribution is

$$\frac{v}{U_0} = \left(\frac{y}{\delta} \right)^{1/7} \qquad (\text{Equation 7.7})$$

$$\delta^* = \frac{1}{U_0} \int_0^\delta \left[U_0 - U_0 \left(\frac{y}{\delta} \right)^{1/7} \right] dy \qquad [7.15]$$

$$\delta^* = \left[y - \frac{7}{8} \left(\frac{1}{\delta} \right)^{1/7} y^{8/7} \right]_0^\delta$$

$$\text{whence } \delta^* = \frac{\delta}{8} \qquad [7.16]$$

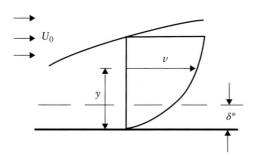

Figure 7.2   Displacement thickness.

7.6 Boundary layers in turbulent pipe flow

In Chapters 4 and 5, the analysis and design of pipelines were demonstrated using the Darcy–Weisbach and Colebrook–White equations. The latter equation

$$\frac{1}{\sqrt{\lambda}} = -2\log\left(\frac{k}{3.7D} + \frac{2.51}{Re\sqrt{\lambda}}\right)$$

owes its origin to the theoretical hypothesis of Prandtl for the general form of the velocity distribution in full pipe flow supported by the experimental work of Nikuradse. Using the concept of the exchange of momentum due to the transverse velocity components in turbulent flow, Prandtl developed his 'mixing theory' expressing the shear stress τ in terms of the velocity gradient dv/dy and 'mixing length' ℓ in the form

$$\tau = \rho\ell^2 \left(\frac{dv}{dy}\right)^2 \tag{7.17}$$

Prandtl further assumed that the mixing length was directly proportional to the distance from the boundary and that the shear stress was constant and hence equal to the boundary shear stress τ_0. Hence,

$$\tau_0 = \rho(\kappa y)^2 \left(\frac{dv}{dy}\right)^2 \tag{7.18}$$

$$\text{whence } \frac{dv}{dy} = \frac{1}{\kappa}\sqrt{\frac{\tau_0}{\rho}}\frac{1}{y} \tag{7.19}$$

The term $\sqrt{\tau_0/\rho}$ has the dimensions of velocity and is called the 'shear velocity' and represented by the symbol v^*. From Equation 7.19,

$$\frac{v}{v^*} = \frac{1}{\kappa}\ln y + C \tag{7.20}$$

$$\text{or } \frac{v}{v^*} = \frac{1}{\kappa}\ln\frac{y}{y'} \tag{7.21}$$

where y' is the 'boundary condition', that is, the distance from the boundary at which the velocity becomes zero (see Figure 7.3).

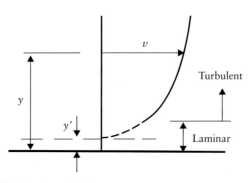

Figure 7.3   Velocity distribution – log law.

Observations of friction head loss and velocity distribution on smooth and artificially roughened pipes, conveying water, by Nikuradse revealed that the forms of the velocity distribution in the smooth and rough turbulent zones were different. For the smooth turbulent zone he assumed y' to be proportional to v/v^*, and Equation 7.21 therefore becomes

$$\frac{v}{v^*} = A \log \left(\frac{v^*y}{v} \right) + B \qquad [7.22]$$

where A and B are constants.

From the experimental data Nikuradse plotted v/v^* against $\log (v^*y/v)$, and a straight-line fit having the equation

$$\frac{v}{v^*} = 5.75 \log \left(\frac{v^*y}{v} \right) + 5.5 \qquad [7.23]$$

verified the hypotheses. The mean velocity is obtained by integration

$$V = \frac{1}{\pi D^2/4} \int_0^{D/2} 2\pi r v \, dr$$

and substitution of v from Equation 7.23 yields

$$\frac{V}{v^*} = 5.75 \log \left(\frac{v^*D}{2v} \right) + 1.75 \qquad [7.24]$$

The equation for the Darcy friction factor can consequently be obtained. Since

$$\tau_0 = \rho g R \frac{h_f}{L} = \rho g \frac{D}{4} \frac{h_f}{L} \quad \text{(see Example 7.5)}$$

$$\text{and} \ \frac{h_f}{L} = \frac{\lambda V^2}{2gD}; \qquad \sqrt{\frac{\tau_0}{\rho}} = V \sqrt{\frac{\lambda}{8}} \qquad [7.25]$$

substitution into Equation 7.25 yields the equation

$$\frac{1}{\sqrt{\lambda}} = 2 \log \left(\frac{VD}{v} \frac{\sqrt{\lambda}}{2.51} \right)$$

$$\text{or} \ \frac{1}{\sqrt{\lambda}} = 2 \log \left(\frac{Re\sqrt{\lambda}}{2.51} \right) \qquad [7.26]$$

Similarly for the rough turbulent zone, Nikuradse found y' to be proportional to k and obtained the velocity distribution in the form

$$\frac{v}{v^*} = 5.75 \log \left(\frac{y}{k} \right) + 8.5 \qquad [7.27]$$

$$\text{and} \ \frac{V}{v^*} = 5.75 \log \left(\frac{D}{2k} \right) + 4.75 \qquad [7.28]$$

$$\text{whence} \ \frac{1}{\sqrt{\lambda}} = 2 \log \left(\frac{D}{2k} \right) + 1.74 \qquad [7.29]$$

$$\text{or} \ \frac{1}{\sqrt{\lambda}} = 2 \log \left(\frac{3.7D}{k} \right) \qquad [7.30]$$

Equations 7.26 and 7.30 are referred to as the Kármán–Prandtl equations. While Equation 7.30 was obtained for artificially roughened pipes, Colebrook and White found that the addition of the Kármán–Prandtl equations produced a 'universal' function which fitted data collected on commercial pipes covering a wide range of Reynolds numbers and relative roughness values. The Colebrook–White equation was expressed as

$$\frac{1}{\sqrt{\lambda}} = -2\log\left(\frac{k}{3.7D} + \frac{2.51}{Re\sqrt{\lambda}}\right) \qquad [7.31]$$

where k now becomes the 'effective roughness size' equivalent to Nikuradse's uniform roughness elements.

7.7 The laminar sub-layer

The λ versus Re curves for artificially roughened pipes are shown in Figure 7.4.

At the lower range of Reynolds numbers the curves for the rough pipes merge into the single smooth pipe curve. Thus, rough pipes can behave like smooth ones at low Reynolds numbers. This phenomenon is explained by the presence of a sub-layer, formed adjacent to the boundary, in which the flow is laminar. The presence of such a layer was also confirmed by the velocity distributions obtained by Nikuradse. At low Reynolds numbers the sub-layer is sufficiently thick to cover the boundary roughness elements so that the turbulent boundary layer is, in effect, flowing over a smooth boundary (Figure 7.5a).

The sub-layer thickness decreases with increasing Reynolds number, and at high Reynolds numbers the roughness elements are fully exposed to the turbulent boundary layer. At intermediate Reynolds numbers, in the transitional turbulent zone, the friction factor is influenced by both the relative roughness and Reynolds number.

Figure 7.6 shows a sub-layer formed on a smooth boundary beneath a turbulent boundary layer.

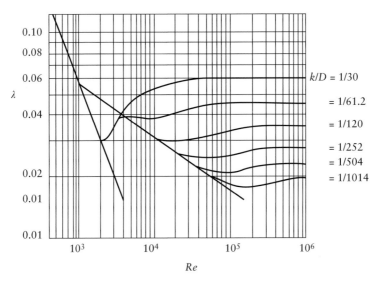

Figure 7.4   Variation of λ with Re for artificially roughened pipes.

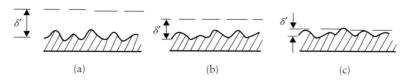

(a)   (b)   (c)

Figure 7.5   Variations in thickness of a laminar sub-layer with Reynolds number: (a) smooth turbulent, (b) transitional turbulent and (c) rough turbulent.

In the laminar sub-layer,

$$\tau = \mu \frac{dv}{dy}$$

$$\text{whence} \quad \frac{dv}{dy} = \frac{\tau}{\rho v} \quad \text{and} \quad v = \frac{\tau y}{\rho v} \tag{7.32}$$

At the upper boundary of the sub-layer, $y = \delta'$, the velocities given by Equations 7.23 and 7.32 are identical and $\tau = \tau_0$.

Whence

$$v^* \left[5.75 \log \left(\frac{v^* \delta'}{v} \right) + 5.5 \right] = \frac{(v^*)^2 \delta'}{v}$$

the solution to which is

$$\delta' = \frac{11.6 v}{v^*} \tag{7.33}$$

substituting $v^* = V\sqrt{\lambda/8}$ and introducing D on both sides yield

$$\frac{\delta'}{D} = \frac{32.8}{Re\sqrt{\lambda}} \tag{7.34}$$

Thus the thickness of the laminar sub-layer decreases with Reynolds number.

For rough surfaces the zone of turbulent flow is clearly related to the ratio of the magnitudes of δ' and k. Since δ'/D is inversely proportional to $Re\sqrt{\lambda}$,

$$\frac{Re\sqrt{\lambda}}{D/2k} \propto \frac{k}{\delta'}$$

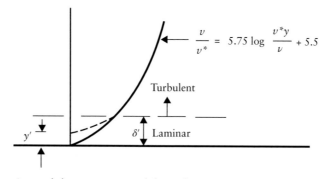

Figure 7.6   Laminar sub-layer on a smooth boundary.

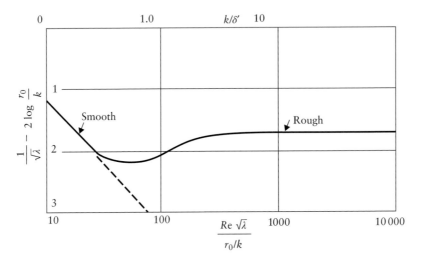

Figure 7.7   The transition zone for artificially roughened pipes.

Each of the curves in Figure 7.4 should therefore deviate from the smooth pipe law at the same value of k/δ'. Thus, superimposing all curves by plotting $(1/\sqrt{\lambda}) - 2\log(D/2k)$ on a base of

$$\frac{Re\sqrt{\lambda}}{D/2k} \quad \text{(Figure 7.7)}$$

shows that the transition from the smooth law begins when δ' is approximately $4k$ and ends when k is approximately $6\delta'$ (see Figure 7.7).

Worked examples

Example 7.1

A thin smooth plate 2 m long and 3 m wide is towed edgewise through water, at 20°C, at a speed of 1 m/s.

(a) Calculate the total drag and the thickness of the boundary layer at the trailing edge.

(b) If the plate were towed with the 3 m side in the direction of flow, what would be the drag?

Solution:

(a)

$$v_{20°C} = 1 \times 10^{-6} \text{ m}^2/\text{s}; \qquad \rho = 1000 \text{ kg/m}^3$$

Assuming the laminar layer to become unstable at $Re_x = 500\,000$, the length of the layer $x_0 = 0.5$ m.

Using Equation 7.11 for the combined drag in the laminar and turbulent layers,

$$F_s = 2 \left(\frac{1.33 x_0}{Re_{x_0}^{1/2}} + \frac{0.074 L}{Re_L^{1/5}} - \frac{0.074 x_0}{Re_{x_0}^{1/5}} \right) B\rho \frac{U_0^2}{2}$$

$$F_s = 2 \left(\frac{1.33 \times 0.5}{707.1} + \frac{0.074 \times 2}{18.2} - \frac{0.074 \times 0.5}{13.8} \right) \frac{3 \times 1000 \times 1}{2}$$

$$F_s = 19.17 \text{ N}$$

Boundary layer thickness: Experiments have shown that when the turbulent boundary layer develops downstream of the laminar layer, the characteristics of the turbulent layer are those of one which develops at the leading edge. Thus,

$$\delta_L = \frac{0.37 L}{Re_L^{1/5}} = 0.041 \text{ m}$$

If the drag were assumed to be entirely due to the turbulent boundary layer,

$$F_s = 2 C_f B L \rho \frac{U_0^2}{2}$$

$$C_f = \frac{0.074}{Re_L^{1/5}} = \frac{0.074}{18.2} = 0.0041$$

$$F_s = 2 \times 0.004 \times 2 \times 3 \times 1000 \times \frac{1}{2} = 24.6 \text{ N}$$

(b)

$$B = 2.0; \qquad L = 3.0$$

From Equation 7.11,

$$F_s = 2 \times 2 \left(\frac{1.33 \times 0.5}{707.1} + \frac{0.074 \times 3}{19.74} - \frac{0.074 \times 0.5}{13.8} \right) 1000 \times \frac{1}{2}$$

$$= 19.01 \text{ N}$$

Example 7.2

A 5 m long smooth model of a ship is towed in fresh water of kinematic viscosity 1×10^{-6} m²/s at 3.5 m/s. The wetted hull area is 1.4 m². What is the skin friction drag?

Solution:

Assuming that a turbulent boundary layer develops at the leading edge,

$$\text{drag} = C_f A \rho \frac{U_0^2}{2}$$

$$C_f = \frac{0.074}{Re_L^{1/5}}; \qquad Re_L = \frac{3.5 \times 5}{1 \times 10^{-6}} = 17.5 \times 10^6$$

whence $C_f = 0.00263$

$$\text{and drag} = 0.00263 \times 1.4 \times 1000 \times \frac{3.5^2}{2}$$

$$= 22.55 \text{ N}$$

Example 7.3

Water enters a 300 mm diameter test section of a water tunnel at a uniform velocity of 15 m/s. Assuming that the boundary layer starts 0.5 m upstream of the test section, estimate the increase in axial velocity at the end of a 3 m test section due to the growth of the boundary layer. Take $v = 1 \times 10^{-6}$ m²/s.

Solution:

$$\text{Length of boundary layer} = 3.5 \text{ m}$$
$$Re_L = 15 \times 3.5 \times 10^6 = 52.5 \times 10^6$$
$$Re_L^{1/5} = 35.0$$
$$\delta_L = \frac{0.37L}{Re_L^{1/5}} = \frac{0.37 \times 3.5}{35} = 0.037 \text{ m}$$

From Section 7.5, assuming that the velocity distribution in the boundary layer is of the form

$$\frac{v}{U_0} = \left(\frac{y}{\delta}\right)^{1/7} \quad \text{(Equation 7.15)}$$

the displacement thickness is expressed by

$$\delta^* = \frac{\delta}{8} \quad \text{(Equation 7.16)}$$
$$\text{or } \delta^* = 0.004625 \text{ m}$$
$$\text{Effective duct diameter} = 300 - 9.25 = 290.75 \text{ mm}$$

Then

$$U_0 A_0 = U_L A_L$$
$$\text{whence } U_L = 15 \left(\frac{300}{290.75}\right)^2 = 15.97 \text{ m/s}$$

Change in pressure due to boundary layer formation: apply Bernoulli's equation, assuming ideal fluid flow:

$$\frac{p_0}{\rho g} + \frac{U_0^2}{2g} = \frac{p_L}{\rho g} + \frac{U_L^2}{2g}$$
$$\frac{p_0 - p_L}{\rho g} = \Delta h = \frac{U_L^2 - U_0^2}{2g}$$
$$\Delta h = \frac{15.97^2 - 15^2}{2g} = 1.53 \text{ m}$$

Example 7.4

Water at 20°C enters the 250 mm square working section of a water tunnel at 20 m/s with a turbulent boundary layer of thickness equal to that from a starting point 0.45 m upstream. Estimate the length of the sides of the divergent duct for constant pressure core flow at 1, 2 and 3 m downstream from the duct entrance.

Solution:

From Equation 7.16, displacement thickness is

$$\delta_x^* = \frac{\delta_x}{8}$$

From Equation 7.7,

$$\delta_x = \frac{0.37x}{Re_x^{1/5}}$$

At 1 m from the entrance of the working section, the length of the boundary layer is 1.45 m.

$$Re_x = \frac{20 \times 1.45}{1 \times 10^{-6}} = 29 \times 10^6$$

$$\text{whence } \delta = 0.01726 \text{ m}$$

$$\text{and } \delta^* = 0.00216 \text{ m} = 2.16 \text{ mm}$$

Thus the section size is 254.32 mm².
   At 2 m from the entrance, the boundary layer is 2.45 m long.

$$\delta^* = 3.29 \text{ mm}$$

$$\Rightarrow \text{section size} = 256.58 \text{ mm}$$

At 3 m from the entrance,

$$\delta^* = 4.32 \text{ m}$$

$$\Rightarrow \text{section size} = 258.64 \text{ mm}$$

Example 7.5

The velocity distribution in the rough turbulent zone is expressed by

$$\frac{v}{\sqrt{\tau_0/\rho}} = 5.75 \ \log \left(\frac{y}{k}\right) + 8.5 \quad \text{(Equation 7.27)} \tag{i}$$

Local axial velocities measured at 25 and 75 mm across a radius from the inner wall of a 150 mm diameter pipe conveying water at 15°C were 0.815 and 0.96 m/s, respectively. Calculate (a) the effective roughness size, (b) the hydraulic gradient and (c) the discharge.

Chapter 7

Solution:

(a) Writing $v^* = \sqrt{\tau_0/\rho}$ and inserting the pairs of values of v and y in the velocity distribution equation,

$$\frac{0.96}{v^*} = 5.75 \log\left(\frac{75}{k}\right) + 8.5 \qquad \text{(ii)}$$

$$\frac{0.815}{v^*} = 5.75 \log\left(\frac{25}{k}\right) + 8.5 \qquad \text{(iii)}$$

Note: y can be expressed in millimetres since y/k is dimensionless; the calculated value of k will then be in millimetres. Hence,

$$\frac{0.96 - 0.815}{v^*} = 5.75 \log\left(\frac{75}{25}\right)$$

$$\text{whence } v^* = 0.0528$$

$$\text{and } \tau_0 = 2.79 \text{ N/m}^2$$

Substituting for v^* in (ii),

$$\frac{0.96}{0.0528} = 5.75 \log\left(\frac{75}{k}\right) + 8.5$$

$$\text{whence } k = 1.55 \text{ mm}$$

(b) The hydraulic gradient ($S_f = h_f/L$) can be related to the boundary shear stress. Consider an element of flow in a pipeline (see Figure 7.8).
  For steady, uniform flow, equate the forces on the element of length δL.

$$(p_1 - p_2)A + \rho g A \delta L \sin\alpha = \tau_0 P \delta L$$

$$\text{i.e. } \frac{p_1 - p_2}{\rho g} + \delta z = \frac{\tau_0 P \delta L}{\rho g A}$$

$$\text{or } h_f = \frac{\tau_0 \delta L}{\rho g R}$$

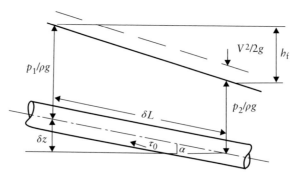

Figure 7.8 Flow analysis in a pipeline.

where $R = A/P = D/4$.

$$\frac{h_f}{L} = S_f = \frac{0.0528^2}{9.806 \times 0.0375} = 0.00758$$

(c) The discharge can now be evaluated using the Darcy–Colebrook–White combination:

$$Q = -2A\sqrt{2gDS_f}\, \log\left[\frac{k}{3.7D} + \frac{2.51\nu}{D\sqrt{2gDS_f}}\right]$$

$$Q = 13.37 \text{ L/s}$$

Mean velocity

$$V = 0.757 \text{ m/s}$$
$$Re = 1.004 \times 10^5$$
$$\frac{k}{D} = \frac{1.55}{150} = 0.0103$$

Referring to the Moody chart, this is in the rough turbulent zone.

Alternatively, as shown in Section 7.6, the mean velocity V can thus be expressed as

$$\frac{V}{\sqrt{\tau_0/\rho}} = 5.75 \log\left(\frac{D}{2k}\right) + 4.75$$

whence $V = 0.762$ m/s   and $Q = 13.45$ L/s

Thus, location of local velocity equals to mean velocity:

$$\frac{v - V}{\sqrt{\tau_0/\rho}} = 5.75 \log\left(\frac{2y}{D}\right) + 3.75 \quad \text{(from Equations 7.28 and 7.29)}$$

i.e. $\log\left(\dfrac{2y}{D}\right) = -0.6522$

$$\frac{2y}{D} = 0.22275$$
$$y = 16.7 \text{ mm}$$

Example 7.6

The velocity distribution in the smooth turbulent zone is given by

$$\frac{v}{v^*} = 5.75 \log\left(\frac{v^* y}{\nu}\right) + 5.5 \quad \text{(Equation 7.23)}$$

The axial velocity at 100 mm from the wall across a radius in a 200 mm perspex pipeline conveying water at 20°C was found to be 1.2 m/s. Calculate the hydraulic gradient and discharge. $\nu = 1.0 \times 10^{-6}$ m²/s.

$$\frac{1.2}{v^*} = 5.75 \log\left(\frac{v^* \times 0.1}{1 \times 10^{-6}}\right) + 5.5$$

(*Note*: y must be expressed here in metres.)

Solve by trial or graphical interpolation.

Solution:

$$v^* = 0.045 \text{ m/s} \ (= \sqrt{\tau_0/\rho})$$

Now

$$S_f = \frac{\tau_0}{\rho g R} \quad \text{(see previous example)}$$

$$\Rightarrow S_f = \frac{0.045^2}{9.806 \times 0.05} = 0.00413$$

Now if the flow is assumed to be in the smooth turbulent zone ($k/D = 0$), the discharge may be calculated from the Darcy–Colebrook–White equation in the form

$$Q = -2A\sqrt{2gDS_f} \ \log\left(\frac{2.51v}{D\sqrt{2gDS_f}}\right)$$

$$= 32.03 \text{ L/s}$$

$$V = 1.02 \text{ m/s}; \qquad Re = 2.04 \times 10^5$$

Example 7.7

Sand grains 0.5 mm in diameter are glued to the inside of a 200 mm diameter pipeline. At what velocity of flow of water at 15°C will the surface roughness (a) cause the flow to depart from the smooth pipe and (b) enter the rough pipe curve?

Solution:

(a) The transition from the smooth law begins when $\delta' = 4k$

$$\text{whence } \delta' = 4 \times 0.5 = 2.0 \text{ mm}$$

Also,

$$\delta' = \frac{32.8 \ D}{Re\sqrt{\lambda}} \quad \text{(Equation 7.34)}$$

$$2.0 = \frac{32.8 \times 200}{Re\sqrt{\lambda}}$$

$$\text{whence } Re\sqrt{\lambda} = 3280.0 \tag{i}$$

The smooth turbulent zone is represented by

$$\frac{1}{\sqrt{\lambda}} = 2\log\left(\frac{Re\sqrt{\lambda}}{2.51}\right)$$

$$\frac{1}{\sqrt{\lambda}} = 2\log\left(\frac{52\,480}{2.51}\right) = 6.232$$

Then from Equation (i),

$$Re = 20\,441 = \frac{VD}{v}$$

$$\text{whence } V = 0.102 \text{ m/s}$$

(b) At $k = 6\delta'$, the flow enters the rough turbulent zone.

$$\delta' = \frac{32.8D}{Re\sqrt{\lambda}} = 0.0833$$

$$Re\sqrt{\lambda} = 78\,751.0 \tag{ii}$$

The rough turbulent zone is represented by

$$\frac{1}{\lambda} = 2\log\frac{3.7\,D}{k} = 2\log\frac{3.7 \times 200}{0.5}$$
$$= 6.341$$

Substituting in Equation (ii),

$$Re = 499\,372.0$$
$$\text{whence } V = 2.5 \text{ m/s}$$

References and recommended reading

Douglas, J. F., Gasiorek, J. M., Swaffield, J. A. and Jack, L. B. (2011) *Fluid Mechanics*, 6th edn, Pearson Prentice Hall, Harlow.

Massey, B. S. and Ward-Smith, J. (2012) *Mechanics of Fluids*, 9th edn, Taylor & Francis, Abingdon, UK.

Problems

1. A pontoon 15 m long and 4 m wide with vertical sides floats to a depth of 0.5 m. The pontoon is towed in sea water at 10°C ($\rho_s = 1024$ kg/m^3, $\mu = 1.31 \times 10^{-3}$ N s/m^2) at a speed of 2 m/s. (a) Determine the viscous resistance and the thickness of the boundary layer at the downstream end. (b) What is the shear stress at the mid-length?

2. Sealed hollow pipes 2 m in external diameter and 6 m long, fitted with rounded nose-pieces to reduce wave-making drag, are towed in a river to a construction site; the pipes float to a depth of 1.5 m. The towing speed is 5 m/s in the water which has a density of 1000 kg/m^3 and dynamic viscosity of 1.2×10^{-3} N s/m^2. Determine the viscous drag of each pipe, assuming that (a) a turbulent boundary layer exists over the entire length and (b) the drag is a combination of that due to the laminar and turbulent boundary layers.

3. Air of density 1.3 kg/m^3 and dynamic viscosity 1.8×10^{-5} N s/m^2 enters the test section 1 m wide and 0.5 m deep of a wind tunnel at a velocity of 20 m/s. Determine the increase in axial velocity 10 m downstream from the entrance to the test section due to the development of the boundary layer, assuming that this forms at the entrance to the test section.

4. Wind velocities over flat grassland were observed to be 3.1 and 3.3 m/s at the height of 3 and 6 m above the ground, respectively. Determine the effective roughness of the surface and estimate the wind velocity at a height of 25 m.

5. The centre line velocity in a 100 mm diameter brass pipeline ($k = 0.0$) conveying water at 20°C was found to be 3.5 m/s. Determine the boundary shear stress, the hydraulic gradient, the discharge and the thickness of the laminar sub-layer.

6. The velocities at 50 and 150 mm from the pipe wall of a 300 mm diameter pipeline convey-ing water at 15°C were 1.423 and 1.674 m/s, respectively. Determine the effective rough-ness size, the hydraulic gradient, the discharge and the Darcy friction factor, and verify that the flow is in the rough turbulent zone.

7. Show that the equation

$$\frac{v_{max} - v}{\sqrt{\tau_0/\rho}} = 5.75 \ \log \frac{D}{2y}$$

applies to full-bore flow in circular pipes in both the rough and smooth turbulent zones.

Chapter 8
Steady Flow in Open Channels

8.1 Introduction

Open channel flow, for example flow in rivers, canals and sewers not flowing full, is characterised by the presence of the interface between the liquid surface and the atmosphere. Therefore, unlike full pipe flow, where the pressure is normally above atmospheric pressure, but sometimes below it, the pressure on the surface of the liquid in open channel flow is always that of the ambient atmosphere.

The energy per unit weight of the liquid flowing in a channel at a section where the depth of flow is y and the mean velocity is V is

$$H = z + y\cos\theta + \frac{\alpha V^2}{2g} \quad \text{(see Figure 8.1)} \tag{8.1}$$

where z is the position energy (or head), θ the bed slope and α the Coriolis coefficient (see also Equation 3.11).

$$\alpha = \frac{1}{AV^3}\int_0^A v^3\,dA$$

The motivating force establishing flow is predominantly the gravity force component acting in parallel with the bed slope, but net pressure forces and inertia forces may also be present. Flow in channels may be unsteady, resulting from changes in inflow such as floods or changes in depth caused by control gate operation. Steady flow can either be uniform or be varied depending upon whether or not the mean velocity is constant with distance. In gradually varied flow there is a gentle change in depth with distance; a common example is the backwater curve (Figure 8.2a). Rapidly varied surface profiles are created by changes in channel geometry, for example flow through a venturi flume (Figure 8.2b).

Steady, uniform flow occurs when the motivating forces and drag forces are exactly balanced over the reach under consideration. This type of flow is analogous with steady pressurised flow in a pipeline of constant diameter. Thus the area of flow in the channel

Nalluri & Featherstone's Civil Engineering Hydraulics: Essential Theory with Worked Examples, Sixth Edition. Martin Marriott.
© 2016 John Wiley & Sons, Ltd. Published 2016 by John Wiley & Sons, Ltd.
Companion Website: www.wiley.com/go/Marriott

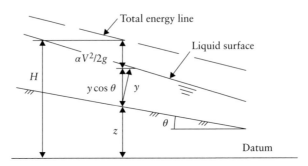

Figure 8.1 Energy components.

must remain constant with distance, a condition requiring the bed slope and channel geometry to remain constant. The liquid surface is parallel to the bed.

8.2 Uniform flow resistance

The nature of the boundary resistance is identical to that of full pipe flow (Chapter 4), and the Darcy–Weisbach and Colebrook–White equations for non-circular sections may be applied (see Section 4.2).

Noting that the energy gradient S_f is equal to the bed slope S_0 in uniform flow:

$$\text{Darcy–Weisbach:} \quad \frac{h_f}{L} = S_f = S_0 = \frac{\lambda V^2}{8gR} \tag{8.2}$$

$$\text{Colebrook–White:} \quad \frac{l}{\sqrt{\lambda}} = -2\log\left(\frac{k}{14.8R} + \frac{2.51\nu}{4RV\sqrt{\lambda}}\right) \tag{8.3}$$

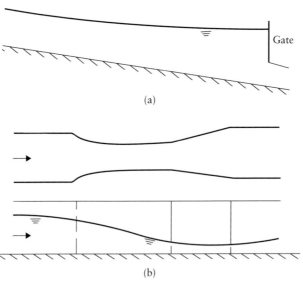

Figure 8.2 Steady, varied-flow surface profiles: (a) gradually varied and (b) rapidly varied through a venturi flume

Table 8.1 Typical values of Manning's n.

Type of surface	n
Concrete	
Culvert, straight and free of debris	0.011
Culvert, with bends, connections and some debris	0.013
Cast on steel forms	0.013
Cast on smooth wood forms	0.014
Unfinished, rough wood form	0.017
Excavated or dredged channels	
Earth, after weathering, straight and uniform	0.022
Earth, winding, clean	0.025
Earth bottom, rubble sides	0.030

For a comprehensive list, see Chow (1959).

Eliminating λ from Equations 8.2 and 8.3 yields

$$V = -\sqrt{32gRS_0}\,\log\left(\frac{k}{14.8R} + \frac{1.255\nu}{R\sqrt{32gRS_0}}\right) \qquad [8.4a]$$

$$\text{where } R = \frac{A}{P} = \frac{\text{area of flow}}{\text{wetted perimeter}}$$

In addition to the Darcy–Weisbach equation, the Manning equation is widely used in open channel water flow computations. This was derived from the Chezy equation $V = C\sqrt{RS_0}$ by writing $C = R^{1/6}/n$, resulting in

$$V = \frac{R^{2/3}}{n}S_0^{1/2} \quad \text{(SI units)} \qquad [8.4b]$$

where n is called the Manning roughness factor and its value is related to the type of boundary surface. If the value of n is taken to be constant regardless of depth, then unlike the Darcy friction factor it does not account for changes in relative roughness; nor does it include the effects of viscosity (see Table 8.1). The Manning equation is in general applicable to shallow flows in rough boundaries, and for a boundary of roughness k, Manning's n may be written as (Strickler's equation)

$$n = \frac{k^{1/6}}{26} \qquad [8.4c]$$

where k is in metres.

8.3 Channels of composite roughness

In applying the Manning formula to channels having different n values for the bed and sides it is necessary to compute an equivalent n value to be used for the whole section. The water area is 'divided' into N parts having wetted perimeters $P_1, P_2, \ldots, P_N$ with associated roughness coefficients $n_1, n_2, \ldots, n_N$. Horton and Einstein assumed that each sub-area has a velocity equal to the mean velocity.

Thus,

$$n = \left(\frac{\sum_{i=1}^{N} P_i n_i^{3/2}}{P} \right)^{2/3} \qquad [8.5]$$

where P is the total wetted perimeter.

Pavlovskij and others equated the sum of the component resisting forces to the total resisting force and thus found

$$n = \left(\frac{\sum_{i=1}^{N} P_i n_i^2}{P} \right)^{1/2} \qquad [8.6]$$

Lotter applied the Manning equation to sub-areas and equated the sum of the individual discharge equations to the total discharge. Thus the equivalent roughness coefficient is

$$n = \frac{PR^{5/3}}{\sum_{i=1}^{N} P_i R_i^{5/3} / n_i} \qquad [8.7]$$

8.4 Channels of compound section

A typical example of a compound section (two-stage channel) is a river channel with flood plains. The roughness of the side channels will be different (generally rougher) from that of the main channel and the method of analysis is to consider the total discharge to be the sum of component discharges computed by the Manning equation.

Thus in the channel shown in Figure 8.3, assuming that the bed slope is the same for the three sub-areas,

$$Q = \left(\frac{A_1}{n_1} R_1^{2/3} + \frac{A_2}{n_2} R_2^{2/3} + \frac{A_3}{n_3} R_3^{2/3} \right) S_0^{1/2}$$

The above assumption leads to large discrepancies between computed and measured discharges under flood flow (above bank-full stages) conditions. The interaction between the slower-moving berm flows and the fast-moving main channel flow significantly increases head losses. As a result, the discharge computed by this conventional method will overestimate the flow. Utilising research data from the Flood Channel Facility at Wallingford, Ackers (1992) has shown that the discrepancy between the conventional calculations and the measured flow is dependent on flood flow levels. He formulated appropriate

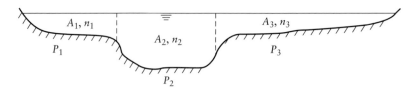

Figure 8.3   Compound channel section.

correction factors for each region of flow; a detailed exposure of the analysis of the research is beyond the scope of this book.

8.5 Channel design

The design of open channels involves the selection of suitable sectional dimensions such that the maximum discharge will be conveyed within the section. The bed slope is sometimes constrained by the topography of the land in which the channel is to be constructed.

In the design of an open channel, a resistance equation, whether Darcy or Chezy or Manning, may be used. However, at least one other equation is required to define the relationship between width and depth. This second series of equations incorporates the design criteria; for example, in rigid-boundary (non-erodible) channels the designer will wish to minimise the construction cost, resulting in what is commonly termed **the most economic section**. In addition, there may be a constraint on the maximum velocity to prevent erosion or on the minimum velocity to prevent settlement of sediment.

In the case of erodible unlined channels (excavated in natural ground, e.g. clay and silts), the design criterion will be that the boundary shear stress exerted by the moving liquid will not exceed the 'critical tractive force' of the bed and side material.

8.5.1 Rigid-boundary channels – best hydraulic or 'economic' section

Using the Darcy-type resistance equation,

$$Q = A\sqrt{\frac{8g}{\lambda}\frac{A}{P}S_0} = \frac{KA^{3/2}}{P^{1/2}}$$

$$A = f(y); \quad P = f(y)$$

Q_{max} is achieved when

$$\frac{dQ}{dy} = 0 \Rightarrow \frac{d}{dy}\left(\frac{A^3}{P}\right) = 0$$

$$\frac{3A^2}{P}\frac{dA}{dy} - \frac{A^3}{P^2}\frac{dP}{dy} = 0$$

$$\text{whence } 3P\frac{dA}{dy} - A\frac{dP}{dy} = 0$$

For a given area, $dA/dy = 0$; then, for Q_{max}, $dP/dy = 0$; in other words, the wetted perimeter is a minimum. For a trapezoidal channel (see Figure 8.4),

$$A = (b + Ny)y$$

$$P = b + 2y\sqrt{1 + N^2}$$

For a given area A,

$$P = \frac{A}{y} - Ny + 2y\sqrt{1 + N^2}$$

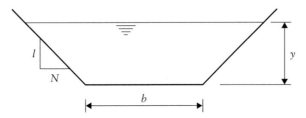

Figure 8.4 Trapezoidal channel.

For Q_{max},

$$\frac{dP}{dy} = -\frac{A}{y^2} - N + 2\sqrt{1 + N^2} = 0$$

$$\Rightarrow \frac{dP}{dy} = -(b + Ny) - Ny + 2y\sqrt{1 + N^2} = 0$$

$$\text{or } b + 2Ny = 2y\sqrt{1 + N^2} \qquad\qquad [8.8]$$

It can be shown that if a semicircle of radius y is drawn with its centre in the liquid surface, it will be tangential to the sides and bed. Thus the most economic section approximates as closely as possible to a circular section which is known to have the least perimeter for a given area.

For a rectangular section, $N = 0$ and $b = 2y$.

8.5.2 Mobile-boundary channels (erodible)

The 'critical tractive force' theory and the 'maximum permissible velocity' concept are commonly used in the design of erodible channels for stability.

8.5.2.1 Critical tractive force theory

The force exerted by the water on the wetted area of a channel is called the *tractive force*. The average **unit tractive force** is the average shear stress given by $\bar{\tau}_0 = \rho g R S_0$. Boundary shear stress is not, however, uniformly distributed; the distribution varies somewhat with channel shape, but not with size. For trapezoidal sections the maximum shear stress on the bed may be taken as $\rho g y S_0$ and on the sides as $0.76 \rho g y S_0$ (see Figure 8.5); however, the shear distribution depends on the channel aspect ratio b/y (see Table 8.2).

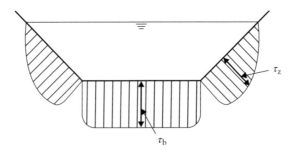

Figure 8.5 Distribution of shear stress on channel boundary.

Table 8.2   Maximum bed/side shear stress.

Aspect ratio b/y	$\tau_{b\,max}/\rho gyS_0$	$\tau_{s\,max}/\rho gyS_0$
2	0.890	0.735
4	0.970	0.750
>8	0.985	0.780

If the shear stresses can be kept below that which will cause the material of the channel boundary to move, the channel will be stable. The critical tractive force of a particular material is the unit tractive force, which will not cause erosion of the material on a horizontal surface. Material on the sides of the channel is subjected, in addition to the shear force due to the flowing water, to a gravity force down the slope. It can be shown that if τ_{cb} is the critical tractive force, the maximum critical shear stress due to the water flow on the sides is

$$\tau_{cs} = \tau_{cb} \sqrt{1 - \frac{\sin^2 \theta}{\sin^2 \phi}}$$

[8.9]

where θ is the slope of the sides to the horizontal and ϕ the angle of repose of the material. Table 8.3 gives some typical values of critical tractive force and permissible velocity.

Table 8.3   Critical tractive force and mean velocity for different bed materials.

Material	Size (mm)	Critical tractive force (N/m²)	Approximate mean velocity (m/s)	The Manning coefficient of roughness
Sandy loam (non-colloidal)		2.0	0.50	0.020
Silt loam (non-colloidal)		2.5	0.60	0.020
Alluvial silt (non-colloidal)		2.5	0.60	0.020
Ordinary firm loam		3.7	0.75	0.020
Volcanic ash		3.7	0.75	0.020
Stiff clay (very colloidal)		1.22	1.15	0.025
Alluvial silts (colloidal)		12.2	1.15	0.025
Shales and hardpans		31.8	1.85	0.025
Fine sand (non-colloidal)	0.062–0.25	1.2	0.45	0.020
Medium sand (non-colloidal)	0.25–0.5	1.7	0.50	0.020
Coarse sand (non-colloidal)	0.5–2.0	2.5	0.60	0.020
Fine gravel	4–8	3.7	0.75	0.020
Coarse gravel	8–64	14.7	1.25	0.025
Cobbles and shingles	64–256	44.0	1.55	0.035
Graded loam and cobbles (non-colloidal)	0.004–64	19.6	1.15	0.300
Graded silts to cobbles (colloidal)	0–64	22.0	1.25	0.300

8.5.2.2 Maximum permissible mean velocity concept

This appears to be a rather uncertain concept since the depth of flow has a significant effect on the boundary shear stress. Fortier and Scobey (1926) published the values in Table 8.3 for well-seasoned channels of small bed slope and depths below 1 m.

8.6 Uniform flow in part-full circular pipes

Circular pipes are widely used for underground storm sewers and wastewater sewers. Storm sewers are usually designed to have sufficient capacity so that they do not run full when conveying the computed surface runoff resulting from a storm of a specified average return period. Under these conditions, 'open channel flow' conditions exist. However, more intense storms may result in the capacity of the pipe, when running full at a hydraulic gradient equal to the pipe slope, being exceeded and pressurised pipe conditions will follow. Wastewater sewers, on the other hand, generally carry relatively small discharges and the design criterion in this case is that the mean velocity under the design flow conditions should exceed a 'self-cleansing velocity' so that sediment will not be permanently deposited.

Although the flow in sewers is rarely steady (and hence non-uniform), some commonly used design methods adopt the assumption of uniform flow at design flow conditions; the rational method for storm sewer design is an example, as is the modified rational method (National Water Council, 1981). Foul sewers are invariably designed under assumed uniform flow conditions. Mathematical simulation and design models of flow in storm sewer networks take account of the unsteady flows using the dynamic equations of flow, but such models often incorporate the steady uniform flow relations in storage–discharge relationships.

Flow equations may be calculated from the geometrical properties derived from Figure 8.6.

$$z = \frac{D}{2} - y$$

$$\theta = \cos^{-1}\left(\frac{2z}{D}\right) \text{ radians}$$

$$A = \frac{D^2}{8}(2\theta - \sin 2\theta)$$

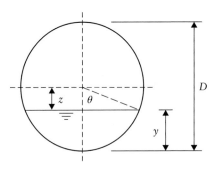

Figure 8.6 Circular channel section.

$$P = D\theta$$

$$R = \frac{A}{P}$$

$$V = -\sqrt{32gRS_0} \log\left(\frac{k}{14.8R} + \frac{1.255v}{R\sqrt{32gRS_0}}\right) \qquad [8.10]$$

$$Q = -A\sqrt{32gRS_0} \log\left(\frac{k}{14.8R} + \frac{1.255v}{R\sqrt{32gRS_0}}\right) \qquad [8.11]$$

8.7 Steady, rapidly varied channel flow energy principles

The computation of non-uniform surface profiles caused by changes of channel section and the like requires the application of energy and momentum principles.

The energy per unit weight of liquid at a section of a channel above some horizontal datum is

$$H = z + y\cos\theta + \frac{\alpha V^2}{2g} \quad \text{(see Figure 8.1)}$$

For mild slopes, $\cos\theta = 1.0$,

$$H = z + y + \frac{\alpha V^2}{2g} \qquad [8.12]$$

Specific energy E is measured relative to the bed:

$$E = y + \frac{\alpha V^2}{2g} \quad \text{or} \quad E = y + \frac{\alpha Q^2}{2gA^2}$$

For a steady fixed inflow into a channel, the specific energy at a particular section can be varied by changing the depth by means of a structure such as a sluice gate with an adjustable opening. Plotting y versus E for a section where the relationship $A = f(y)$ is known and Q is fixed results in the 'specific energy curve'; Figure 8.7 shows that at a given energy level two alternative depths are possible.

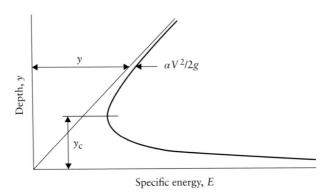

Figure 8.7   Specific energy curve.

Specific energy becomes a minimum at a certain depth called the *critical depth* (y_c). For this condition,

$$\frac{dE}{dy} = 1 - \frac{\alpha Q^2}{A^3 g}\frac{dA}{dy} = 0$$

Now $dA/dy = B$, the surface width

$$\text{whence}\quad \frac{\alpha Q^2 B}{A^3 g} = 1 \quad \text{or} \quad \frac{Q}{\sqrt{g/\alpha}} = A\sqrt{\frac{A}{B}} \qquad [8.13]$$

For the special case of a rectangular channel and where $\alpha = 1.0$,

$$\frac{V^2}{gy} = 1$$

that is, the Froude number is unity and

$$y_c = \sqrt[3]{\frac{Q^2}{gb^2}} = \sqrt[3]{\frac{q^2}{g}} \qquad [8.14]$$

where q is the discharge per unit width.
At the critical depth,

$$E_{min} = y_c + \frac{V_c^2}{2g} \quad \text{and} \quad V_c = \sqrt{gy_c}$$

$$E_{min} = y_c + \frac{y_c}{2}$$

In a rectangular channel, the depth of flow at the critical flow is two-thirds of the specific energy at critical flow.

The velocity corresponding to the critical depth is called the **critical velocity**.

At depths below the critical the flow is called **supercritical**, and at depths above the critical the flow is **subcritical** or **tranquil**.

From the specific energy equation we can write

$$Q = A\sqrt{\frac{2g}{\alpha}(E - y)}$$

which shows that $Q = f(y)$ for a constant E. This exhibits a maximum discharge occurring at a depth equal to critical depth (see Figure 8.21).

8.8 The momentum equation and the hydraulic jump

The hydraulic jump is a stationary surge and occurs in the transition from a supercritical to subcritical flow (Figure 8.8).

A smooth transition is not possible: if this were to occur, the energy would vary according to the route ABC on the specific energy curve. At B the energy would be less than that at C, corresponding to the downstream depth y_s. Therefore a rapid depth change occurs corresponding to the route AC on the curve. The depth at which the jump starts is called the **initial depth**, y_i, and the downstream depths the **sequent depth**, y_s. For a given channel and discharge, there is a unique relationship between y_i and y_s which requires application of the momentum equation.

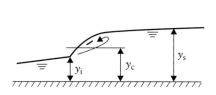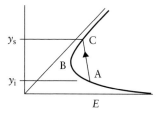

Figure 8.8 The hydraulic jump.

The steady-state momentum equation may be applied to the situation shown in Figure 8.9.

Assuming hydrostatic pressure distribution,

$$\rho g A_1 \bar{y}_1 + \rho Q(\beta_1 V_1 - \beta_2 V_2) - \rho g A_2 \bar{y}_2 + \rho g \bar{A} S_0 - \rho g \bar{A} S_f = 0$$

where $\bar{y}$ is the depth of the centre of area of the cross section and $\bar{A} = (A_1 + A_2)/2$.

$$\Rightarrow A_1 \bar{y}_1 + \frac{Q}{g}(\beta_1 V_1 - \beta_2 V_2) - A_2 \bar{y}_2 + \bar{A}(S_0 - S_f) = 0 \qquad [8.15]$$

Equation 8.15 may be rewritten as $M_1 = M_2$, where

$$M = A\bar{y} + \frac{Q}{g}\beta V = f(y)$$

The M function (specific force) exhibits a minimum value at critical depth. When applied to the analysis of the hydraulic jump, the term $(S_0 - S_f)$ may be neglected. In the special case of a rectangular channel, the above equation reduces to a quadratic which can be solved for either y_i or y_s. Taking $\beta_1 = \beta_2$ (β = Boussinesq coefficient = $(1/AV^2)\int_0^A v^2\, dA$, often taken for practical purposes to equal 1.0),

$$y_i = \frac{y_s}{2}\left(\sqrt{1 + 8\beta Fr_s^2} - 1\right) \qquad [8.16]$$

$$\text{or} \quad y_s = \frac{y_i}{2}\left(\sqrt{1 + 8Fr_i^2} - 1\right) \qquad [8.17]$$

$$\text{where } Fr_s = \frac{V_s}{\sqrt{gy_s}} \quad \text{and} \quad Fr_i = \frac{V_i}{\sqrt{gy_i}} \quad \text{(Froude numbers)}$$

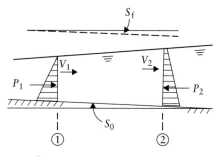

Figure 8.9 Reference diagram for momentum equation.

The energy loss through the jump ($E_L = E_1 - E_2$) can also be shown as

$$E_L = \frac{(y_2 - y_1)^3}{4y_1y_2}$$ [8.18]

The length of the jump is a function of the approach flow Froude number, and for $Fr_1 > 9$, the length is approximately equal to $7y_2$. The excess kinetic energy of the downstream flow over a control structure (such as a spillway) is often destroyed by the formation of a hydraulic jump (energy dissipator) over a confined solid structure known as a **stilling basin**.

8.9 Steady, gradually varied open channel flow

This condition occurs when the motivating and drag forces are not balanced, with the result that the depth varies gradually along the length of the channel (Figure 8.10).

The dynamic equation of gradually varied flow is obtained by differentiating the energy equation $H = z + y\cos\theta + \alpha V^2/2g$ with respect to distance along the channel bed (x-direction):

$$\frac{dH}{dx} = \frac{dz}{dx} + \cos\theta\frac{dy}{dx} + \frac{d}{dx}\left(\frac{\alpha V^2}{2g}\right)$$

$$\text{Now } S_f = -\frac{dH}{dx}; \qquad S_0 = \sin\theta = -\frac{dz}{dx}$$

$$\frac{dy}{dx} = \frac{S_0 - S_f}{\cos\theta + d\left(\alpha V^2/2g\right)/dy}$$

$$\frac{d}{dy}\left(\frac{V^2}{2g}\right) = \frac{d}{dy}\left(\frac{Q^2}{2gA^2}\right) = -\frac{2Q^2}{2gA^3}\frac{dA}{dy}$$

and $dA/dy = T$, the width at the liquid surface.

$$\frac{d}{dy}\left(\frac{V^2}{2g}\right) = -\frac{Q^2T}{A^3g}$$

Since channel slopes are usually small, $\cos\theta = 1.0$

$$\text{whence } \frac{dy}{dx} = \frac{S_0 - S_f}{1 - \alpha Q^2T/A^3g}$$ [8.19]

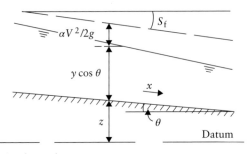

Figure 8.10   Varied flow in channel.

which gives the slope of the water surface relative to the bed. Various types of gradually varied surface profiles can occur, depending on whether the bed slope is mild, steep, critical, horizontal or adverse, and on the depth of flow, as shown in Example 8.24. Further treatment and illustration of this topic can be found in texts such as Chow (1959) and Chadwick *et al.* (2013).

8.10 Computations of gradually varied flow

The gradually varied flow (Equation 8.19) may be solved using numerical methods, which have superseded earlier approaches by direct and graphical integration. Two such methods are described in the following sections.

Computations of gradually varied surface profiles should proceed upstream from the control section in subcritical flow and downstream from the control section in supercritical flow.

8.11 The direct step method

The direct step method is a simple method applicable to prismatic channels. As in the graphical integration method, depths of flow are specified and the distances between successive depths calculated.

Consider an element of the flow (Figure 8.11).

Equating total heads at 1 and 2,

$$S_0 \Delta x + y_1 + \alpha_1 \frac{V_1^2}{2g} = y_2 + \alpha_2 \frac{V_2^2}{2g} + S_f \Delta x$$

$$\text{i.e. } \Delta x = \frac{E_2 - E_1}{S_0 - S_f} \qquad [8.20]$$

where E is the specific energy.

In the computations S_f is calculated for depths y_1 and y_2 and the average taken denoted by $\bar{S}_f$.

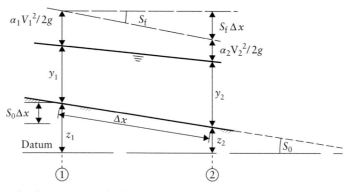

Figure 8.11 The direct step method.

8.12 The standard step method

The standard step method is applicable to non-prismatic channels and therefore to natural rivers. The station positions are predetermined and the objective is to caculate the surface elevations, and hence the depths, at the stations. A trial-and-improvement method is employed (see Figure 8.12).

$$y_1 + \frac{\alpha V_1^2}{2g} + h_f = S_0 \Delta x + y_2 + \frac{\alpha V_2^2}{2g}$$

$$Z_1 = y_1; \qquad Z_2 = S_0 \Delta x + y_2$$

and assuming $\alpha_1 = \alpha_2 = \alpha$,

$$Z_1 + \frac{\alpha V_1^2}{2g} + h_f = Z_2 + \frac{\alpha V_2^2}{2g} \qquad [8.21]$$

Writing $H_1 = Z_1 + \alpha V_1^2/2g$ and $H_2 = Z_2 + \alpha V_2^2/2g$, Equation 8.21 becomes $H_1 + h_f = H_2$.

Proceeding upstream (in subcritical flow), for example, H_1 is known and Δx is predetermined. Z_2 is estimated, for example, by adding a small amount to Z_1; y_2 is obtained from $y_2 = Z_2 - z_2$. The area and wetted perimeter, and hence hydraulic radius corresponding to y_2, are obtained from the known geometry of the section.

Calculate $S_{f,2} = n^2 V_2^2 / R_2^{4/3}$ and $\bar{S}_f = (S_{f,2} + S_{f,1})/2$.
Calculate $\alpha V_2^2/2g$ and $H_{(2)} = Z_2 + \alpha V_2^2/2g$.
Calculate $H_{(1)} = H_1 + \bar{S}_f \Delta x$.

Compare $H_{(2)}$ and $H_{(1)}$; if the difference is not within prescribed limits (e.g. 0.001 m), re-estimate Z_2 and repeat until agreement is reached; Z_2, y_2 and $H_{(2)} = H_2$ are then recorded and Z_2 and H_2 become Z_1 and H_1 for the succeeding station.

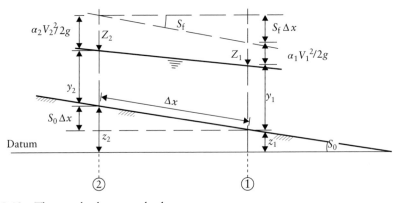

Figure 8.12   The standard step method.

8.13 Canal delivery problems

When a channel is connected to two reservoirs its discharge capacity depends upon inlet (upstream) and outlet (downstream) conditions imposed by the water levels in the reservoirs. The reservoir–canal–reservoir interaction depends upon the channel characteristics such as its boundary roughness, its slope, the length between reservoirs and the state of water levels in the reservoirs.

8.13.1 Case 1: Upstream reservoir water level is constant

For a given boundary characteristic the discharge rate in the channel depends on whether it is a long or short channel with a mild or critical or steep bed slope. A long channel delivers water with no interference from the downstream water levels (i.e. with no downstream control); only the inlet controls the flow rate. This suggests that any water surface profiles likely to develop with the available downstream water levels will not be long enough to reach the inlet, thus allowing the inlet to discharge freely. On the other hand, if the channel is short, the water surface profiles could submerge the inlet and this submergence affects the flow rate. The types of profile and their appropriate lengths would depend on the channel slope.

(a) *Long channel with mild slope* (see Figure 8.20): The flow on entering the inlet establishes uniform conditions from a short distance downstream of the entry. Thus at entry we have two simultaneous equations to compute the flow depth y_0 and discharge Q_0 or velocity:

> Energy equation: upstream water level above inlet, $H = y_0 + \alpha V^2/2g + K\alpha V^2/2g$
> Uniform flow resistance equation (say Manning's), $V = (1/n)R^{2/3}S^{1/2}$

See Example 8.7 for the solution of these two equations.

(b) *Long channel with critical slope:* The flow now establishes with its normal flow depth equal to the critical depth from the inlet, thus allowing maximum possible discharge through the channel. We now have at the channel entry two equations enabling the computations of Q and y:

$$\text{Energy equation at inlet: } H = y_c + \alpha V_c^2/2g + \alpha K V_c^2/2g$$

Either the critical depth criterion $\alpha Q^2 B/gA^2 = 1$ or the appropriate uniform flow resistance equation.

(c) *Long channel with steep slope:* The flow depth at the entry is critical, the channel delivering maximum possible discharge, and if the channel is sufficiently long uniform flow will establish further downstream of the entry. The flow up to this point will be non-uniform with the development of an S_2 profile which asymptotically merges with the uniform flow depth (see Example 8.8).

The problems are much more complicated if the channels are short; in other words, any downstream control or disturbance (e.g. downstream water level variations) extends its influence right up to the entry, thus submerging the entry and changing the delivery capacities of the channel. Such problems are solved iteratively by computer.

8.13.2 Case 2: Downstream water level is constant and upstream level varies

Long channel with mild slope: here the discharge gradually increases with increasing upstream level (y_1) with the formation of M_1 profiles and attains uniform flow conditions ($Q = Q_0$) when $y_1 = y_2$, the downstream level. Further increases in y_1 produce M_2 profiles, ultimately delivering a maximum discharge whose critical depth is equal to y_2. Any further rise in y_1 would develop an M_2 profile terminating with its corresponding critical depth, now greater than y_2; this necessitates a corresponding increase in y_2.

8.13.3 Case 3: Both water levels varying (mild slope)

For a constant Q, the levels y_1 and y_2 are fluctuating, thus leading to a number of possible surface profiles. With $y_1 = y_2$, uniform flow is established.

However, for water levels above uniform flow depth M_1, profiles develop with the upper limit occurring when $y_2 = y_1 + S_0 L$, L being the length of the channel between reservoirs. For water levels below uniform depth M_2, profiles develop with the minimum depth of flow occurring when $y_2 = y_c$, the critical depth corresponding to the given discharge.

8.14 Culvert flow

Highway cross drainage is normally provided with culverts, bridges and dips. Culverts are structures buried under a high-level embankment (see Figure 8.13). The culvert consists of a pipe barrel (conveyance part, i.e. the channel) with protection works at its entrance and exit. It creates a backwater effect to the approach flow, causing a pondage of water above the culvert entrance. The hydraulic design of the culvert is based upon the characteristics of the barrel flow (free surface flow, orifice flow or pipe flow) conditions which depend on its length, roughness, gradient and upstream and downstream water levels.

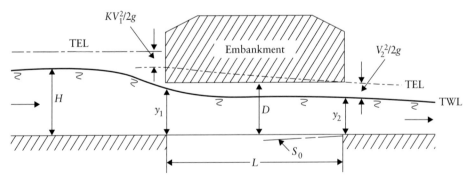

Figure 8.13 Culvert flow with free entrance. L, length of culvert; D, height of culvert; S_0, bed slope; y_1, depth at entrance; y_2, depth at exit; y_0, uniform flow depth; y_c, critical depth; K, entry loss coefficient; TEL, total energy line; TWL, tail water level.

Free entrance conditions:

1. $H/D < 1.2$; $y_0 > y_c < y_2 < D$; any length; mild slope: open channel subcritical flow
2. $H/D < 1.2$; $y_0 > y_c > y_2 < D$; any length; mild slope: open channel subcritical flow
3. $H/D < 1.2$; $y_0 < y_c > y_2 < D$; any length; steep slope: open channel supercritical flow; critical depth at inlet
4. $H/D < 1.2$; $y_0 < y_c < y_2 < D$; any length; steep slope: open channel supercritical flow; formation of hydraulic jump in barrel
    Submerged entrance conditions:
5. $H/D > 1.2$; $y_2 < D$; short; any slope: orifice flow
6. $H/D > 1.2$; $y_2 < D$; long; any slope: pipe flow
7. $H/D > 1.2$; $y_2 > D$; any length; any slope: pipe flow

See Example 8.27 for a complete analysis of culvert flow.

8.15 Spatially varied flow in open channels

Spatially varied flow (SVF) is represented by the discharge variation along the length of the channel due to lateral inflow (side spillway channel) or outflow (side weir or bottom racks).

8.15.1 Increasing flow (q_*, inflow rate per unit length)

In this case there exists a considerable amount of turbulence due to the addition of the incoming flow, and the energy equation is not of much use. With the usual assumptions introduced in the development of non-uniform flow equations, and assuming that the lateral inflow has no x-momentum added to the channel flow, we can deduce an equation for the surface slope as

$$\frac{dy}{dx} = \frac{S_0 - S_f - 2\beta Q q_*/gA^2}{1 - \beta Q^2 T/gA^3} \qquad [8.22]$$

In the case of subcritical flow all along the channel, the control (critical depth) of the profile is at the downstream end of the channel. For all other flow situations the establishment of the control point is essential to initiate the computational procedures.

In a rectangular channel ($T = B$, the bed width), the location of the control point x_c may be written approximately (Henderson, 1966) as

$$x_c = \frac{8q_*^2\beta^3}{gB^2\left(S_0 - gP/C^2B\right)^3} \qquad [8.23]$$

where C is the Chezy coefficient and P the wetted perimeter of the channel. The control point in a channel will exist only if the channel length $L > x_c$. In a given length of the channel, for a control section to exist, its slope S_0 should have a minimum value given by

$$S_0 > \frac{gP}{C^2B} + \frac{2\beta}{B}\left(\frac{q_*^2B}{gL}\right)^{1/3} \qquad [8.24]$$

and the flow upstream of the control is subcritical.

8.15.2 Decreasing flow (q_*, outflow rate per unit length) – side weir

Assuming the energy loss (due to diversion of the water) in the parent channel is zero, the water surface slope equation can be deduced as

$$\frac{dy}{dx} = \frac{S_0 - S_f - \alpha Q q_*/gA^2}{1 - \alpha Q^2 T/gA^3} \qquad [8.25]$$

Equation 8.25 can be extended to be applicable to a side weir of short length with $S_0 = S_f = 0$ and $\alpha = 1$. In the case of a rectangular channel, Equation 8.25 is rewritten as

$$\frac{dy}{dx} = \frac{Qy(-dQ/dx)}{gB^2y^3 - Q^2} \qquad [8.26]$$

The outflow per unit length, $q_* = -dQ/dx$, is given by the weir equation

$$-\frac{dQ}{dx} = \frac{2}{3}C_M\sqrt{2g}(y - s)^{3/2} \qquad [8.27]$$

where C_M is the De Marchi discharge coefficient; s, sill height; and y, flow depth in the channel. If the specific energy in the channel, E, is assumed constant, the discharge in the channel at any section is given by

$$Q = By\sqrt{2g(E - y)} \qquad [8.28]$$

Combining Equations 8.26, 8.27 and 8.28 and integrating, we obtain

$$x = \frac{3B}{2C_M}\phi_M(y, E, s) + \text{Constant} \qquad [8.29]$$

in which

$$\phi_M(y, E, S) = \frac{2E - 3s}{E - s}\sqrt{\frac{E - y}{y - s}} - 3\sin^{-1}\sqrt{\frac{E - y}{y - s}} \qquad [8.30]$$

The weir length L between two sections is then given by

$$L = x_2 - x_1 = \frac{3}{2}\frac{B}{C_M}(\phi_{M2} - \phi_{M1}) \qquad [8.31]$$

The De Marchi coefficient, C_M, for a rectangular sharp-crested side weir is given by

$$C_M = 0.81 - 0.60Fr_1 \qquad [8.32]$$

for both subcritical and supercritical approach flows, Fr_1 being the flow Froude number. For a broad-crested side weir, the discharge coefficient is given by

$$C_M = (0.81 - 0.60Fr_1)K \qquad [8.33]$$

where K is a parameter depending on the crest length W, and for a 90° branch channel is given by

$$K = 1.0 \quad \text{for } \frac{y_1 - s}{W} > 2.0 \qquad [8.34]$$

and

$$K = 0.80 + 0.10\left(\frac{y_1 - s}{W}\right) \quad \text{for } \frac{y_1 - s}{W} < 2.0 \qquad [8.35]$$

8.15.3 Decreasing flow (bottom racks)

The flow over bottom racks (e.g. kerb openings) is spatially varied with the surface slope given by

$$\frac{dy}{dx} = \frac{2\varepsilon C\sqrt{y(E-y)}}{3y - 2E}$$ [8.36]

in which ε is the void ratio (opening area to total rack area); E, the specific energy (constant); and C, a coefficient of discharge depending on the configuration of openings. Further treatment of these topics can be found, for example, in French (1994, 2007).

Worked examples

Example 8.1

Measurements carried out on the uniform flow of water in a long rectangular channel 3.0 m wide and of bed slope 0.001 revealed that at a depth of flow of 0.8 m, the discharge of water at 15°C was 3.6 m³/s. Estimate the discharge of water at 15°C when the depth is 1.5 m using (a) the Manning equation and (b) the Darcy equation, and state any assumptions made.

Solution:

From the flow measurement, the value of n and the effective roughness size (k) can be found:

(a)

$$Q = \frac{A}{n} R^{2/3} S_0^{1/2} \quad \text{(Equation 8.4)}$$ (i)

$$n = \frac{A}{Q} R^{2/3} S_0^{1/2}$$

$$= 0.0137$$

$$\text{when } y = 1.5\text{m}; \quad Q = 8.60\,\text{m}^3/\text{s}$$

(b)   Using Equation 8.11,

$$Q = -A\frac{\sqrt{32gRS_0}}{2.303}\ln\left(\frac{k}{14.8R} + \frac{1.255v}{R\sqrt{32gRS_0}}\right)$$ (ii)

(Note the conversion from log (base 10) to ln (base e).)

$$\text{whence } \frac{k}{14.8R} + \frac{1.255v}{R\sqrt{32gRS_0}} = \exp\left(-\frac{Q \times 2.303}{A\sqrt{32gRS_0}}\right)$$

$$k = \left[\exp\left(-\frac{Q \times 2.303}{A\sqrt{32gRS_0}}\right) - \frac{1.255v}{R\sqrt{32gRS_0}}\right] \times 14.8 \times R = 0.00146 \text{ m}$$

Substitution in Equation (ii) with $y = 1.5$ m and $k = 0.00146$ m yields $Q = 8.44$ m³/s.

Example 8.2

A concrete-lined trapezoidal channel has a bed width of 3.5 m, side slopes at 45° to the horizontal, a bed slope 1 in 1000 and the Manning roughness coefficient of 0.015. Calculate the depth of uniform flow when the discharge is 20 m³/s.

Solution:

See Figure 8.14.

$$A = (b + Ny)y = (3.5 + y)\,y$$
$$P = b + 2y\sqrt{1 + N^2} = 3.5 + 2\sqrt{2}y$$
$$R = \frac{A}{P} = \frac{(3.5 + y)y}{3.5 + 2\sqrt{2}y}$$

Manning equation: $Q = \dfrac{A}{n} R^{2/3} S^{1/2}$

$$\text{i.e. } Q = \frac{(3.5 + y)y}{0.015}\left[\frac{(3.5 + y)y}{3.5 + 2\sqrt{2}y}\right]^{2/3}(0.001)^{1/2} \qquad \text{(i)}$$

Setting $Q = 20$ m³/s, Equation (i) may be solved for y by trial or by graphical interpolation from a plot of discharge against depth for a range of y values substituted into Equation (i) (see Figure 8.15 and the table below).
   At 20 m³/s, depth of uniform flow is 1.73 m.
   Of course, the graph (Figure 8.15) will enable the depth at any other discharge to be determined.

Depth y (m)	A (m²)	P (m)	R (m)	Q (m³/s)
1.0	4.50	6.33	0.711	7.56
1.2	5.64	6.89	0.818	10.40
1.4	6.86	7.46	0.920	13.67
1.6	8.16	8.02	1.017	17.39
1.8	9.54	8.59	1.110	21.57
2.0	11.00	9.16	1.200	26.21

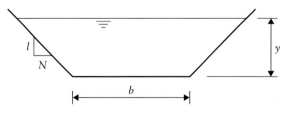

Figure 8.14   Flow through trapezoidal channel.

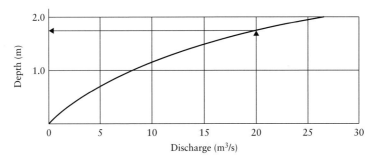

Figure 8.15 Normal depth versus discharge.

Example 8.3

Assuming that the flow in a river is in the rough turbulent zone, show that in a wide river a velocity measurement taken at 0.6 of the depth of flow will approximate closely to the mean velocity in the vertical.

Solution:

See Figure 8.16.
   In Chapter 7 it was shown that the velocity distribution in the turbulent boundary layer formed in the fluid flow past a rough surface is

$$\frac{v}{\sqrt{\tau_0/\rho}} = 5.75 \log \frac{y}{k} + 8.5 \tag{i}$$

$$\text{or} \quad \frac{v}{\sqrt{\tau_0/\rho}} = 5.75 \log \frac{30y}{k} \tag{ii}$$

Noting that the local velocity given by Equation (i) is reduced to zero at $y' = k/30$ from the boundary, the mean velocity in the vertical is obtained from

$$V = \frac{1}{y_0} \int_{y'}^{y_0} v \, dy$$

$$\text{whence } V = \frac{5.75\sqrt{\tau_0/\rho}}{2.303 y_0} \int_{y'}^{y_0} \ln \frac{30y}{k} dy$$

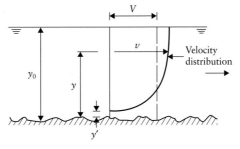

Figure 8.16 Velocity profile in rough turbulent zone.

$$= \frac{5.75\sqrt{\tau_0/\rho}}{2.303y_0}\left[y_0 \ln\frac{30}{k} + y_0 \ln y_0 - y_0 - \left(y' \ln\frac{30}{k} + y' \ln y' - y'\right)\right]$$

$$= \frac{5.75\sqrt{\tau_0/\rho}}{2.303y_0}\left(y_0 \ln\frac{30y_0}{k} - y_0 + y'\right)$$

i.e. $V = 5.75\sqrt{\frac{\tau_0}{\rho}}\left(\log\frac{30y_0}{k} - \frac{1}{2.303}\right)$ (iii)

ignoring the single y' term.

The distance above the bed at which the mean velocity coincides with the local velocity is obtained by equating (ii) and (iii):

$$\log\frac{30y}{k} = \log\frac{30y_0}{k} - 0.434$$

$$\text{or } y = 0.37y_0 \simeq 0.4y_0$$

This verifies the field practice of taking current meter measurements at 0.6 of depth to obtain a close approximation to the mean velocity in the depth.

Example 8.4

Derive Chezy's resistance equation for uniform flow in open channels, and show that the Chezy coefficient C is a function of the flow Reynolds number and the channel's relative roughness and is given by

$$C = 5.75\sqrt{g}\log\left(\frac{12R}{k + \delta'/3.5}\right)$$

where δ' is the sub-layer thickness given by Equation 7.33.

Solution:

Balancing the gravity and resisting forces along a reach length L of the channel (for uniform flow), we obtain

$$\rho g A L S_0 = \tau P L$$ (i)

giving the uniform boundary shear stress

$$\tau = \rho g R S_0$$ [8.37]

In turbulent flows, $\tau \propto V^2$ (V is the mean velocity of flow), and we can hence write

$$V = C\sqrt{RS_0}$$ [8.38]

Comparing Equation 8.38 (Chezy's) with the Darcy–Weisbach equation (Equation 8.2), we obtain

$$C = \sqrt{\frac{8g}{\lambda}}$$ [8.39]

In pipe flow, $\lambda = f(Re, k/D)$, and extending this to open channel flow ($D_e = 4R$),

$$\lambda = f(4RV/\nu, k/4R) \qquad [8.40]$$

Equation 8.40 is represented by the same Moody diagram constructed for pipe flow.
   Evaluation of the Chezy coefficient:
   The velocity distributions (two-dimensional flows) for smooth and rough boundaries given by Equations 7.23 and 7.27 can be written as

$$\frac{u}{u^*} = 5.75 \, \log \left(\frac{9u^*y}{\nu} \right) \quad \text{and} \quad \frac{u}{u^*} = 5.75 \, \log \left(\frac{30y}{k} \right) \qquad \text{(ii)}$$

In turbulent two-dimensional flows, $u = V$ at $y = 0.4y_0$ (see Example 8.3), y_0 being the flow depth. Also, Equation 7.33 suggests that $u_* = 11.6\nu/\delta'$. Combining these with (ii) and (iii) and replacing y_0 by R yield

$$V = 5.75u_* \, \log \left(\frac{12R}{k + \delta'/3.5} \right) \qquad \text{(iii)}$$

By writing $u_* = (gRS_0)^{1/2}$ in Equation (iv) and comparing with the Chezy equation (Equation 8.38), we can deduce

$$C = 5.75 \sqrt{g} \log \left(\frac{12R}{k + \delta'/3.5} \right) \qquad [8.41]$$

Example 8.5

A trapezoidal channel with side slopes 1:1 and bed slope 1:1000 has a 3 m wide bed composed of sand ($n = 0.02$) and sides of concrete ($n = 0.014$). Estimate the discharge when the depth of flow is 2.0 m.

Solution:

See Figure 8.17.

$P_1(=P_3) = 2.828 \, \text{m};$ $P_2 = 3.0 \, \text{m};$ $P = 8.656 \, \text{m}$ (on solid surface only)
$A_1(=A_3) = 2.0 \, \text{m}^2;$ $A_2 = 6.0 \, \text{m}^2;$ $A = 10.0 \, \text{m}^2$
$R_1(=R_3) = 0.7072 \, \text{m};$ $R_2 = 2.0 \, \text{m};$ $R = 1.155 \, \text{m}$

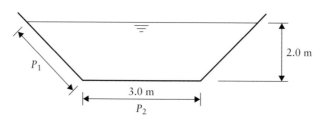

Figure 8.17   The equivalent Manning, n in channel of variable boundary roughness.

Evaluate composite roughness:

Horton and Einstein: $n = \left(\dfrac{\sum P_i n_i^{1.5}}{P} \right)^{2/3}$

$$n = \left(\frac{2 \times 2.828 \times 0.014^{1.5} + 3 \times 0.02^{1.5}}{8.656} \right)^{2/3} = 0.0162$$

Pavlovskij: $n = \left(\dfrac{\sum P_i n_i^2}{P} \right)^{1/2}$

$$n = \left(\frac{2 \times 2.828 \times 0.014^2 + 3 \times 0.02^2}{8.656} \right)^{1/2} = 0.0163$$

Lotter: $n = \dfrac{P R^{5/3}}{\sum_{i=1}^{N} P_i R_i^{5/3} / n_i} = 0.0157$

With $y = 2.0\,\mathrm{m}$,

$$\text{Discharge (Horton and Einstein)} = \frac{A}{n} R^{2/3} S_0^{1/2} = 21.49 \ \mathrm{m^3/s}$$
$$\text{Discharge (Pavlovskij)} = 21.36 \ \mathrm{m^3/s}$$
$$\text{Discharge (Lotter)} = 22.17 \ \mathrm{m^3/s}$$

Example 8.6

The cross section of the flow in a river during a flood was as shown in Figure 8.18. Assuming the roughness coefficients for the side channel and main channel to be 0.04 and 0.03, respectively, estimate the discharge.

Bed slope = 0.005
Area of main channel (bank-full) = 280 m²
Wetted perimeter of main channel = 54 m
Area of flow in side channel = 152.25 m²

Wetted perimeter of side channel around the solid boundary only (excluding the interfaces X–X between the main and side channel flows) = 104.24 m. Area of main channel

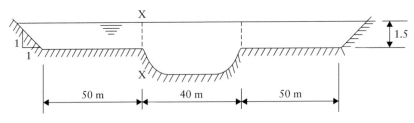

Figure 8.18   Two-stage (compound) channel.

component $= 280 + 40 \times 1.5 = 340$ m^2.

$$\text{Discharge} = \frac{340}{0.03}\left(\frac{340}{54}\right)^{2/3}\sqrt{0.005} + \frac{152.25}{0.04}\left(\frac{152.25}{104.24}\right)^{2/3}\sqrt{0.005}$$

$$= 3079 \, \text{m}^3/\text{s}$$

Note that the treatment of this problem by the equivalent roughness methods of Horton and Pavlovskij will produce large errors in the computed discharge due to the inherent assumptions. However, the Lotter method should produce a similar result to that computed above since it basically uses the same method.

$$\text{Lotter equivalent roughness, } n = \frac{PR^{5/3}}{\sum_{i=1}^{N} P_i R_i^{5/3}/n_i}$$

$$n = 0.0241$$

$$\text{and } Q = \frac{492.25}{0.0241}\left(\frac{492.25}{158.242}\right)^{2/3}\sqrt{0.005}$$

$$= 3077 \, \text{m}^3/\text{s}$$

Example 8.7

A long rectangular concrete-lined channel ($k = 0.3$ mm), 4.0 m wide, bed slope 1:500, is fed by a reservoir via an uncontrolled inlet. Assuming that uniform flow is established a short distance from the inlet and that entry losses are equal to $0.5V^2/2g$, determine the discharge and depth of uniform flow in the channel when the level in the reservoir is 2.5 m above the bed of the channel at inlet.

Figure 8.19 is an example of natural channel control; the discharge is affected both by the resistance of the channel and by the energy available at the inlet.

Two simultaneous equations therefore need to be solved:

1. Apply the energy equation to sections 1 and 2, assuming negligible velocity head at 1:

$$2.5 = y + \frac{V^2}{2g} + h_L = y + \frac{V^2}{2g} + \frac{0.5V^2}{2g} = y + \frac{Q^2}{(by)^2 2g}(1 + 0.5) \qquad \text{(i)}$$

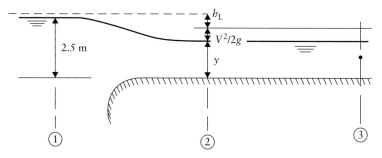

Figure 8.19   Channel inlet.

$$\text{or } Q_2 = by\sqrt{2g\frac{(2.5 - y)}{1.5}} \tag{ii}$$

2. Resistance equation applied downstream of section 2:

$$Q_3 = -A\sqrt{32gRS_0}\log\left(\frac{k}{14.8R} + \frac{1.255v}{R\sqrt{32gRS_0}}\right) \tag{iii}$$

Solution:

Equation (ii) could be incorporated in Equation (iii) to yield an implicit equation by y, which could then be found iteratively. However, a graphical solution can be obtained by generating curves for Q versus y from Equations (ii) and (iii).

y (m)	0.4	0.8	1.2	1.6	2.0	2.4	2.5
Q_2 (m³/s)	8.38	15.09	19.79	21.95	20.45	10.98	0
Q_3 (m³/s)	2.14	5.94	10.50	15.51	20.80	26.30	27.70

Q_2 and Q_3 are plotted against y in Figure 8.20 whence discharge $= 20.5$ m³/s at a uniform flow depth of 1.98 m, given by the point of intersection of the two curves.

   Note: Care must be taken in treating this method of solution as a universal case. For example, if the channel slope is steep the flow may be supercritical and the plots of Equations (ii) and (iii) would appear as in Figure 8.21.

   The solution is now not the point of intersection of the two curves. The depth passes through the critical depth at inlet and this condition controls the discharge, given by Q_c. Channel resistance no longer controls the flow and the depth of uniform flow corresponds with Q_c on the curve of Equation (iii).

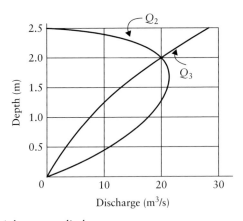

Figure 8.20 Depth at inlet versus discharge.

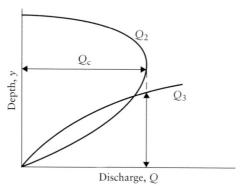

Figure 8.21   Q curves in steep-sloped channel.

Example 8.8

Using the data of Example 8.7 but with a channel bed slope of 1:300, calculate the discharge and depth of uniform flow.

Solution:

The discharge-versus-depth curve using the inlet energy relationship, Equation (ii) of Example 8.7, is unaffected by the bed slope. Q_3 is recomputed from Equation (iii) of Example 8.7 with $S_0 = 1/300$.

y (m)	0.4	0.8	1.2	1.6	2.0	2.4	2.5
Q_2 (m³/s)	8.38	15.06	19.79	21.95	20.45	10.98	0
Q_3 (m³/s)	3.94	10.91	19.27	28.44	38.14	48.20	50.77

The plotted curves of Q_2 versus y and of Q_3 versus y appear as in Figure 8.21. The flow at the channel inlet is critical, thus controlling the discharge ($=Q_c$). Downstream, the uniform depth of flow is supercritical.

Summary: Discharge $= 21.95 \text{ m}^3/\text{s}$;   depth in channel $= 1.31 \text{ m}$

Example 8.9

Determine the dimensions of a trapezoidal channel, lined with concrete ($k = 0.15$ mm) with side slopes at 45° to the horizontal and bed slope 1: 1000 to discharge 20 m³/s of water at 15°C under uniform flow conditions such that the section is the most economic.

Solution:

See Figure 8.4.

$$N = 1.0$$
$$b + 2Ny = 2y\sqrt{1 + N^2}$$

$$b + 2y = 2\sqrt{2}y$$
$$\text{or } b = 0.828y$$
$$\text{Then } A = 1.828y^2; \qquad P = 3.656y$$

$$Q = -A\sqrt{32gRS_0} \, \log\left(\frac{k}{14.8R} + \frac{1.255v}{R\sqrt{32gRS_0}}\right)$$

This must be solved by trial, calculating Q for a series of y.

y (m)	0.5	1.0	1.5	1.8	1.9	2.0
Q (m³/s)	0.54	3.30	9.50	15.26	17.56	20.06

Adopt $y = 2.0$ m, then

$$b = 1.414 \text{ m}$$

Example 8.10

A trapezoidal irrigation channel excavated in silty sand having a critical tractive force on the horizontal of 2.4 N/m² and angle of friction 30° is to be designed to convey a discharge of 10 m³/s on a bed slope of 1:10 000. The side slopes will be 1 (vertical) : 2 (horizontal) ($n = 0.02$).

Solution:

The channel bed is almost horizontal and the critical tractive force on the bed may therefore be taken as 2.4 N/m².

The limiting tractive force on the sides is

$$\tau_{cs} = \tau_{cb}\sqrt{1 - \frac{\sin^2\theta}{\sin^2\phi}} \quad \text{(see Section 8.5.1)}$$

$$= 2.4\sqrt{1 - \frac{\sin^2(26.565°)}{\sin^2(30°)}}$$

$$= 2.4\sqrt{1 - 0.8} = 1.073 \text{ N/m}^2$$

$$\Rightarrow 0.76\rho g y S_0 \not> 1.073 \text{ N/m}^2$$

$$\Rightarrow y \not> \frac{1.073}{0.76 \times 1000 \times 9.81 \times 0.0001}$$

$$y \not> 1.44 \text{ m}$$

Now $Q = \dfrac{A}{n}R^{2/3}S_0^{1/2}$

$$10 = \frac{(b+2y)y}{n}\left(\frac{(b+2y)y}{b+2y\sqrt{5}}\right)^{2/3} S_0^{1/2}$$

$$10 = \frac{(b+2.88) \times 1.44}{0.02} \left(\frac{(b+2.88) \times 1.44}{b+6.44} \right)^{2/3} \times \sqrt{0.0001}$$

Solving by trial (graphical interpolation) for series of values of b,

b	1.0	2.0	4.0	8.0	10.0	12.0
RHS	2.31	3.11	4.78	8.27	10.05	11.84

$$\text{required } b = 9.95 \, \text{m}$$
$$V = 0.54 \, \text{m/s}$$

which agrees reasonably with the maximum mean velocity criterion (Table 8.3).

Example 8.11 (Maximum mean velocity criterion)

Using the data of the previous example, determine the channel dimensions such that the mean velocity does not exceed 0.5 m/s when conveying the discharge of 10 m³/s.

Solution:

$$Q = AV; \quad 10 = A \times 0.5$$
$$\text{whence } A = 20 \, \text{m}^2$$
$$\text{and } A = (b+2y)y$$
$$\text{whence } b = (20/y) - 2y$$
$$Q = \frac{A}{n} R^{2/3} S_0^{1/2} = \frac{A^{5/3}}{nP^{2/3}} S^{1/2}$$
$$\text{i.e. } 10 = \frac{20^{5/3} \times \sqrt{0.0001}}{0.02 \times P^{2/3}}$$
$$\text{whence } P = 20 \, \text{m}$$
$$\text{and } P = b + 2y\sqrt{5} = \frac{20}{y} + 2y\left(\sqrt{5}-1\right)$$
$$\text{that is,} \quad 2.472y^2 - 20y + 20 = 0$$
$$\text{whence } y = 1.169 \, \text{m}$$
$$\text{and } b = 14.77 \, \text{m}$$

Example 8.12

Check the proposed design of a branch of a wastewater sewerage system receiving the flow from 300 houses. The pipe is 150 mm in diameter with a proposed slope of 1 in 100. The roughness of the sewer is 1.5 mm. The average flow, known as dry weather flow (DWF), is calculated assuming a population of 3 person per house, a flow of 200 L/day per person and an allowance of 10% for additional infiltration into the sewer. A peak flow of at least twice the average (2 DWF) is generally achieved each day, and this flow value is to be used to check the minimum self-cleansing velocity criterion of 0.75 m/s. A flow value of 6 DWF is to be used to check the pipe discharge capacity.

Solution:

From Equations 8.10 and 8.11 with kinematic viscosity, $v = 1.14 \times 10^{-6}$ m²/s, or by using appropriate hydraulic charts or tables (HR Wallingford, 1990), the full pipe velocity and discharge may be found to be

$$V_{full} = 0.88 \text{ m/s}$$

$$Q_{full} = 15.5 \text{ L/s}$$

To check the pipe discharge capacity, 6 DWF is calculated:

$$6\,DWF = \frac{6 \times 300 \times 3 \times 200 \times 1.1}{24 \times 3600} = 13.8\,L/s$$

This is satisfactory since it does not exceed the Q_{full} value of 15.5 L/s.

The velocity V at $Q = 2$ DWF = 4.6 L/s must also be checked.

A chart showing proportional depth (depth divided by diameter D) against proportional velocity (V/V_{full}) and proportional discharge (Q/Q_{full}) is useful for this purpose. This is shown in Figure 8.22, with the proportional discharge as the upper line and proportional velocity as the lower line. The proportional discharge in this example is 4.6/15.5 = 0.30. Using this value to enter the figure as indicated, or by using calculated tables, it is found

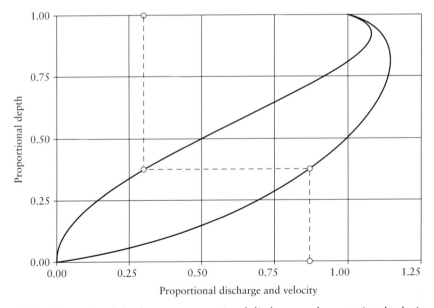

Figure 8.22   Proportional depth versus proportional discharge and proportional velocity.

that the proportional depth is 0.38 and the corresponding proportional velocity is 0.87. The velocity at 2 DWF is therefore given by

$$V = \left(\frac{V}{V_{full}}\right)V_{full} = 0.87 \times 0.88 \, \text{m/s} = 0.77 \, \text{m/s}$$

The design is therefore satisfactory since this exceeds the required self-cleansing velocity of 0.75 m/s.

Example 8.13

Design a branch within a storm sewer network which has a length of 100 m, a bed slope of 1 in 125 and a roughness size of 0.6 mm and which receives the storm runoff from 3.5 ha of impermeable surface using the rational method. In designing the upstream pipes, the maximum 'time of concentration' at the head of the pipe has been found to be 6.2 min. Use the relationship between rainfall intensity and average storm duration tabulated below.

Storm duration (min)	6.0	7.0	8.0
Average rainfall intensity (mm/h)	61.4	57.2	53.5

Notes: The rational method gives the peak discharge (Q_p) from an urbanised catchment in the form

$$Q_p = \frac{1}{360}Ai \,(\text{m}^3/\text{s})$$

where A is the impermeable area (ha) assuming 100% runoff; and i is the average rainfall intensity (mm/h) during the storm.

Since the average rainfall intensity of storms of a given average return period decreases with increase in storm duration, the critical design storm is that which has a duration equal to the 'time of concentration of the catchment', t_c. t_c is the longest time of travel of a liquid element to the point in question in the catchment and includes the times of overland and pipe flow; the time of pipe flow is based on full-bore velocities. For further coverage of design rainfall data see, for example, Mansell (2003).

Solution:

The selection of the appropriate pipe diameter is by trial and improvement. Try $D = 500$ mm.

Full-bore conditions are found in a similar way to the previous example:

$$V_{full} = 1.94 \, \text{m/s}$$

$$Q_{full} = 0.38 \, \text{m}^3/\text{s}$$

$$\text{Travel time along pipe} = \frac{100}{1.94 \times 60} = 0.9 \, \text{min}$$

$$\text{Thus } t_c = 6.2 + 0.9 = 7.1 \text{ min};$$

$$\text{whence } i = 56.8 \text{ mm/h}$$

$$\Rightarrow Q_p = \frac{3.5 \times 56.8}{360} = 0.55 \text{ m}^3/\text{s}$$

This is greater than the full-bore discharge (Q_{full}) of the 500 mm diameter pipe which is therefore too small. Try 600 mm diameter pipe.

$$V_{full} = 2.18 \text{ m/s}$$

$$Q_{full} = 0.62 \text{ m}^3/\text{s}$$

$$\text{Travel time along pipe} = \frac{100}{2.18 \times 60} = 0.8 \text{ min}$$

$$t_c = 7.0 \text{ min}; \quad \text{whence } i = 57.2 \text{ mm/h}$$

$$Q_p = \frac{3.5 \times 57.2}{360} = 0.56 \text{ m}^3/\text{s}$$

This does not exceed the full-bore discharge capacity, and so a 600 mm diameter sewer is a satisfactory design without surcharging.

Example 8.14

A rectangular channel 5 m wide laid to a mild bed slope conveys a discharge of 8 m³/s at a uniform flow depth of 1.25 m.

(a) Determine the critical depth.
(b) Neglecting the energy loss, show how the height of a streamlined sill constructed on the bed affects the depth upstream of the sill and the depth at the crest of the sill.
(c) Show that if the flow at the crest becomes critical, the structure can be used as a flow-measuring device using only an upstream depth measurement.

Solution:

(a)

$$y_c = \sqrt[3]{\frac{q^2}{g}}; \qquad q = \frac{8}{5} = 1.6 \text{ m}^3/(\text{s m})$$

$$y_c = \sqrt[3]{\frac{1.6^2}{9.81}} = 0.639 \text{ m}$$

(b) Neglecting losses between 1 and 2 (see Figure 8.23),

$$E_1 = E_2 + z$$

In the case of a uniform rectangular channel the specific energy curve is the same for any section, and if this is drawn for the specified discharge it can be used to show the variation of y_1 and y_2.

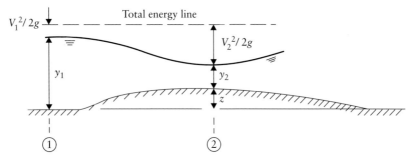

Figure 8.23 Flow over hump.

y (m)	V = Q/by (m/s)	V²/2g (m)	E = y + V²/2g (m)
0.2	8.00	3.262	3.462
0.3	5.33	1.450	1.750
0.4	4.00	0.815	1.215
0.6	2.67	0.362	0.962
0.8	2.00	0.204	1.004
1.0	1.60	0.130	1.130
1.2	1.33	0.091	1.291
1.4	1.14	0.067	1.467
1.6	1.00	0.051	1.651

For small values of z (crest height) and assuming that the upstream depth is the uniform flow depth (y_n) (see Figure 8.24), the equation

$$E_1 = E_2 + z$$

can be evaluated (for y_2) by entering the diagram with y_1 (=y_n) moving horizontally to meet the E curve (at x) setting off z to the left to meet the E curve again at w, which corresponds with the depth at the crest y_2.

This procedure can be repeated for all values of z up to z_c at which height the flow at the crest will just become 'critical'. Within this range of crest heights, the upstream depth (the uniform flow depth, y_n) remains unaltered.

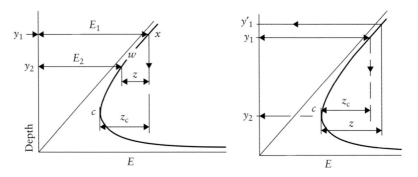

Figure 8.24 Specific energy curves with hump.

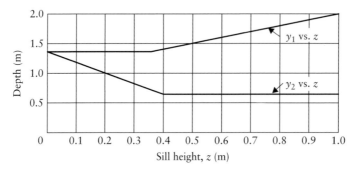

Figure 8.25 Sill height versus depths upstream and over hump.

Of course, the solution can also be obtained from the equation

$$y_1 + \frac{V_1^2}{2g} = y_2 + \frac{V_2^2}{2g} + z$$

but it is important to realise that if z exceeds z_c, then y_1 will not remain equal to y_n.

If z exceeds z_c, the E curve can still be used to predict the surface profile; y_2 remains equal to y_c. Set off z to the right from c (right-hand diagram of Figure 8.24).

Note that y_1 has increased (to y_1') to give the increased energy to convey the discharge over the crest.

The solution using a numerical method of solution of the energy equation for greater accuracy is tabulated: the graphical method described gives similar values. See Figure 8.25 and the table here:

z (m)	0.1	0.2	0.3	0.4	0.5	0.6	0.7	0.8	0.9
y_1 (m)	1.250	1.250	1.250	1.280	1.390	1.500	1.610	1.715	1.820
y_2 (m)	1.130	1.000	0.850	0.639	0.639	0.639	0.639	0.639	0.639

(c)   See Figure 8.26.

$$H_1 = y_c + \frac{V_c^2}{2g}$$

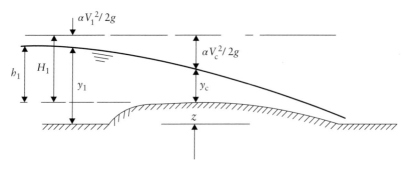

Figure 8.26 Hump as flow-measuring structure.

$$\text{Now } y_c = \frac{2}{3}H_1 \quad \text{(see Section 8.7)}$$

$$H_1 = \frac{2H_1}{3} + \frac{V_c^2}{2g}$$

$$V_c = \sqrt{\frac{2g}{3}}H_1^{1/2}$$

$$\text{and } Q = by_cV_c = \frac{2}{3}bH_1\sqrt{\frac{2g}{3}}H_1^{1/2}$$

$$\text{i.e. } Q = \frac{2b}{3}\sqrt{\frac{2g}{3}}H_1^{3/2}$$

Note that H_1 is the upstream energy measured relative to the crest of the sill. In practice, the upstream depth above the crest (h_1) would be measured and the velocity head $\alpha V_1^2/2g$ is allowed for by a coefficient C_v and energy losses by C_d:

$$Q = \frac{2}{3}b\sqrt{\frac{2g}{3}}C_vC_dh_1^{3/2}$$

(see BS 3680 Part 4A, 1981).

See also Example 8.18, which illustrates the effect of downstream conditions on the existence of critical flow over the sill.

Example 8.15

Venturi flume: A rectangular channel 2.0 m wide is contracted to a width of 1.2 m. The uniform flow depth at a discharge of 3 m³/s is 0.8 m. (a) Calculate the surface profile through the contraction, assuming that the profile is unaffected by downstream conditions. (b) Determine the maximum throat width such that critical flow in the throat will be created.

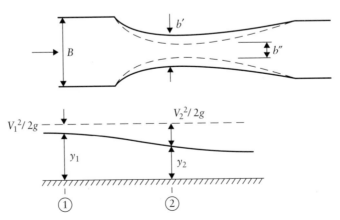

Figure 8.27   Channel contraction – venturi flume.

Solution:

See Figure 8.27.

In principle, the system, and the method of solution, is similar to that of Example 8.14. However, the specific energy diagrams for any of the contracted sections are no longer identical with that for section 1.

$$E_1 = E_2 = y_2 + \frac{V_2^2}{2g}$$

Entering with $y_1 (=y_n)$ to meet the E curve for section 1 (width B) at X and moving vertically to meet the curve for width b' yields y_2', the depth in the throat (see Figure 8.28).

If the throat is further contracted to b'', the vertical through X no longer intercepts the E curve for width b''; this means that the energy at 1 is insufficient and must rise to meet the specific energy at 2 corresponding to the critical depth at 2 $(E_{2\,min})$.

For a given discharge a minimum degree of contraction is required to establish critical flow at the throat; this is b_c corresponding to the specific energy curve which is just tangential to the vertical through X. Provided it is recognised that if $b < b_c, y_1$ will be greater than y_n, the problem can be solved numerically.

$$\text{At uniform flow } E_1 = y_n + \frac{Q^2}{2gB^2y_n^2}$$

$$= 0.8 + \frac{3^2}{19.62 \times 2^2 \times 0.8^2}$$

$$= 0.979 \text{ m}$$

If flow at the throat were critical, the minimum specific energy would be $y_c + y_c/2 = 1.5y_c$

$$\text{and } y_c = \sqrt[3]{\frac{q^2}{g}} = \sqrt[3]{\frac{Q^2}{gb^2}}$$

$$y_c = \sqrt[3]{\frac{3^2}{9.81 \times 1.2^2}} = 0.86 \text{ m}$$

$$\text{and } E_{min} = 1.291 \text{ m}$$

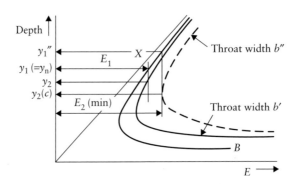

Figure 8.28   Specific energy curves at throat.

This is greater than the energy upstream at uniform flow; thus, the flow in the throat will be critical.

Therefore, $y_2 = 0.86$ m and y_1 is obtained from

$$1.291 = y_1 + \frac{Q^2}{2gB^2y_1^2}$$

$$1.291 = y_1 + \frac{3^2}{19.62 \times 2^2 \times y_1^2}$$

whence $y_1 = 1.215$ m

(b) Let b_c be the throat width to create critical flow at the specified discharge. Critical flow can just be achieved with the upstream energy of 0.979 m.

$$0.979 = y_c + \frac{Q^2}{2gb_c^2y_c^2} \tag{i}$$

Since 0.979 m is also the energy corresponding to the critical flow at the throat,

$$y_c = 2/3 \times 0.979 = 0.653 \text{ m}$$

whence from Equation (i)

$$0.326 = \frac{3^2}{19.62 \times b_c^2 \times 0.653^2}$$

The solution to which is $b_c = 1.816$ m.

Note: In a similar manner to that of Example 8.14(c), it can be shown that if the flow in the throat is critical, the discharge can be calculated from the theoretical equation

$$Q = \frac{2}{3}\sqrt{\frac{2g}{3}}bH_1^{3/2} \quad \left(\text{Practical form: } Q = \frac{2}{3}\sqrt{\frac{2g}{3}}bC_vC_db_1^{3/2}\right)$$

where H_1 is the upstream energy $(h_1 + V_1^2/2g)$ and h_1 the upstream depth. In practice, the throat would be made narrower than that calculated in the example above in order to create supercritical flow conditions in the expanding section downstream of the throat followed by a hydraulic jump in the downstream channel. The reader is referred to BS 3680 Part 4C (1981) and also to Example 8.19.

Example 8.16

A vertical sluice gate with an opening of 0.67 m produces a downstream jet depth of 0.40 m when installed in a long rectangular channel 5.0 m wide conveying a steady discharge of 20.0 m³/s. Assuming that the flow downstream of the gate eventually returns to the uniform flow depth of 2.5 m,

(a) verify that a hydraulic jump occurs. Assume $\alpha = \beta = 1.0$.
(b) calculate the head loss in the jump.
(c) if the head loss through the gate is $0.05V_j^2/2g$, calculate the depth upstream of the gate and the force on the gate.
(d) if the downstream depth is increased to 3.0 m, analyse the flow conditions at the gate.

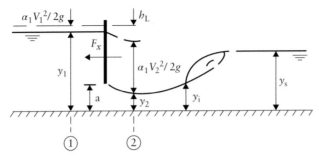

Figure 8.29   Sluice gate control and hydraulic jump.

Solution:

(a)   See Figure 8.29.

If a hydraulic jump is to form the required initial depth, (y_i) must be greater than the jet depth.

$$y_i = \frac{y_s}{2}\left(\sqrt{1+8F_s^2}-1\right); \qquad Fr_s = \frac{V_s}{\sqrt{gV_s}} = \frac{20}{5.0\times 2.5\sqrt{9.8\times 2.5}}$$

That is, $Fr_s = 0.323$ and hence $y_i = 0.443$ m (Equation 8.16). Therefore, a jump will form.

(b)   Head loss at jump:

$$h_L = \left(y_i + \frac{V_i^2}{2g}\right) - \left(y_s + \frac{V_i^2}{2g}\right)$$

$$= 0.443 - 2.5 + \frac{1}{2g}\left[\left(\frac{4}{0.443}\right)^2 - \left(\frac{4}{2.5}\right)^2\right]$$

$$= 1.97\,\mathrm{m}$$

(c)   (i) Apply the energy equation to 1 and 2:

$$y_1 + \frac{V_1^2}{2g} = y_2 + \frac{V_2^2}{2g} + 0.05\frac{V_2^2}{2g}$$

$$V_2 = \frac{4}{0.4} = 10\,\mathrm{m/s}; \qquad \frac{V_2^2}{2g} = 5.099\,\mathrm{m}$$

$$y_1 + \frac{q^2}{2gy_1^2} = 5.752\,\mathrm{m}$$

whence $y_1 = 5.73$ m

(ii) F_x = gate reaction per unit width

Apply momentum equation to element of water between 1 and 2:

$$\frac{\rho g y_1^2}{2} + \rho q(V_1 - V_2) - \frac{\rho g y_2^2}{2} - F_x = 0$$

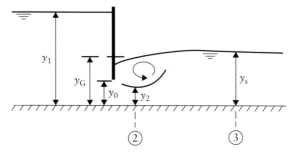

Figure 8.30 Submerged hydraulic jump.

(*Note*: The force due to the friction head loss through the gate is implicitly included in the above equation since this affects the value of y_1.)

$$1000 \left[\frac{9.806}{2}(5.73^2 - 0.4^2) + 4(0.693 - 10) \right] - F_x = 0$$

whence $F_x = 123$ kN/m width

(d) With a sequent depth of 3.0 m, the initial depth required to sustain a jump is 0.327 m (following the procedure of (a)). Therefore the jump will be submerged (see Figure 8.30), since the depth at the vena contracta is 0.4 m.

Apply the momentum equation to 2 and 3, neglecting friction and gravity forces:

$$\frac{\rho g y_G^2}{2} + \rho q(V_2 - V_s) - \frac{\rho g y_s^2}{2} = 0$$

$$y_G^2 - y_s^2 + \frac{2q^2}{g}\left(\frac{1}{y_2} - \frac{1}{y_s}\right) = 0$$

$$\text{whence } y_G = y_s \sqrt{1 + 2Fr_s^2 \left(1 - \frac{y_s}{y_2}\right)}$$

$$\text{where } Fr_s = \frac{V_s}{\sqrt{g y_s}}; \quad y_s = 3.0 \text{ m}; \quad y_2 = 0.4 \text{ m}$$

$$\Rightarrow y_G = 1.39 \text{ m}$$

Applying the energy equation to 1 and 2,

$$y_1 + \frac{V_1^2}{2g} = y_G + \frac{V_2^2}{2g} + 0.05\frac{V_2^2}{2g}$$

$$y_1 + \frac{V_1^2}{2g} = 6.74 \text{ m}$$

whence the upstream depth, y_1, is now 6.73 m.

Example 8.17

A sluice gate is discharging water freely (modular flow) under a head of 5 m (upstream of the gate) with a gate opening of 1.5 m. Compute the discharge rate per unit width of the gate. If the water depth immediately downstream of the gate is 2 m (drowned or non-modular flow), determine the discharge rate.

Solution:

Referring to Figure 8.29 for the modular/free flow case,

$$q = C_d a \sqrt{2gy_1} \tag{8.42}$$

where

$$C_d = \frac{C_c}{\sqrt{1 + C_c a/y_1}} \tag{8.43}$$

The contraction coefficient C_c $(=y_2/a)$ is a function of a/y_1; a reasonable constant value of $C_c = 0.60$ may be assumed for most conditions. Equation 8.42 can also be written as

$$q = C'_d a \sqrt{2g(y_1 - C_c a)} \tag{8.44}$$

where

$$C'_d = \frac{C_c}{\sqrt{1 - (C_c a/y_1)^2}} \tag{8.45}$$

For the submerged (non-modular) flow condition, the discharge

$$q_s = C'_{ds} a \sqrt{2g(y_1 - y_G)} \tag{8.46}$$

where $C'_{ds} = C'_d$ with $C_c = 0.60$. The depth of submergence, y_G, downstream of the gate (see Figure 8.30) is computed by the momentum equation (see Example 8.16) as

$$\frac{y_G}{y_s} = \sqrt{1 + Fr_s^2 \left(1 - \frac{y_s}{C_c a}\right)} \tag{8.47}$$

(i) Modular flow:

$$\frac{a}{y_1} = 0.3$$
$$\Rightarrow C_d = 0.552 \quad \text{(Equation 8.43)}$$
$$\text{hence } q = 8.2 \text{ m}^3/(\text{s m}) \quad \text{(Equation 8.42)}$$

(ii) Non-modular flow:

$$C'_{ds} = C'_d = 0.610 \quad \text{(Equation 8.45)}$$
$$\Rightarrow q = 0.61 \times 1.5 \times \sqrt{2g(5 - 2)} = 7.08 \text{ m}^3/(\text{s m})$$

Note: If we assume that the flow condition immediately downstream of the gate remains unaffected by submergence, we can obtain V_1 by the energy and continuity equations.

$$5 + \frac{V_1^2}{2g} = 2 + \frac{V_2^2}{2g}; \qquad 5V_1 = 0.6 \times 1.5 V_2$$
$$\text{as } V_1 = 1.4 \text{ m/s}; \quad \text{hence } q = y_1 \times V_1 = 7.0 \text{ m}^3/(\text{s m})$$

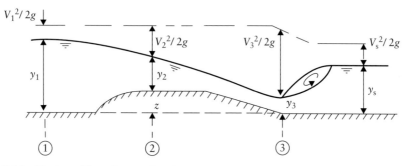

Figure 8.31   Design of broad-crested weir.

Example 8.18

A broad-crested weir is to be constructed in a long rectangular channel of mild bed slope for discharge monitoring by single upstream depth measurement.

Bed width = 4.0 m. Discharge measurement ranges from 3.0 to 20.0 m³/s. Depth–discharge (uniform flow) rating curve for the channel:

Depth (m)	0.5	1.0	1.5	2.0	2.5
Discharge (m³/s)	3.00	8.15	14.22	20.80	27.70

Select a suitable crest height for the weir.

Solution:

See Figure 8.31.

Ideally the design criterion is that a hydraulic jump should form downstream of the sill. From the table the depth of uniform flow at 20 m³/s is 1.95 m (y_n). Required initial depth to sustain a hydraulic jump of sequent depth 1.95 m (y_n) is

$$y_i = \frac{y_s}{2}\left(\sqrt{1 + 8Fr_s^2} - 1\right)$$

$$y_i = 0.913\,\text{m} \; (=y_3)$$

$$= y_3 + \frac{Q^2}{2g(by_3)^2} = 0.913 + 1.529 = 2.442\,\text{m}$$

For critical flow conditions at the crest of the sill,

$$y_2 = y_c = \sqrt[3]{\frac{q^2}{g}} = 1.37\,\text{m} \quad \text{and} \quad \frac{V_2^2}{2g} = \frac{V_c^2}{2g} = \frac{Q^2}{2g(by_c)^2} = 0.679\,\text{m}$$

Thus the specific energy at critical flow at the crest of the sill is

$$E_{2(\text{crit})} = y_c + \frac{V_c^2}{2g} = 1.37 + 0.679 = 2.049\,\text{m}$$

To find the minimum sill height, equate energy at sections 2 and 3:

$$\Rightarrow z + E_{2(\text{crit})} = E_3$$
$$\Rightarrow z + 2.049 = 2.442;$$
whence $z = 0.393$ (say 0.4 m)

The upstream depth can be calculated by equating E_1 to E_3 neglecting losses.

$$\Rightarrow y_1 + \frac{Q^2}{2g(by_1)^2} = 2.442$$
whence $y_1 = 2.17$ m

At the lower discharge of 3.0 m³/s the depth of uniform flow is 0.5 m. Required initial depth for a hydraulic jump of sequent depth 0.5 m is 0.29 m ($=y_3$). The minimum specific energy required to convey the discharge of 3 m³/s over the sill is that corresponding to critical flow conditions.

$$y_c = \sqrt[3]{\frac{q^2}{g}} = 0.386 \text{ m}; \qquad \frac{V_c^2}{2g} = 0.193 \text{ m}$$
$$E_c = 0.579 \text{ m}$$

With the established crest height of 0.4 m the minimum total energy is $0.4 + 0.579 = 0.979$ m.

Since this is much greater than 0.5 m, the downstream uniform flow depth, the flow at the crest is certainly critical, provided there are no downstream constraints.

Check the existence of a hydraulic jump:

$$E_3 = 0.979 \text{ m} \quad \text{(neglecting losses)}$$
$$y_3 + \frac{Q^2}{2g(by_3)^2} = 0.979 \text{ m}; \quad \text{whence } y_3 = 0.191 \text{ m}$$

Since this is less than that required for a jump to form (0.29 m), a hydraulic jump will form in the channel downstream of section 3. The design is therefore satisfactory.

Example 8.19 (The 'critical depth flume')

Using the data of Example 8.15, determine the minimum width of the throat of the venturi flume such that a hydraulic jump will be formed in the downstream channel with a sequent depth equal to the depth of uniform flow. Determine the upstream depth under these conditions.

Solution:

See Figure 8.32.

$$\text{Downstream depth } (=\text{sequent depth}) = \text{depth of uniform flow}$$
$$= 0.8 \text{ m}$$

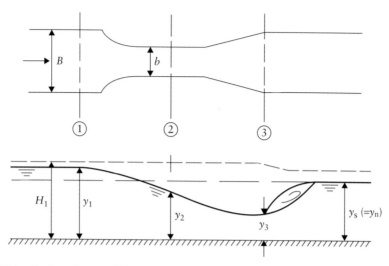

Figure 8.32 Design of venturi flume.

$$Q = 3 \text{ m}^3/\text{s}; \quad \text{channel width} = 2.0 \text{ m}$$

$$V_s = \frac{3}{2 \times 0.8} = 1.875 \text{ m/s}; \quad Fr_s = \frac{V_s}{\sqrt{gy_s}} = 0.67$$

Required initial depth for a hydraulic jump to form in the channel, with a sequent depth of 0.8 m,

$$y_i = \frac{y_s}{2}\left(\sqrt{1 + 8Fr_s^2} - 1\right) = 0.456 \text{ m}$$

Thus the maximum value of

$$y_3 = 0.456 \text{ m} \quad \text{and} \quad E_3 = y_3 + \frac{Q^2}{2g(by_3)^2} = 1.007 \text{ m}$$

Conditions in the throat will be critical; if b is the throat width,

$$y_{2,c} = \sqrt[3]{\frac{Q^2}{gb^2}}$$

Equating energies at 2 and 3,

$$y_{2,c} + \frac{V_{2,c}^2}{2g} = E_3 = 1.007 \text{ m}$$

that is, $\sqrt[3]{\dfrac{Q^2}{gb^2}} + \dfrac{Q^2}{2gb^2\left(Q^2/gb^2\right)^{2/3}} = 1.007$

that is, $1.5\left(\dfrac{Q^2}{gb^2}\right)^{1/3} = 1.007 \text{ m}$

whence $b = 1.74 \text{ m}$

(*Note:* This is narrower than that in Example 8.15, which specifically stated that, in that case, downstream controls did not affect the flow profile.)

The upstream depth can be calculated, neglecting losses, from $E_1 = E_2 = E_3$.

$$\text{that is, } y_1 + \frac{Q^2}{2gB^2y_1^2} = 1.007 \text{ m}$$

$$\text{whence } y_1 = 0.98 \text{ m}$$

(*Note:* This is greater than the uniform flow depth.)

Example 8.20

A trapezoidal concrete-lined channel has a constant bed slope of 0.0015, a bed width of 3 m and side slopes 1:1. A control gate increases the depth immediately upstream to 4.0 m when the discharge is 19.0 m³/s. Compute the water surface profile to a depth 5% greater than the uniform flow depth. Take $n = 0.017$ and $\alpha = 1.1$.

Notes: The energy gradient at each depth is calculated as though uniform flow existed at that depth. For hand calculation the Manning equation is much simpler than the Darcy–Colebrook–White equations which could, however, be incorporated in a computer program.

Calculations: Using the Manning equation the depth of uniform flow at 19.0 m³/s is 1.75 m. The systematic calculations are shown in tabular form below.

$$\Delta x = \int_{y_1}^{y_2} \frac{dx}{dy}\, dy = x_2 - x_1 \quad \text{where} \quad \frac{dx}{dy} = \frac{1 - (\alpha Q^2 T / A^3 g)}{S_0 - S_f}$$

y (m)	B (m)	A (m²)	R (m)	dx/dy	Δx (m)	x (m)
4.0	11.0	28.0	1.956	677.71		0
3.9	10.8	26.91	1.918	679.10	67.8	67.8
3.8	10.6	25.84	1.880	680.65	68.0	135.8
3.7	10.4	24.79	1.840	682.47	68.2	204.0
3.6	10.2	23.76	1.800	684.59	68.3	272.3
3.5	10.0	22.75	1.760	687.92	68.6	340.9
3.4	9.8	21.76	1.725	690.00	68.9	409.8
3.3	9.6	20.79	1.685	693.45	69.2	472.0
3.2	9.4	19.84	1.646	697.58	69.5	548.5
3.1	9.2	18.91	1.607	702.54	70.0	618.5
3.0	9.0	18.00	1.567	708.56	70.6	689.1
2.9	8.8	17.11	1.527	715.94	71.2	760.3
2.8	8.6	16.24	1.487	725.10	72.0	832.3
2.7	8.4	15.39	1.447	736.64	73.1	905.4
2.6	8.2	14.56	1.406	751.43	74.4	979.8
2.5	8.0	13.75	1.365	770.80	76.1	1056

2.4	7.8	12.96	1.324	796.90	78.4	1134
2.3	7.6	12.19	1.282	833.35	81.5	1216
2.2	7.4	11.44	1.240	886.86	86.0	1302
2.1	7.2	10.71	1.198	971.33	92.9	1395
2.0	7.0	10.00	1.155	1120.92	104.6	1499
1.9	6.8	9.31	1.111	1448.07	128.5	1628
1.8	6.6	8.64	1.068	2667.73	205.8	1834

Note: The surface profile is illustrated by plotting y versus x on the channel bed.

Example 8.21

Using the data of Example 8.20, compute the surface profile using the direct step method.

$$y_0 = 4.0 \text{ m}; \qquad S_0 = 0.0015; \qquad Q = 19 \text{ m}^3/\text{s}$$

$$n = 0.017; \qquad \alpha = 1.1$$

Solution:

At the control section, depth = 4.0 m, $A = 28.0 \text{ m}^2$, $R = 1.956$ m

$$\text{Specific energy} \left(y + \frac{Q^2}{2gA^2} \right) = 4.0 + \frac{1.1 \times 19^2}{19.62 \times 28^2} = 4.026 \text{ m}$$

$$S_f = \frac{Q^2 n^2}{A^2 R^{4/3}} = 5.44 \times 10^{-5}$$

y (m)	A (m²)	R (m)	E (m)	ΔE (m)	S_f	$\bar{S}_f$	Δx (m)	x (m)
4.0	28.0	1.956	4.026	—	5.44×10^{-5}	0	0	0
3.9	26.91	1.918	3.928	0.098	6.05×10^{-5}	5.74×10^{-5}	67.84	67.84
3.8	25.84	1.880	3.830	0.098	6.74×10^{-5}	6.39×10^{-5}	67.98	135.82
3.7	24.79	1.840	3.733	0.097	7.52×10^{-5}	7.13×10^{-5}	68.16	203.98
↓	↓	↓	↓	↓	↓	↓	↓	↓
1.8	8.64	1.068	2.0712	0.0623	1.28×10^{-3}	1.16×10^{-3}	184.94	1809.3

The surface profile is very similar to that calculated by the integration method (Example 8.20).

Example 8.22

Using the standard step method, compute the surface profile using the data of Example 8.20.

The solution is shown in the next table, the intermediate iterations where $H_{(1)} \neq H_{(2)}$ have not been included. It is noted that the result is almost identical with the numerical

Chapter 8

Example 8.22

x	Z	y	A	V	$\alpha V^2/2g$	$H_{(1)}$	R	S_f	$\bar{S}_f$	Δx	b_f	$H_{(2)}$
0.00	4.000	4.000	28.00	0.679	0.026	4.026	1.900					
100.00	4.002	3.582	26.39	0.720	0.029	4.031	1.843	0.000064	0.000059	100	0.0059	4.032
200.00	4.005	3.705	24.84	0.765	0.033	4.038	1.786	0.000075	0.000069	100	0.0069	4.038
300.00	4.008	3.558	23.33	0.814	0.037	4.045	1.729	0.000088	0.000082	100	0.0082	4.046
400.00	4.012	3.412	21.88	0.868	0.042	4.054	1.673	0.000105	0.000097	100	0.0097	4.055
500.00	4.017	3.267	20.47	0.928	0.048	4.065		0.000125	0.000115	100	0.0115	4.066
→	→	→	→	→	→	→	→	→	→	→	→	→
1300.00	4.164	2.196	11.41	1.665	0.155	4.302	1.239	0.000602	0.000548	100	0.0548	4.302
1400.00	4.188	2.188	10.62	1.788	0.179	4.367	1.193	0.000731	0.000666	100	0.0666	4.368
1500.00	4.242	1.992	9.94	1.910	0.205	4.447	1.152	0.000874	0.000802	100	0.0802	4.448
1600.00	4.311	1.911	9.38	2.024	0.230	4.541	1.117	0.001022	0.000948	100	0.0948	4.541
1700.00	4.397	1.847	8.95	2.122	0.253	4.650	1.088	0.001162	0.001093	100	0.1093	4.650
1800.00	4.500	1.800	8.64	2.120	0.271	4.771	1.068	0.001280	0.001221	100	0.1221	4.772
1900.00	4.618	1.768	8.43	2.254	0.285	4.903	1.054	0.001369	0.001325	100	0.1325	4.903

$Q = 19.0 \text{ m}^3/\text{s}$, $n = 0.017$, $S_0 = 0.0015$, $\alpha = 1.1$, $b = 3.0$ m, side slopes $= 1.1$, $y_n = 1.75$ m.

integration and direct step methods. However, unless a computer is used the calculations in the standard step method are laborious, and for prismatic channels with constant bed slopes the other methods would be quicker. The standard step method is particularly suited to natural channels in which the channel geometry and bed elevation at spatial intervals, which are not necessarily equal, have been measured. Variations in roughness coefficient, n, along the channel can also be incorporated.

Example 8.23

A vertical sluice gate situated in a rectangular channel of bed slope 0.005, width 4.0 m and Manning's $n = 0.015$ has a vertical opening of 1.0 m and $C_c = 0.60$. Taking $\alpha = 1.1$ and $\beta = 1.0$, determine the location of the hydraulic jump when the discharge is 20 m³/s and the downstream depth is regulated to 2.0 m.

Solution:

See Figure 8.33.
   Note: $S_f = Q^2 n^2 / A^2 R^{4/3}$.

Depth at vena contracta $= 1.0 \times 0.6 = 0.6$ m
Depth of uniform flow (from the Manning equation) $= 1.26$ m

Critical depth, $y_c = 1.366$ m   (Equation 8.14)

Initial depth at jump $= 0.884$ m   (Equation 8.16)

Note: $y_s = 2.0$ m.
   Proceeding downstream (in supercritical flow) from the control section, $y = 0.6$ m, and using the direct step method with $\Delta y = 0.04$ m, the calculations are shown in the next table.

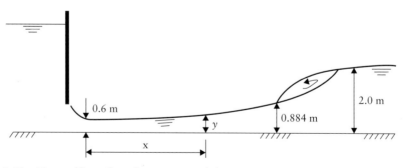

Figure 8.33   Non-uniform flow downstream of sluice gate.

y	A	R	E	ΔE	S_f	$\bar{S}_f$	Δx	x
0.60	2.4	0.461	4.495	—	0.0438	—	—	0
0.64	2.56	0.485	4.063	0.4317	0.0361	0.040	12.36	12.36
0.68	2.72	0.507	3.712	0.3510	0.0300	0.0330	12.50	24.87
0.72	2.88	0.529	3.425	0.2876	0.0253	0.0278	12.67	37.54
0.76	3.04	0.551	3.188	0.2372	0.0216	0.0235	12.85	50.39
0.80	3.20	0.571	2.991	0.1967	0.0185	0.0200	13.07	63.46
0.84	3.36	0.591	2.827	0.1637	0.0160	0.0173	13.31	76.11
0.88	3.52	0.611	2.691	0.1365	0.0140	0.0150	13.61	90.38

Example 8.24

Identify the types of water surface profiles behind (upstream) and between the gate and hydraulic jump using the data of Example 8.23.

Solution:

The solution here is generalised by rewriting the basic water surface slope equation (Equation 8.19) as

$$\frac{dy}{dx} = \frac{S_0(1 - S_f/S_0)}{1 - \alpha Q^2 B/gA^3} \tag{i}$$

If the conveyance of a channel is K and its section factor is Z, we can write $Q^2 = K^2 S_f = K_0^2 S_0$, $Z^2 = A^3/B$ and $Z_c^2 = \alpha Q^2/g$; Equation (i) now becomes

$$\frac{dy}{dx} = \frac{S_0 \left[1 - (K_0/K)^2\right]}{1 - (Z_c/Z)^2} \tag{ii}$$

Equation (ii) for the case of a wide rectangular channel with the Manning resistance equation reduces to

$$\frac{dy}{dx} = \frac{S_0 \left[1 - (y_0/y)^{10/3}\right]}{1 - (y_c/y)^3} \tag{iii}$$

Equation (iii) is very convenient to identify the types of surface profile once the normal depth, y_0, and critical depth, y_c, are established in a channel of given slope, S_0.

Case (i): Mild channel
In this channel, $S_0 < S_c$ (critical slope) and hence $y_0 > y_c$.

For $y > y_0 > y_c$ (zone 1): dy/dx is positive (i.e. increasing depths along x–M_1 profile)
For $y < y_0 > y_c$ (zone 2): dy/dx is negative (i.e. decreasing depths along x–M_2 profile)
For $y < y_0 < y_c$ (zone 3): dy/dx is positive (i.e. increasing depths along x–M_3 profile)

Case (ii): Steep channel
Here $S_0 > S_c$ and hence $y_0 < y_c$, and depending on the level of y, again three profiles (S_1, S_2 and S_3) exist.

Case (iii): Critical slope channel
Here $S_0 = S_c$ and $y_0 = y_c$; zone 2 is absent, and two profiles (C_1 and C_3) exist.

Case (iv): Horizontal channel
Here $S_0 = 0$ and y_0 will not exist and hence zone 1 is absent; again two profiles (H_2 and H_3) exist.

Case (v): Adverse slope channel
Now S_0 is negative with no y_0 and two profiles (A_2 and A_3) exist in this case. The discharge $Q = 20$ m³/s and by the Manning resistance equation $y_0 = 1.26$ m. The critical depth in rectangular channel $y_c = (\alpha Q^2/b^2 g)^{1/3}$. Hence $y_c = 1.366$ m $> y_0$; hence, steep channel with gate opening below critical depth. The depth upstream of the gate (by energy balance) $= 4.05$ m $> y_c > y_0$; zone 1 is on steep slope; S_1 profile exists immediately upstream of the gate.

The flow downstream of the gate is supercritical (zone 3 of the steep channel) merging with the controlled subcritical flow with the formation of hydraulic jump. Here the S_3 profile forms between the depths, 0.6 m (just d/s of gate) and 0.884 m, sequential to the controlled depth of 2 m.

Example 8.25

Discharge from a natural lake occurs through a very long rectangular channel of bed width 3 m, Manning's $n = 0.014$ and the bed slope $= 0.001$. The maximum level of the water surface in the lake above the channel bed at the lake outlet is 3 m. Calculate the discharge in the channel. If the channel slope were to be 0.008, compute the discharge. Also, determine the uniform flow depth and the minimum length of the channel for the uniform flow to establish. Ignore entrance losses.

Solution:

Slope $= 0.001$; first establish whether this is a mild or steep slope. If the slope were to be assumed critical, the channel will have inlet control with critical depth; critical depth in a rectangular channel is $2H/3$, H being the energy head (lake level above channel inlet) available.

$$\text{Therefore the critical depth at inlet, } y_c = \frac{2}{3} \times 3 = 2 \text{ m}$$

Since the channel is long (with no downstream control), uniform flow with depth $y_0 = y_c$ establishes at the inlet itself; using the Manning resistance equation, the corresponding critical slope, S_c, is computed.

At critical depth the discharge is maximum and is computed from $y_c = (q^2/g)^{1/3}$ (from $\alpha Q^2 B/gA^3 = 1$). Hence, $Q = qb = 26.58$ m³/s, and from $Q = An^{-1}R^{2/3}S^{1/2}$ the slope, $S = S_c = 0.0047$.

If the channel is other than rectangular in cross section, two simultaneous equations (i) the energy equation and (ii) the critical depth criterion, $\alpha Q^2 B/gA^3 = 1$, must be solved for computing y_c.

Since the bed slope $S_0 < S_c$, the channel is of mild slope.

Two equations at the inlet to solve two unknowns, depth and velocity, are required:

(i) Energy equation, $H = y_0 + V^2/2g$

(ii) The Manning resistance equation, $V = n^{-1}R^{2/3}S^{1/2}$

Both are applicable at the inlet (long channel; uniform flow establishes from the inlet). Simultaneous solution of (i) and (ii) gives

$$y_0 = 2.75 \text{ m} \quad \text{and} \quad V = 2.213 \text{ m/s}$$
$$\text{Thus the discharge, } Q_0 = 18.26 \text{ m}^3/\text{s}$$

The slope $S_0 = 0.008 > S_c = 0.0047$; the channel is a steep-sloped one, and the inlet controls the flow. The discharge is maximum ($= 26.58\text{m}^3/\text{s}$) with critical depth at the inlet ($y_c = 2$ m). The corresponding uniform flow depth (y_0) from the Manning resistance equation

$$26.58 \times 3 \times y_0 \times (1/0.014) \times [(3 \times y_0)/(3 + 2y_0)]^{2/3} \times (0.008)^{1/2}$$

is computed. Thus $y_0 = 1.64$ m; this unform flow establishes in the channel if its length is at least equal to the length of the surface profile (non-uniform flow) that exists between the inlet depth, y_c, and the uniform flow depth, y_0.

The channel is of steep slope, and the flow between these two depths corresponds to region 2 and hence an S_2 profile develops whose length can be computed by any appropriate method. By the step method we obtain $L \approx 80$ m between the two depths; the calculations should commence at a depth slightly less than the critical depth and terminate at a depth slightly higher than the normal (uniform) depth.

Example 8.26

A rectangular channel ($b = 15$ m, length $= 10$ km, slope $= 1/10\,000$, Manning's $n = 0.015$) fed by an upstream lake is discharging into a downstream lake. If the upstream and downstream lake levels are 1.5 and 2 m (above the channel bed), respectively, determine the discharge rate in the channel.

Solution:

The channel delivery depends upon the following considerations:

(i) Is the channel long (i.e. no downstream control)?

(ii) Is the slope mild or critical or steep?

If we assume a long and mild channel, two equations at its inlet are applicable (i.e. the energy and resistance equations). If we assume a long and critical sloped channel, the two equations are the energy equation and the critical depth criterion. Here it is convenient to assume initially a long and critical sloped channel. Since the channel is of rectangular cross section, $y_c = 2H/3 = 1$ m. Therefore, from the Manning resistance equation (uniform flow depth $= y_c$), the critical slope, $S_c = 0.00256 > S_0 = 0.0001$ (see Example 8.25). Hence the channel is of mild slope; if we still assume that it is long, we now can obtain the corresponding uniform flow rate, Q_0, from the energy and resistance equations.

(iii) Is the channel long enough to satisfy the above assumptions?

Identify the type of water surface profile based on the discharge, uniform flow depth and downstream lake level and compute its length. If the surface profile length is shorter than channel length, the channel is long enough and uniform flow does exist at the inlet (i.e. free inlet). If the profile length is longer than the channel length, the inlet will be drowned (short channel) and the discharge rate is reduced. To compute the actual discharge rate in a short channel, the following iterative procedure is to be followed:

(a) Assume $Q < Q_0$ (since the inlet is drowned) and compute the corresponding flow depth at the inlet by the energy equation.
(b) Compute the new surface profile length corresponding to this discharge and verify whether it fits between the inlet depth and the downstream lake level.
(c) Repeat (a) and (b) until the profile length matches the channel length.

The problem is best approached on a computer for executing these iterative procedures.

Example 8.27

A concrete twin-box-type culvert is proposed to discharge a design flood of 13.5 m³/s. The following data refer to each opening:

Manning's n	= 0.013
Height	= 0.75 m
Width	= 1.5 m
Length	= 30 m
Slope	= 1/100
Entrance conditions	= square edge, loss coefficient, $K = 0.5$
Downstream conditions	= free jet

Establish the rating curve (discharge vs. headwater elevation above the invert at the entrance) for rising head conditions over a discharge range from 0 to 13.5 m³/s. Neglect the velocity of approach. Determine the minimum elevation of the road surface assuming a free board of 300 mm to avoid any flooding of the highway.

Solution:

The culvert behaviour is dependent on the headwater level H, the height of culvert D, slope S_o and length L; here the outlet is to discharge freely and has no effect on the type of flow through the culvert.

(i) For $H/D \leq 1.2$, open channel flow. If entrance control exists, the depth at the inlet is critical (i.e. the slope is either critical or steep). Assuming entrance control, $y_c = (2/3) \times H$ (rectangular channel; inlet with no entrance losses) and $V_c = \sqrt{gy_c}$; hence, Q can be computed. Also from the Manning resistance equation, the critical slope, S_c, can be computed and checked against the proposed slope of the culvert; if the slope is then found to be mild, the depth and discharge calculations must be computed by the energy and resistance equations (see Example 8.25).

The energy equation at the inlet gives $H = 1.75y_c$, assuming entrance control with a loss coefficient of 0.5 (i.e. $S_0 \geq S_c$).

For $H = 0.1$ m, $y_c = 0.057$ m and $V_c = 0.748$ m/s. From the Manning equation, $S_c = 0.00476$, and as $S_0 (= 0.01) > S_c (= 0.00476)$, entrance control exists. The discharge through the culvert of width b can be written as $Q = by_c\sqrt{gy_c}$ for one box. Various values for y_c and hence the headwater levels $(H = 1.75y_c)$ are assumed until $H = 1.2D (=0.9$ m, the upper limit for open channel flow); a check for $S_0 > S_c$ is necessary for all values.

Headwater level H (m)	Discharge Q (m³/s) (two boxes)
0.175	0.297
0.525	1.544
0.700	2.377
0.900	3.465

(ii) For $H/D \geq 1.2$, the culvert entry behaves like an orifice (constriction); if the normal depth in the barrel corresponding to the orifice discharge is less than D, the flow downstream of the inlet is free. The orifice flow equation

$$Q = C_d \times b \times D \times \left[2g\left(H - \frac{D}{2}\right)\right]^{1/2} \quad \text{for one box}$$

with the discharge coefficient $C_d = 0.62$ (assumed). Computations between $H = 1.2D$ and the value at which $y_0 = D$ are as below:

Headwater level H (m)	Discharge Q (m³/s) (two boxes)
0.900	4.477
1.300	5.943
1.700	7.113
2.100	8.116
2.500	9.007

(iii) For $H/D > 1.2$ and $y_0 > D$, pipe flow exists in the culvert. The energy equation between the inlet and outlet of the culvert gives

$$H + S_0L = D + \frac{(1 + K)V^2}{2g} + S_fL$$

where S_f is the friction slope given by the Manning equation, as $S_f = (Vn)^2/R^{4/3}$. The above reduce to

$$Q = 3.41(H - 0.45)^{1/2} \quad \text{for one box}$$

pipe flow condition with the following:

Headwater level H (m)	Discharge (m³/s) (two boxes)
2.500	9.774
2.900	10.685
3.100	11.112
3.500	11.921
3.900	12.679
4.200	13.219
4.368	13.500 (design discharge)

Elevation of the road surface with a free board of 300 mm $= 4.368 + 0.300 = 4.668$ m above the culvert invert at its inlet.

Example 8.28

A lateral spillway channel, 120 m long and trapezoidal in section, is designed to carry a discharge which increases at a rate of 3.7 m³/s per metre length. The cross section has a bed width of 3 m with side slopes of 0.5 (horizontal): 1 (vertical). The bed slope is 0.15 and Manning's $n = 0.015$. Compute the water surface profile of the design discharge assuming uniform velocity distribution.

Solution:

The type of surface profile depends upon the length of the channel and whether the channel characteristics would permit the existence of a control section (i.e. a section where the depth is critical). The existence of a critical depth section and its longitudinal location are to be examined first; this is achieved by a trial-and-improvement process using Equation 8.23 and the critical depth criterion, $Q^2B/gA^3 = 1$. The following table of results is self-explanatory:

x (m)	$Q = qx$ (m³/s)	y_c (m)	A (m²)	P (m)	B (m)	$R = A/P$ (m)	$C = R^{1/6}/n$ (m$^{1/2}$/s)	x (m)
60.0	222.0	5.97	35.73	16.37	8.97	2.18	76.00	43.0
43.0	159.1	4.90	26.70	13.98	7.90	1.91	74.26	56.0
56.0	207.2	5.57	32.24	15.48	8.57	2.08	75.34	47.6
47.0	173.9	5.22	29.28	14.69	8.22	1.99	74.79	51.6
52.0	192.4	5.52	31.79	15.36	8.52	2.07	75.26	48.0
48.0	177.6	5.29	29.86	14.85	8.29	2.01	74.90	51.0
51.0	188.7	5.46	31.28	15.23	8.46	2.05	75.16	48.7
49.0	181.3	5.35	30.36	14.98	8.35	2.03	75.00	50.0
50.0	185.0	5.40	30.78	15.09	8.40	2.04	75.07	49.0
49.5	183.1	5.38	30.61	15.05	8.38	2.03	75.04	49.6

From the table, $x_c = 49.5$ m $< L$ ($=120$ m). As the length available is greater than x_c, the water surface profile upstream of the control section is in subcritical flow, while that of the downstream part is in supercritical flow. The surface profile computations are generally carried out by numerical integration combined with trial and improvement; the procedures are laborious and well described in textbooks of open channel hydraulics (see French, 1994, 2007; also see Example 15.3).

The non-uniform flow (GVF and SVF) computations may be carried out using the advanced numerical methods. The surface slope dy/dx, given by Equations 8.19, 8.22 and 8.25, is a function of x and y and can be written as $dy/dx = F(x, y)$; and the standard fourth-order Runge–Kutta (SRK) method uses the following operation:

$$y_{i+1} = y_i + \frac{1}{6}(K_1 + 2K_2 + 2K_3 + K_4) \tag{8.48}$$

in which

$$K_1 = \Delta x F(x_i, y_i)$$

$$K_2 = \Delta x F\left(x_i + \frac{\Delta x}{2}, y_i + \frac{K_1}{2}\right)$$

$$K_3 = \Delta x F\left(x_i + \frac{\Delta x}{2}, y_i + \frac{K_2}{2}\right)$$

$$K_4 = \Delta x F(x_i + \Delta x, y_i + K_3)$$

The solution is easily achieved with the help of a computer.

Example 8.29

A rectangular channel of bed width 2 m, Manning's $n = 0.014$, is laid on a slope of 1/1000. A side weir is to be designed at a section such that it comes into operation when the discharge in the channel exceeds 0.6 m³/s. A lateral outflow of 0.15 m³/s is expected to be delivered by the side weir when the channel discharge is 0.9 m³/s. Compute the elements of the weir.

Solution:

The sill (crest) height of the side weir is decided by the flow depth corresponding to 0.6 m³/s. The normal depths in the channel (by the Manning resistance equation):

For $Q = 0.6$ m³/s, $y_0 = 0.33$ m
For $Q = 0.9$ m³/s, $y_0 = 0.44$ m

The crest height is therefore (for the weir to come into operation) $s = 0.33$ m. For $Q = 0.9$ m³/s, the critical depth in the channel $y_{c1} = (q^2/g)^{1/3} = 0.274$ m. Now the sill height $s > y_{c1}$ and $y_0 > y_{c1}$, the dy/dx of the flow profile over the weir is positive; and the flow is subcritical. The De Marchi equation assumes $y_1 \approx y_0 = 0.44$ m. With the assumption $E_1 = E_2$, we can write

$$E_1 = 0.493 = y_2 + \frac{Q_2^2}{(by_2)^2 2g}; \quad Q_2 = 0.90 - 0.15 = 0.75 \text{ m}^3/\text{s}$$

and hence $y_2 = 0.46$ m (subcritical flow). The De Marchi functions $\phi_1 = -1.84$ and $\phi_2 = -1.42$. The De Marchi coefficient $C_m = 0.81 - 0.60 Fr_1$ (assuming a sharp-crested weir), where $Fr_1 = V_1/\sqrt{(2g)} = 0.49$. Hence, $C_m = 0.516$, and by Equation 8.31, the weir length, $L = 2.442$ m.

References and recommended reading

Ackers, P. (1992) Hydraulic design of two-stage channels. *Proceedings of the Institution of Civil Engineers, Water, Maritime & Energy,* 96, 247–257, and discussion in March 1994, 106, 99–101.

Balkham, M., Fosbeary, C., Kitchen, A. and Rickard, C. (2010) *Culvert Design and Operation Guide,* C689, CIRIA, London.

British Standards Institution (1981) *Methods of Measurement of Liquid Flow in Open Channels – Thin Plate Weirs,* BS 3680, Part 4A, British Standards Institution, London.

British Standards Institution (1981) *Methods of Measurement of Liquid Flow in Open Channels – Flumes,* BS 3680, Part 4C, British Standards Institution, London.

Chadwick, A., Morfett, J. and Borthwick, M. (2013) *Hydraulics in Civil and Environmental Engineering,* 5th edn, Taylor & Francis, Abingdon, UK.

Chanson, H. (2004) *The Hydraulics of Open Channel Flow,* 2nd edn, Elsevier Butterworth-Heinemann, Oxford.

Chow, V. T. (1959) *Open-Channel Hydraulics,* McGraw-Hill, New York.

Fortier, S. and Scobey, F. C. (1926) Permissible canal velocities. *Transactions of the American Society of Civil Engineers,* 89, 940–984.

French, R. H. (1994) *Open Channel Hydraulics,* 2nd edn, McGraw-Hill, New York.

French, R. H. (2007) *Open Channel Hydraulics,* Water Resources Publications, Highlands Ranch, CO.

Henderson, F. M. (1966) *Open Channel Flow,* Macmillan, New York.

HR Wallingford (1990) *Charts for the Hydraulic Design of Channels and Pipes,* 6th edn, Thomas Telford, London.

HR Wallingford (1990) *Tables for the Hydraulic Design of Pipes and Sewers,* 5th edn, Thomas Telford, London.

HR Wallingford and Barr, D. I. H. (2006) *Tables for the Hydraulic Design of Pipes, Sewers and Channels,* 8th edn, Thomas Telford, London.

Mansell, M. G. (2003) *Rural and Urban Hydrology,* Thomas Telford, London.

National Water Council (1981) *Design and Analysis of Urban Storm Drainage, the Wallingford Procedure, Volume 4: The Modified Rational Method,* National Water Council Standing Technical Committee Report No. 31, National Water Council, London.

Novak, P., Moffat, A. I. B., Nalluri, C. and Narayanan, R. (2007) *Hydraulic Structures,* 4th edn, Taylor & Francis, Abingdon, UK.

Ramsbottom, D., Day, R. and Rickard, C. (1997) *Culvert Design Mannual,* Report 168, CIRIA, London.

Water UK/WRc (2012) *Sewers for Adoption – A Design and Construction Guide for Developers,* 7th edn, WRc, Swindon.

Problems

1. Water flows uniformly at a depth of 2 m in a rectangular channel of width 4 m and bed slope 1:2000. What is the mean shear stress on the wetted perimeter?

2. (a) At a measured discharge of 40 m^3/s, the depth of uniform flow in a rectangular channel 5 m wide and with a bed slope of 1:1000 was 3.05 m. Determine the mean effective roughness size and the Manning roughness coefficient.

   (b) Using (i) the Darcy–Weisbach equation (together with the Colebrook–White equation) and (ii) the Manning equation, predict the discharge at a depth of 4 m.

Chapter 8

3. Determine the depth of uniform flow in a trapezoidal concrete-lined channel of bed width 3.5 m and bed slope 0.0005 with side slopes at 45° to the horizontal when conveying 36 m³/s of water. The Manning roughness coefficient is 0.014.

4. Determine the rate of uniform flow in a circular section channel 3 m in diameter of effective roughness 0.3 mm, laid to a gradient of 1:1000 when the depth of flow is 1.0 m. What are the mean velocity and the mean boundary shear stress?

5. A circular storm water sewer 1.5 m in diameter and effective roughness size 0.6 mm is laid to a slope of 1:500. Determine the maximum discharge which the sewer will convey under uniform open channel conditions. If a steady inflow from surface runoff exceeds the maximum open channel capacity by 20%, show that the sewer will become pressurised (surcharged) and calculate the hydraulic gradient necessary to convey the new flow.

6. Assuming that a rough turbulent velocity distribution having the form

$$\frac{v}{\sqrt{\tau_0/\rho}} = 5.75 \ \log \frac{30y}{k}$$

    exists in a wide river, show that the average of current meter measurements taken at 0.2 and 0.8 of the depth from the surface approximates to the mean velocity in a vertical section.

7. A long, concrete-lined trapezoidal channel with a bed slope 1:1000, bed width 3.0 m, side slopes at 45° to the horizontal and Manning roughness 0.014 receives water from a reservoir. Assuming an energy loss of $0.25\,V^2/2g$ calculate the steady discharge and depth of uniform flow in the channel when the level in the reservoir is 2.0 m above the channel bed at inlet.

8. A trapezoidal channel with a bed slope of 0.005, bed width 3 m and side slopes 1:1.5 (vertical:horizontal) has a gravel bed ($n = 0.025$) and concrete sides ($n = 0.013$). Calculate the uniform flow discharge when the depth of flow is 1.5 m using (a) the Einstein, (b) the Pavlovskij and (c) the Lotter methods.

9. Figure 8.34 shows the cross section of a river channel passing through a flood plain. The main channel has a bank-full area of 300 m², a top width of 50 m, a wetted perimeter of 65 m and the Manning roughness coefficient of 0.025. The flood plains have a Manning roughness of 0.035 and the gradient of the main channel and plain is 0.00125. Determine the depth of flow over the flood plain at a flood discharge of 2470 m³/s.

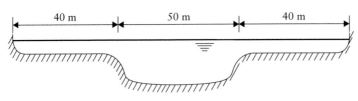

Figure 8.34 Compound channel.

10. A concrete-lined rectangular channel is to be constructed to convey a steady maximum discharge of 160 m^3/s to a hydropower installation. The bed slope is 1:5000 and Manning's n appropriate to the type of surface finish is 0.015. Determine the width of the channel and the depth of flow for the 'most economic' section. Give reasons why the actual constructed depth would be made greater than the flow depth.

11. A concrete-lined trapezoidal channel with a bed slope of 1:2000 is to be designed to convey a maximum discharge of 75 m^3/s under uniform flow conditions. The side slopes are at 45° and Manning's $n = 0.014$. Determine the bed width and depth of flow for the 'most economic' section.

12. A channel with a bed slope of 1:2000 is to be constructed through a stiff clay formation. Compare the relative costs of the alternative design of best rectangular concrete-lined and trapezoidal unlined channels to convey 60 m^3/s if the cost of the 100 mm thick lining per square metre is twice the cost of excavation per cubic metre. Manning's n for concrete lining is 0.014 and for the unlined channel is 0.025. Side slopes (stable) are 1:1.5.

13. An unlined irrigation channel of trapezoidal section is to be constructed through a sandy formation at a bed slope of 1:10 000 to convey a discharge of 40 m^3/s. The side slopes are at 25° to the horizontal. The angle of internal friction of the material is 35° and the critical tractive force is 2.5 N/m^2; Manning's n is 0.022. Assuming that the maximum boundary shear stress exerted on the bed, due to the water flow, is 0.98 $\rho g y S_0$ and that on the sides is 0.75 $\rho g y S_0$, determine the bed width and flow depth for a non-eroding channel design.

14. A vertical sluice gate in a long rectangular channel 5 m wide is lowered to produce an opening of 1.0 m. Assuming that free flow conditions exist at the vena contracta downstream of the gate verify that the flow in the vena contracta is supercritical when the discharge is 15 m^3/s, and determine the depth just upstream of the gate. $C_v = 0.98$; $C_c = 0.6$. Take the upstream velocity energy coefficient (Coriolis) to be 1.0 and that at the vena contracta to be 1.2.

15. A sill is to be constructed on the bed of a rectangular channel conveying a specific discharge of 5 m^3/s per metre width. The depth of uniform flow is 2.5 m. Neglecting energy losses,
    (a) Determine the variation of the depths upstream of the sill (y_1) and over the sill (y_2) for a range of sill heights (z) from 0.1 to 0.8 m. Take the Coriolis coefficient to be 1.2.
    (b) Determine the critical depth (y_c).
    (c) Determine the minimum sill height (z_c) to create critical flow conditions at the sill.

16. A venturi flume with a throat width of 0.5 m is constructed in a rectangular channel 1.5 m wide. The depth of uniform flow in the channel at a discharge of 1.6 m^3/s is 0.92 m; $\alpha = 1.1$. Assuming that downstream conditions do not influence the natural flow profile through the contraction and neglecting losses, verify that the flume acts as a 'critical depth flume' and determine the upstream and throat depths.

Show that for critical flow conditions in the throat, the discharge can be obtained from

$$Q = \frac{2}{3}\sqrt{\frac{2g}{3}}bH_1^{3/2}$$

where b is the throat width and $H_1 = y_1 + (\alpha V_1^2/2g)$.

Calculate the discharge when the upstream depth is 1.0 m. (Verify that critical flow conditions are maintained in the throat.)

17. A vertical sluice gate in a long rectangular channel 4 m wide has an opening of 1.0 m and a coefficient of contraction of 0.6. At a discharge of 25 m³/s, the depth of uniform flow (y_n) is 3.56 m. Assuming that a hydraulic jump were to occur in the channel downstream with a sequent depth equal to y_n and taking the Boussinesq coefficient to be 1.2, what would be the initial depth of the jump? Hence verify that a hydraulic jump will occur. Determine the depth upstream of the gate and the hydrodynamic force on the gate, assuming $C_v = 0.98$ and $\alpha = 1.2$.

18. If in Problem 17 the gate is raised to give an opening of 1.5 m, determine whether or not a hydraulic jump will form. Calculate the depths immediately upstream of the gate and at the position of the vena contracta and the force on the gate.

19. A long rectangular channel 8 m wide, bed slope 1:5000 and the Manning roughness 0.015 conveys a steady discharge of 40 m³/s. A sluice gate raises the depth immediately upstream to 5.0 m. Taking the Coriolis coefficient α to be 1.1, determine the uniform flow depth and the distance from the gate at which this depth is exceeded by 10%. What is the depth 5000 m from the gate?

20. A rectangular channel having a bed width of 4 m, a bed slope of 0.001 and Manning's $n = 0.015$ conveys a steady discharge of 25 m³/s. A barrage creates a depth upstream of 4.0 m. Compute the water surface profile taking $\alpha = 1.1$.

21. A long rectangular channel, 2.5 m wide, bed slope 1:1000 and the Manning roughness coefficient 0.02, discharges 4.5 m³/s freely to atmosphere at the downstream end. Taking $\alpha = 1.1$ and noting that at a free overfall the depth approximates closely to the critical depth, compute the surface profile to within approximately 10% of the uniform flow depth.

22. A long trapezoidal channel of bed width 3.5 m, side slopes at 45°, bed slope 0.0003 and Manning's $n = 0.018$ conveys a steady flow of 50 m³/s. A control structure creates an upstream depth of 5.0 m. Taking $\alpha = 1.1$, determine the distance upstream at which the depth is 4.2 m.

23. Two reserviors are connected by a wide rectangular channel of length 1500 m, where Manning's n is 0.02, the bed slope is 4×10^{-4}, the channel entry loss coefficient K is 0.02, the channel invert elevation (u/s) is 101.00 m above ordnance datum (AOD) and the water level in the u/s reservoir is 104.00 m AOD (constant).
    (a) Determine the limiting downstream reservoir level to cause uniform flow in the channel.
    (b) If the downstream reservoir level is 103.50 m AOD, examine whether or not it will affect the uniform flow rate (submerged inlet and reduced flow).

24. A culvert is proposed under a highway embankment where the design flood is 15 m³/s, the width of the highway is 30 m and the natural drainage slope is 0.015. The available pipe barrels are corrugated pipes of diameter in multiples of 250 mm, with Manning's n equal to 0.024 and the entry loss coefficient 0.9.

   (a) Compute the proposed culvert barrel size if the maximum permissible headwater level is 4 m above the invert, with the barrel discharging free at its outlet.

   (b) If a fare-edged entry (loss coefficient = 0.25) is chosen, calculate the required barrel diameter for the conditions in (a).

Chapter 9
Dimensional Analysis, Similitude and Hydraulic Models

9.1 Introduction

Hydraulic engineering structures or machines can be designed using (i) pure theory, (ii) empirical methods, (iii) semi-empirical methods, which are mathematical formulations based on theoretical concepts supported by suitably designed experiments, (iv) physical models or (v) mathematical models.

The purely theoretical approach in hydraulic engineering is limited to a few cases of laminar flow, for example the Hagen–Poiseuille equation for the hydraulic gradient in the laminar flow of an incompressible fluid in a circular pipeline. Empirical methods are based on correlations between observed variables affecting a particular physical system. Such relationships should only be used under circumstances similar to those under which the data were collected. Due to the inability to express the physical interaction of the parameters involved in mathematical terms, some such methods are still in use. One well-known example is in the relationship between wave height, fetch, wind speed and duration for the forecasting of ocean wave characteristics.

A good example of a semi-empirical relationship is the Colebrook–White equation for the friction factors in turbulent flow in pipes (see Chapters 4 and 7). This was obtained from theoretical concepts and experiments designed on the basis of dimensional analysis; it is universally applicable to all Newtonian fluids.

Dimensional analysis also forms the basis for the design and operation of physical scale models which are used to predict the behaviour of their full-sized counterparts called **prototypes**. Such models, which are generally geometrically similar to the prototype, are used in the design of aircraft, ships, submarines, pumps, turbines, harbours, breakwaters, river and estuary engineering works, spillways and so on.

While mathematical modelling techniques have progressed rapidly due to the advent of high-speed digital computers, enabling the equations of motion coupled with semi-empirical relationships to be solved for complex flow situations such as pipe network analysis, pressure transients in pipelines and unsteady flows in rivers and estuaries, there

Nalluri & Featherstone's Civil Engineering Hydraulics: Essential Theory with Worked Examples,
Sixth Edition. Martin Marriott.
© 2016 John Wiley & Sons, Ltd. Published 2016 by John Wiley & Sons, Ltd.
Companion Website: www.wiley.com/go/Marriott

are many cases, particularly where localised flow patterns cannot be mathematically modelled, when physical models are still needed.

Without the technique of dimensional analysis, experimental and computational progress in fluid mechanics would have been considerably retarded.

9.2 Dimensional analysis

The basis of dimensional analysis is to condense the number of separate variables involved in a particular type of physical system into a smaller number of non-dimensional groups of the variables.

The arrangement of the variables in the groups is generally chosen so that each group has a physical significance.

All physical parameters can be expressed in terms of a number of basic dimensions; in engineering, the basic dimensions such as mass (M), length (L) and time (T) are sufficient for this purpose. For example, velocity = distance/time ($=LT^{-1}$); discharge = volume/time ($=L^3T^{-1}$). Force is expressed using Newton's law of motion (force = mass × acceleration); hence, force = MLT^{-2}.

A list of some physical quantities with their dimensional forms can be seen below.

Physical quantity	Symbol	Dimensional form
Length	ℓ	L
Time	t	T
Mass	m	M
Velocity	V	LT^{-1}
Acceleration	a	LT^{-2}
Discharge	Q	L^3T^{-1}
Force	F	MLT^{-2}
Pressure	p	$ML^{-1}T^{-2}$
Power	P	ML^2T^{-3}
Density	ρ	ML^{-3}
Dynamic viscosity	μ	$ML^{-1}T^{-1}$
Kinematic viscosity	v	L^2T^{-1}
Surface tension	σ	MT^{-2}
Bulk modulus of elasticity	K	$ML^{-1}T^{-2}$

9.3 Physical significance of non-dimensional groups

The main components of force which may act on a fluid element are those due to viscosity, gravity, pressure, surface tension and elasticity. The resultant of these components is called the inertial force, and the ratio of this force to each of the force components indicates the relative importance of the force types in a particular flow system.

For example, the ratio of inertial force to viscous force is

$$\frac{F_i}{F_\mu} = \frac{\rho L^3 LT^{-2}}{\tau L^2}$$

$$\text{Now } \tau = \mu \frac{dy}{dy} = \mu LT^{-1}L^{-1}$$

$$\text{whence } \frac{F_i}{F_\mu} = \frac{\rho L^2 T^{-1}}{\mu} = \frac{\rho L V}{\mu} = \frac{\rho \ell V}{\mu}$$

where ℓ is a typical length dimension of the particular system.

The dimensionless term $\rho \ell V / \mu$ is in the form of the Reynolds number.

Low Reynolds numbers indicate a significant dominance of viscous forces in the system, which explains why this non-dimensional parameter may be used to identify the regime of flow (i.e. whether laminar or turbulent).

Similarly, it can be shown that the **Froude number** is the ratio of inertial force to gravity force in the form

$$Fr = \frac{V^2}{g\ell} \quad \left(\text{but usually expressed as } Fr = \frac{V}{\sqrt{g\ell}} \right)$$

The **Weber number**, We, is the ratio of inertial to surface tension force and is expressed by $V / \sqrt{\sigma / \rho \ell}$.

9.4 The Buckingham π theorem

This states that the n quantities $Q_1, Q_2, \ldots, Q_n$ involved in a physical system can be arranged in $(n - m)$ non-dimensional groups of the quantities, where m is the number of basic dimensions required to express the quantities in dimensional form.

Thus $f_1(Q_1, Q_2, \ldots, Q_n) = 0$ can be expressed as $f_2(\pi_1, \pi_2, \ldots, \pi_{n-m})$, where f means 'a function of ...'. Each π term basically contains m repeated quantities which together contain the m basic dimensions together with one other quantity. In fluid mechanics $m = 3$, and therefore each π term basically contains four of the quantity terms.

Further information on the historical development of this approach is given in the recommended reading, and a matrix method suited to computer application is presented in Chadwick *et al.* (2013).

9.5 Similitude and model studies

Similitude, or dynamic similarity, between two geometrically similar systems exists when the ratios of inertial force to the individual force components in the first system are the same as the corresponding ratios in the second system at the corresponding points in space. Hence for absolute dynamic similarity, the Reynolds, Froude and Weber numbers must be the same in the two systems. If this can be achieved the flow patterns will be geometrically similar (i.e., kinematic similarity exists).

In using physical scale models to predict the behaviour of prototype systems or designs, it is rarely possible (except when only one force type is relevant) to achieve simultaneous equality of the various force ratios. The 'scaling laws' are then based on equality of the predominant force; strict dynamic similarity is thus not achieved, resulting in 'scale effect'.

Reynolds modelling is adopted for studies of flows without a free surface such as pipe flow and flow around submerged bodies (e.g. aircraft, submarines, vehicles and buildings).

The Froude number becomes the governing parameter in flows with a free surface since gravitational forces are predominant. Hydraulic structures, including spillways, weirs and stilling basins, rivers and estuaries, hydraulic turbines and pumps and wave-making resistance of ships, are modelled according to the Froude law.

Worked examples

Example 9.1

Obtain an expression for the pressure gradient in a circular pipeline, of effective roughness k conveying an incompressible fluid of density ρ, dynamic viscosity μ, at a mean velocity V, as a function of non-dimensional groups.

By comparison with the Darcy–Weisbach equation, show that the friction factor is a function of relative roughness and the Reynolds number.

Solution:

In full pipe flow, gravity and surface tension forces do not influence the flow. Let $\Delta p = $ pressure drop in a length L. Then

$$f_1(\Delta p, L, \rho, V, D, \mu, k) = 0 \tag{i}$$

The repeating variables will be ρ, V and D. Δp clearly is not to be repeated since this variable is required to be expressed in terms of the other variables. If μ or k were to be repeated, the relative effect of the parameter would be hidden.

$$f_2(\pi_1, \pi_2, \pi_3, \pi_4) = 0 \tag{ii}$$

Then

$$\pi_1 = \rho^\alpha D^\beta V^\gamma \Delta p \tag{iii}$$

where α, β and γ are indices to be evaluated.

In dimensional form,

$$\pi_1 = (\mathrm{ML^{-3}})^\alpha \mathrm{L}^\beta (\mathrm{LT^{-1}})^\gamma \mathrm{ML^{-1}T^{-2}}$$

The sum of the indices of each dimension must be zero.

Thus for M, $0 = \alpha + 1$, whence $\alpha = -1$
for T, $0 = -\gamma - 1$, whence $\gamma = -2$
and for L, $0 = -3\alpha + \beta + \gamma - 1$, whence $\beta = 0$

$$\Rightarrow \pi_1 = \frac{\Delta p}{\rho V^2} \tag{iv}$$

$$\pi_2 = \rho^\alpha D^\beta V^\gamma L \tag{v}$$

The π terms are dimensionless, and since D and L have the same dimensions the solution is

$$\pi_2 = \frac{L}{D} \tag{vi}$$

Similarly,

$$\pi_3 = \frac{k}{D} \tag{vii}$$

$$\pi_4 = \rho^\alpha D^\beta V^\gamma \mu \tag{viii}$$

$$\pi_4 = (ML^{-3})^\alpha L^\beta (LT^{-1})^\gamma ML^{-1}T^{-1}$$

Indices of M: $0 = \alpha + 1$; $\alpha = -1$

Indices of T: $0 = -\gamma - 1$; $\gamma = -1$

Indices of L: $0 = -3\alpha + \beta + \gamma - 1$; $\beta = -1$

$$\Rightarrow \pi_4 = \frac{\mu}{\rho D V} \tag{ix}$$

$$\Rightarrow f_2\left(\frac{\Delta p}{\rho V^2}, \frac{L}{D}, \frac{k}{D}, \frac{\mu}{\rho D V}\right) = 0 \tag{x}$$

The π terms can be multiplied or divided, and since the pressure gradient is required Equation (x) may thus be reformed as

$$f_2\left(\frac{\Delta p\, D}{L\, \rho V^2}, \frac{k}{D}, \frac{\mu}{\rho D V}\right) = 0$$

$$\text{whence } \frac{\Delta p}{L} = \frac{\rho V^2}{D}\, \phi\left[\frac{k}{D}, Re\right]$$

where ϕ means 'a function of …' the form of which is to be obtained experimentally.
    The hydraulic gradient,

$$\frac{\Delta h}{L} = \frac{\Delta p}{\rho g L}$$

$$\text{whence } \frac{\Delta h}{L} = \frac{V^2}{gD}\, \phi\left[\frac{k}{D}, Re\right] \tag{xi}$$

Comparing Equation (xi) with the Darcy–Weisbach equation,

$$\frac{h_f}{L} = \frac{\lambda V^2}{2gD}$$

it is seen that λ is dimensionless and that

$$\lambda = \phi\left[\frac{k}{D}, Re\right]$$

This relationship enabled experiments to be designed (as described in Chapter 7), which eventually led to the Colebrook–White equation.

Example 9.2

Show that the discharge of a liquid through a rotodynamic pump having an impeller of diameter D and width B, running at speed N when producing a total head H, can be expressed in the form

$$Q = ND^3\phi\left[\frac{D}{B}, \frac{N^2D^2}{gH}, \frac{\rho ND^2}{\mu}\right]$$

Solution:

$$f_1(N, D, B, Q, gH, \rho, \mu) = 0$$

Note that the presence of g represents the transformation of pressure head to velocity energy; it is convenient, but not essential, to combine g and H instead of treating them separately.

$$f_2(\pi_1, \pi_2, \pi_3, \pi_4) = 0$$

Using ρ, N and D as the recurring variables,

$$\pi_1 = \rho^\alpha N^\beta D^\gamma B$$
$$\pi_1 = \frac{B}{D}$$
$$\pi_2 = \rho^\alpha N^\beta D^\gamma Q$$
$$\pi_2 = (ML^{-3})^\alpha (T^{-1})^\beta L^\gamma L^3 T^{-1}$$

For M, $0 = \alpha$; $\alpha = 0$
For T, $0 = -\beta - 1$; $\beta = -1$
For L, $0 = -3\alpha + \gamma + 3$; $\gamma = -3$

$$\pi_2 = \frac{Q}{ND^3}$$
$$\pi_3 = \rho^\alpha N^\beta D^\gamma (gH)$$

$$\pi_3 = (ML^{-3})^\alpha (T^{-1})^\beta L^\gamma L^2 T^{-2}$$

whence $\pi_3 = \dfrac{gH}{N^2D^2}$

$$\pi_4 = \rho^\alpha N^\beta D^\gamma \mu$$

$$\pi_4 = (ML^{-3})^\alpha (T^{-1})^\beta L^\gamma ML^{-1} T^{-1}$$

whence $\pi_4 = \dfrac{\mu}{\rho ND^2}$

$$\Rightarrow f_1\left(\frac{B}{D}, \frac{Q}{ND^3}, \frac{gH}{N^2D^2}, \frac{\mu}{\rho ND^2}\right) = 0$$

whence $Q = ND^3\phi\left[\dfrac{D}{B}, \dfrac{N^2D^2}{gH}, \dfrac{\rho ND^2}{\mu}\right]$

Note that the π terms may be inverted for convenience and that $\rho ND^2/\mu$ is a form of the Reynolds number and N^2D^2/gH a form of the square of the Froude number.

Example 9.3

Show that the discharge Q of a liquid of density ρ, dynamic viscosity μ and surface tension σ, over a V-notch under a head H, may be expressed in the form

$$Q = g^{1/2}H^{5/2}\phi\left[\frac{\rho g^{1/2}H^{3/2}}{\mu}, \frac{\rho g H^2}{\sigma}, \theta\right]$$

where θ is the notch angle, and hence define the parameters upon which the discharge coefficient of such weirs is dependent.

Solution:

$$f_1(\rho, g, H, Q, \mu, \sigma, \theta) = 0$$

in other words,

$$f_2(\pi_1, \pi_2, \pi_3, \pi_4) = 0$$

Using ρ, g and H as the repeating variables,

$$\pi_1 = \rho^\alpha g^\beta H^\gamma Q$$
$$\pi_1 = (ML^{-3})^\alpha (LT^{-2})^\beta L^\gamma L^3 T^{-1}$$

For M,	$0 = \alpha$;	$\alpha = 0$
For T,	$0 = -2\beta - 1$;	$\beta = -1/2$
For L,	$0 = -3\alpha + \beta + \gamma + 3$;	$\gamma = -5/2$

$$\pi_1 = \frac{Q}{g^{1/2}H^{5/2}}$$
$$\pi_2 = \rho^\alpha g^\beta H^\gamma \mu$$
$$\pi_2 = (ML^{-3})^\alpha (LT^{-2})^\beta L^\gamma ML^{-1}T^{-1}$$
$$\pi_2 = \frac{\mu}{\rho g^{1/2}H^{3/2}}$$
$$\pi_3 = \rho^\alpha g^\beta H^\gamma \sigma$$
$$\pi_3 = (ML^{-3})^\alpha (LT^{-2})^\beta L^\gamma ML^{-2}$$
$$\pi_3 = \frac{\sigma}{\rho g H^2}$$
$$\pi_4 = \rho^\alpha g^\beta H^\gamma \theta$$
$$\pi_4 = \theta \quad \text{(since } \theta \text{ itself is dimensionless)}$$

$$\Rightarrow f_2\left[\frac{Q}{g^{1/2}H^{5/2}}, \frac{\mu}{\rho g^{1/2}H^{3/2}}, \frac{\sigma}{\rho g H^2}, \theta\right] = 0$$

Rearranging,

$$Q = g^{1/2}H^{5/2}\phi\left[\frac{\rho g^{1/2}H^{3/2}}{\mu}, \frac{\rho g H^2}{\sigma}, \theta\right]$$

From energy considerations the discharge over a V-notch is expressed as

$$Q = \frac{8}{15}\sqrt{2g}\, C_d \tan\frac{\theta}{2}H^{5/2}$$

Comparing the above two forms it is seen that

$$C_d = f\left(\frac{\rho g^{1/2}H^{3/2}}{\mu}, \frac{\rho g H^2}{\sigma}\right)$$

The group $\rho g^{1/2}H^{3/2}/\mu$ has the form of the Reynolds number Re, and the group $\rho g H^2/\sigma$ relates to the square of the Weber number, We. Hence

$$C_d = f(Re, We)$$

Surface tension effects represented by the Weber number may become significant at low discharges.

Example 9.4

Derive an expression for the discharge per unit crest length of a rectangular weir over which a fluid of density ρ and dynamic viscosity μ is flowing with a head H.

The crest height is P. By comparison with the discharge equation obtained from energy considerations,

$$q = \frac{2}{3}\sqrt{2g}\, C_d H^{3/2}$$

state the parameters on which the discharge coefficient depends for a given crest profile.

Solution:

$$f_1(q, g, H, \rho, \mu, \sigma, P) = 0$$
$$f_2(\pi_1, \pi_2, \pi_3, \pi_4) = 0$$

With ρ, g and H as the repeating variables,

$$\pi_1 = \rho^\alpha g^\beta H^\gamma q$$
$$\pi_1 = \frac{q}{g^{1/2}H^{3/2}}$$
$$\pi_2 = \rho^\alpha g^\beta H^\gamma \mu$$
$$\pi_2 = \frac{\mu}{\rho g^{1/2}H^{3/2}}$$
$$\pi_3 = \rho^\alpha g^\beta H^\gamma \sigma$$
$$\pi_3 = \frac{\sigma}{\rho g H^2}$$
$$\pi_4 = \rho^\alpha g^\beta H^\gamma P$$
$$\pi_4 = \frac{P}{H}$$
$$\Rightarrow f_2\left(\frac{q}{g^{1/2}H^{3/2}}, \frac{\mu}{\rho g^{1/2}H^{3/2}}, \frac{\sigma}{\rho g H^2}, \frac{P}{H}\right) = 0$$

$$\text{or } q = g^{1/2}H^{3/2}\phi\left[\frac{\rho g^{1/2}H^{3/2}}{\mu}, \frac{\rho g H^2}{\sigma}, \frac{P}{H}\right]$$

Hence

$$C_d = f\left[Re, We, \frac{P}{H}\right]$$

In addition, of course, the discharge coefficient will depend on the crest profile, and the influence of this factor together with that of the non-dimensional groups in the above expression can only be found from experiments.

Example 9.5

A spun iron pipeline 300 mm in diameter and with 0.3 mm effective roughness is to be used to convey oil of kinematic viscosity 7.0×10^{-5} m^2/s at a rate of 80 L/s. Laboratory tests on a 30 mm pipeline conveying water at 20°C ($v = 1.0 \times 10^{-6}$ m^2/s) are to be carried out to predict the hydraulic gradient in the oil pipeline.

Determine the effective roughness of the 30 mm pipe, the water discharge to be used and the hydraulic gradient in the oil pipeline at the design discharge.

Solution:

It was shown in Example 9.1 that the hydraulic gradient in a pressure pipeline is expressed by

$$S_f = \frac{h_f}{L} = \frac{V^2}{gD}\phi\left[\frac{k}{D}, Re\right] \tag{i}$$

For geometrical similarity, the relative roughness k/D must be the same in both systems. Using subscript 'o' for oil and 'w' for water,

$$\left(\frac{k}{D}\right)_o = \frac{0.3}{300} = 0.001$$

$\Rightarrow$ roughness of water pipe $= 0.001 \times 30 = 0.03$ mm

(An unplasticised polyvinyl chloride (PVCu) pipeline with chemically cemented joints could be used.)

For dynamic similarity the Reynolds numbers must be the same.

$$\left(\frac{VD}{v}\right)_w = \left(\frac{VD}{v}\right)_o$$

$$V_o = 1.132 \text{ m/s}; \qquad Re_o = \frac{1.132 \times 0.3}{7.0 \times 10^{-5}} = 4851.0$$

$$\Rightarrow V_w = \frac{4851.0 \times 1.0 \times 10^{-6}}{0.03} = 0.1617 \text{ m/s}$$

The velocity 0.1617 m/s for water is called the **corresponding speed** for dynamic similarity.

$$\Rightarrow \text{Water discharge} = 0.114 \text{ L/s}$$

From Equation (i),

$$\frac{S_{f,o}}{S_{f,w}} = \frac{(V^2/gD)_o}{(V^2/gD)_w}$$

since $\phi[k/D, Re]$ is the same for the two systems at the corresponding speeds.

$$\Rightarrow S_{f,o} = \frac{(1.332^2/0.3)}{(0.1617^2/0.3)} \times 0.0017 = 0.00833$$

Example 9.6

A V-notch is to be used for monitoring the flow of oil of kinematic viscosity 8.0×10^{-6} m^2/s. Laboratory tests using water at 15°C over a geometrically similar notch were used to predict the calibration of the notch when used for oil flow measurement. At a head of 0.15 m, the water discharge was 12.15 L/s. What is the corresponding head when measuring oil flow and what is the corresponding oil discharge?

Solution:

In Example 9.3 it was shown that the discharge over a V-notch is given by

$$Q = g^{1/2}H^{5/2}\phi\left[\frac{\rho g^{1/2}H^{3/2}}{\mu}, \frac{\rho g H^2}{\sigma}, \theta\right] \tag{i}$$

where θ is the same for both notches; and for dynamic similarity the groups $\rho g^{1/2}H^{3/2}/\mu$ and $\rho g H^2/\sigma$ should be the same, respectively, for the water and oil systems. That is

$$Re_o = Re_w \quad \text{and} \quad We_o = We_w$$

using subscript 'o' for oil and 'w' for water.

However, it will be realised that these two relationships will lead to two different scaling laws, but since the surface tension effect will become significant only in relation to the viscous and gravity forces at very low heads this effect can be neglected. The scaling law is therefore obtained by equality of the Reynolds numbers.

$$\left(\frac{H^{3/2}}{\nu}\right)_o = \left(\frac{H^{3/2}}{\nu}\right)_w$$

whence $H_o = 0.15 \left(\dfrac{8 \times 10^{-6}}{1.13 \times 10^{-6}}\right)^{2/3} = 0.553$ m

Using Equation (i), and since the $\phi[\ ]$ terms are the same for both systems,

$$\frac{Q_o}{Q_w} = \left(\frac{H_o}{H_w}\right)^{5/2}$$

whence $Q_o = 12.15 \left(\dfrac{0.553}{0.15}\right)^{5/2}$

$$Q_o = 0.3171 \text{ m}^3/\text{s } (317.1 \text{ L/s})$$

Example 9.7

(a) Show that the net force acting on the liquid flowing in an open channel may be expressed as

$$F = \rho V^2 \ell^2 \phi \left[Re, Fr, We, \frac{k}{\ell} \right] \qquad (ii)$$

where ℓ is a typical length dimension.

(b) A 1:50 scale model of part of a river is to be constructed to investigate channel improvements. A steady discharge of 420 m³/s was measured in the river at a section where the average width was 105 m and water depth 3.5 m.

Determine the corresponding depth, velocity and discharge to be reproduced in the model. Check that the flow in the model is in the turbulent region, and discuss how the boundary resistance in the model could be adjusted to produce a geometrical similarity of surface profiles.

Solution:

(a)

$$f(\rho, V, \ell, g, \mu, \sigma, k) = 0$$

With ρ, V and ℓ as the repeating variables, dimensional analysis yields

$$F = \rho V^2 \ell^2 \phi \left[\frac{V \ell \rho}{\mu}, \frac{V^2}{g\ell}, \frac{\rho V^2 \ell}{\sigma}, \frac{k}{\ell} \right]$$

i.e. $F = \rho V^2 \ell^2 \phi \left[Re, Fr, We, \dfrac{k}{\ell} \right]$

(b) For dynamic similarity the non-dimensional groups should be equal in the model and prototype. Surface tension effects will be negligible in the prototype and its effect must be minimised in small-scale models. Although the boundary resistance, as reflected in the Re and k/ℓ terms, and the gravity force are both significant, open channel models are operated according to the Froude law,

$$\left(\frac{V}{\sqrt{gy}} \right)_m = \left(\frac{V}{\sqrt{gy}} \right)_p \qquad (i)$$

where the length parameter is the depth y. Subscript 'm' relates to the model and 'p' to the prototype.

$$\frac{y_m}{y_p} = \frac{1}{50}$$

$$\text{whence } y_m = \frac{3.5}{50} = 0.07 \text{ m}$$

$$V_p = \frac{Q}{A} = \frac{420}{105 \times 3.5} = 1.14 \text{ m/s}$$

From Equation (i),

$$V_m = V_p \sqrt{\frac{y_m}{y_p}} = 1.14 \sqrt{\frac{1}{50}} = 0.161 \text{ m/s}$$

$$\frac{Q_m}{Q_p} = \frac{V_m A_m}{V_p A_p} = \frac{V_m (by)_m}{V_p (by)_p}$$

$$= \sqrt{\lambda_y}\, \lambda_x \lambda_y = \lambda_y^{3/2} \lambda_x$$

where λ_y is the vertical scale and λ_x is the horizontal scale.

Note that the term λ is commonly used to indicate a scaling ratio in model studies. It is not to be confused with the Darcy friction factor.

In this case $\lambda_y = \lambda_x$ (undistorted model), whence

$$\frac{Q_m}{Q_p} = \sqrt{\frac{1}{50}} \left(\frac{1}{50}\right)^2$$

$$Q_m = 420 \times \left(\frac{1}{50}\right)^{5/2}$$

$$= 0.02376 \text{ m}^3/\text{s}$$

$$= 23.76 \text{ L/s}$$

The Reynolds number in the model for testing the flow regime is best expressed in terms of the hydraulic radius R_m:

$$b_m = \frac{105}{50} = 2.1 \text{ m} \quad (b_m = \text{average width in model} = b_p \times \lambda_x)$$

Wetted perimeter, $P_m = 2.1 + 2 \times 0.07 = 2.24 \text{ m}$

$$R_m = \frac{2.1 \times 0.07}{2.24} = 0.0656 \text{ m}$$

$$Re_m = \frac{0.161 \times 0.0656}{1 \times 10^{-6}} = 10\,561.6$$

which indicates a turbulent flow.

The discharge in the model has been determined, in relation to the geometrical scale, to correspond with the correct scaling of the gravity forces. In order to scale correctly the viscous resistance forces, in the model the discharge ratio should comply with the Reynolds law with the geometrically relative roughness.

In the Froude scaled model, therefore, the resistance forces would be underestimated if the boundary roughness were to be modelled to the geometrical scale, and in practice, roughness elements consisting of concrete blocks, wire mesh or vertical rods are installed and adjusted until the surface profile in the model, when operating at the appropriate scale discharge, is geometrically similar to the observed prototype surface profile.

Example 9.8

(a) Explain why distorted scale models of rivers are commonly used.

(b) A river model is constructed to a vertical scale of 1:50 and a horizontal scale of 1:200. The model is to be used to investigate a flood alleviation scheme. At the design flood of 450 m³/s, the average width and depth of flow are 60 and 4.2 m, respectively. Determine the corresponding discharge in the model and check the Reynolds number of the model flow.

Solution:

(a) The size of a river model is determined by the laboratory space available (although in some cases special buildings are constructed to house a particular model, notably the Eastern Schelde model in the Delft Hydraulics Laboratory, the Netherlands). In the case of a long river reach, a natural model scale may result in such small flow depths that depth and water elevations cannot be measured with sufficient accuracy, the flow in the model may become laminar, surface tension effects may become significant and sediment studies may be precluded because of the low tractive force.

To avoid these problems, geometrical distortion wherein the vertical scale λ_y is larger than the horizontal scale λ_x is used, typical vertical distortions being in the range of 5–100 with a vertical scale not smaller than 1:100.

(b) Velocity in prototype $= \dfrac{450}{60 \times 4.2} = 1.786$ m/s

The Froude scaling law is based on the vertical scale ratio

$$V_m = V_p\sqrt{\lambda_y} = 1.786\sqrt{\frac{1}{50}} = 0.25 \text{ m/s}$$

$$\frac{Q_m}{Q_p} = \frac{V_m \, (xy)_m}{V_p \, (by)} = \lambda_y^{3/2}\lambda_x$$

$$= \left(\frac{1}{50}\right)^{3/2} \times \frac{1}{200} = 1.414 \times 10^{-5}$$

$$\Rightarrow Q_m = 450 \times 1.414 \times 10^{-5} \text{ m}^3/\text{s}$$

$$= 6.36 \text{ L/s}$$

Average dimensions of model channel:

$$\text{Width} = \frac{60}{200} = 0.30 \text{ m}$$

$$\text{Depth} = \frac{4.2}{50} = 0.084 \text{ m}$$

$$\text{Hydraulic radius} \simeq \frac{0.3 \times 0.084}{0.468} = 0.0538$$

$$\text{Reynolds number} = \frac{VR}{v} = \frac{0.25 \times 0.0538}{1 \times 10^{-6}} = 13\,450$$

The flow therefore will be turbulent.

Example 9.9

An estuary model is built to a horizontal scale of 1:500 and vertical scale of 1:50. Tidal oscillations of amplitude 5.5 m and a tidal period of 12.4 h are to be reproduced in the model. What are the corresponding tidal characteristics in the model?

Solution:

The speed of propagation, or celerity, of a gravity wave in which the wavelength is very large in relation to the water depth y, as in the case of tidal oscillations, is given by $c = \sqrt{gy}$. Thus, estuary models must be operated according to the Froude law.

The tidal range is modelled according to the vertical scale.

$$H_m = H_p \times \frac{1}{50} = \frac{5.5}{50} = 0.11 \text{ m}$$

$$\text{Tidal period, } T = \frac{L}{c}$$

where L is the wavelength. Hence

$$\frac{T_m}{T_p} = \frac{L_m}{L_p} \frac{c_p}{c_m} = \lambda_x \sqrt{\frac{y_p}{y_m}} = \frac{\lambda_x}{\sqrt{\lambda_y}}$$

$$\Rightarrow T_m = \frac{12.4 \times 1/500}{\sqrt{1/50}} = 0.1754 \text{ h}$$

$$= 10.52 \text{ min}$$

Example 9.10

The discharge Q from a rotodynamic pump developing a total head H when running at N revolutions per minute is given by

$$Q = ND^3 \phi \left[\frac{D}{B}, \frac{N^2 D^2}{gH}, \frac{\rho ND^2}{\mu} \right] \quad \text{(see Example 9.2)} \qquad \text{(i)}$$

(a) Obtain an expression for the specific speed of a rotodynamic pump and show how to predict the pump characteristics when running at different speeds.

(b) The performance of a new design of rotodynamic pump is to be tested in a 1:5 scale model. The pump is to run at 1450 rev/min.

The model delivers a discharge of 2.5 L/s of water and a total head of 3 m, with an efficiency of 65% when operating at 2000 rev/min. What are the corresponding discharge head and power consumption in the prototype? Determine the specific speed of the pump and hence state the type of impeller.

Solution:

(a) For geometrically similar machines operating at high Reynolds numbers, the term $\rho ND^2/\mu$ becomes unimportant and Equation (i) may be rewritten as

$$f\left(\frac{Q}{ND^3}, \frac{D}{B}, \frac{N^2 D^2}{gH} \right) = 0$$

The term D/B is automatically satisfied by the geometrical similarity, and the terms Q/ND^3 and N^2D^2/gH should have identical values in the model and prototype.

$$\Rightarrow \left(\frac{Q}{ND^3}\right)_m = \left(\frac{Q}{ND^3}\right)_p \qquad \text{(ii)}$$

$$\text{and} \left(\frac{N^2D^2}{gH}\right)_m = \left(\frac{N^2D^2}{gH}\right)_p \qquad \text{(iii)}$$

The scale

$$\frac{D_m}{D_p} = \frac{N_p}{N_m}\left(\frac{H_m}{H_p}\right)^{1/2} \qquad \text{(from Equation (iii))}$$

whence from Equation (ii)

$$\frac{N_m}{N_p} = \sqrt{\frac{Q_p}{Q_m}}\left(\frac{H_m}{H_p}\right)^{3/4} \qquad \text{(iv)}$$

If H_m and Q_m are made equal to unity, then Equation (iv) becomes

$$N_m = \frac{N_p\sqrt{Q_p}}{H_p^{3/4}} \qquad \text{(v)}$$

The term $N\sqrt{Q}/H^{3/4}$ is called the **specific speed** and is interpreted as the speed at which a geometrically scaled model would run in order to deliver unit discharge when generating unit head. All geometrically similar machines have the same specific speed.

In the case of the same pump running at different speeds, $D_m = D_p$ and Equation (ii) becomes

$$\frac{Q_1}{N_1} = \frac{Q_2}{N_2} \quad \text{or} \quad Q_2 = Q_1\left(\frac{N_2}{N_1}\right)$$

and Equation (iii) becomes

$$\frac{N_1}{H_1^2} = \frac{N_2}{H_2^2} \quad \text{or} \quad H_2 = H_1\sqrt{\frac{N_2}{N_1}}$$

(b)   From Equation (ii)

$$Q_p = Q_m\frac{(ND^3)_p}{(ND^3)_m}$$

$$\Rightarrow Q_p = 2.5 \times 5^3 \times \frac{1450}{2000} = 226 \text{ L/s}$$

From Equation (iii)

$$H_p = H_m\left(\frac{N_p}{N_m}\frac{D_p}{D_m}\right)^2$$

$$= 3.0\left(\frac{1450}{2000} \times 5\right)^2 = 39.42 \text{ m}$$

$$\text{Power input required} = \frac{9.81 \times 0.226 \times 39.42}{0.65} = 134.5 \text{ kW}$$

$$\text{Specific speed, } N_s = \frac{N\sqrt{Q}}{H^{3/4}} = \frac{1450\sqrt{226}}{(39.42)^{3/4}} = 1385$$

The impeller is of the centrifugal type (see Chapter 6).

Example 9.11

A model of a proposed dam spillway was constructed to a scale of 1:25. The design flood discharge over the spillway is 1000 m³/s. What discharge should be provided in the model? What is the velocity in the prototype corresponding with a velocity of 1.5 m/s in the model at the corresponding point?

Solution:

Example 9.4 showed that the discharge per unit crest length of a rectangular weir could be expressed as

$$q = g^{1/2}H^{3/2}\phi\left[\frac{\rho g^{1/2}H^{3/2}}{\mu}, \frac{\rho g H^2}{\sigma}, \frac{P}{H}\right] \tag{i}$$

The governing equation for spillways and weirs is identical. Flood discharges over spillways will result in very high Reynolds numbers, and since surface tension effects are also negligible the only factor affecting the discharge coefficient is P/H. In modelling dam spillways, therefore, if the ratios of P/H in the model and prototype are identical and the crest geometry is correctly scaled,

$$\frac{q}{g^{1/2}H^{3/2}} = \text{Constant}$$

Therefore, spillway models are operated according to the Froude law and are made sufficiently large that viscous and surface tension effects are negligible.

$$\left(\frac{V}{\sqrt{g\ell}}\right)_m = \left(\frac{V}{\sqrt{g\ell}}\right)_p \tag{ii}$$

$$Q = V\ell^2 \tag{iii}$$

$$\Rightarrow \left(\frac{Q}{\ell^{5/2}}\right)_m = \left(\frac{Q}{\ell^{5/2}}\right)_p \tag{iv}$$

whence

$$Q_m = 1000\left(\frac{1}{25}\right)^{5/2} = 0.32 \text{ m}^3/\text{s}$$

From Equation (ii)

$$V_p = V_m\sqrt{\frac{\ell_p}{\ell_m}} = V_m\sqrt{25}$$

$$\Rightarrow V_p = 1.5 \times 5 = 7.5 \text{ m/s}$$

References and recommended reading

Allen, J. (1952) *Scale Models in Hydraulic Engineering*, Longman, London.
Chadwick, A., Morfett, J. and Borthwick, M. (2013) *Hydraulics in Civil and Environmental Engineering*, 5th edn, Taylor & Francis, Abingdon, UK.
Novak, P. and Cabelka, J. (1981) *Models in Hydraulic Engineering*, Pitman, London.
Webber, N. B. (1971) *Fluid Mechanics for Civil Engineers*, Spon, London.
Yalin, M. S. (1971) *Theory of Hydraulic Models*, Macmillan, London.

Problems

1. The head loss of water of kinematic viscosity 1×10^{-6} m^2/s in a 50 mm diameter pipeline was 0.25 m over a length of 10.0 m at a discharge of 2.0 L/s. What is the corresponding discharge and hydraulic gradient when oil of kinematic viscosity 8.5×10^{-6} m^2/s flows through a 250 mm diameter pipeline of the same relative roughness?

2. Find the pressure drop at the corresponding speed in a pipe 25 mm in diameter and 30 m long conveying water at 10°C if the pressure head loss in a 200 mm diameter smooth pipe 300 m long in which air is flowing at a velocity of 3 m/s is 10 mm of water. Density of air $= 1.3$ kg/m^3; dynamic viscosity of air $= 1.77 \times 10^{-5}$ N s/m^2; dynamic viscosity of water $= 1.3 \times 10^{-3}$ N s/m^2.

3. A 50 mm diameter pipe is used to convey air at 4°C (density $= 1.12$ kg/m^3 and dynamic viscosity $= 1.815 \times 10^{-5}$ N s/m^2) at a mean velocity of 20 m/s. Calculate the discharge of water at 20°C for dynamic similarity, and obtain the ratio of the pressure drop per unit length in the two cases.

4. If, in modelling a physical system, the Reynolds and Froude numbers are to be the same in the model and prototype, determine the ratio of kinematic viscosity of the fluid in the model to that in the prototype.

5. The sequent depth y_s of a hydraulic jump in a rectangular channel is related to the initial depth y_i, the discharge per unit width, q, g and p. Express the ratio y_s/y_i in terms of a non-dimensional group and compare with the equation developed from momentum principles:

$$y_s = \frac{y_i}{2}\left(\sqrt{1 + Fr_i^2} - 1\right)$$

6. A 60° V-notch is to be used for measuring the discharge of oil having a kinematic viscosity 10 times that of water. The notch was calibrated using water. When the head over the notch was 0.1 m the discharge was 2.54 L/s. Determine the corresponding head and discharge when the notch is used for oil flow measurement.

7. The airflow and wind effects on a bridge structure are to be studied on a 1:25 scale model in a pressurised wind tunnel in which the air density is eight times that of air at atmospheric pressure and at the same temperature. If the bridge structure is subjected to wind speeds of 30 m/s, what is the corresponding wind speed in the wind tunnel? What force on the prototype corresponds with a 1400 N force on the model? (Note: The dynamic viscosity of air is unaffected by pressure changes provided the temperature remains constant.)

8. A rotodynamic pump is designed to operate at 1450 rev/min and to develop a total head of 60 m when discharging 250 L/s.

   The following characteristics of a 1:4 scale model were obtained from tests carried out at 1800 rev/min.

Q_m (L/s)	0	2	4	6	8
H_m (m)	8	7.6	6.4	4.2	1.0

   Obtain the corresponding characteristics of the prototype, and state whether or not it meets its design requirements.

9. **(a)** Show that the power output P of a hydraulic turbine expressed in terms of non-dimensional groups in the form

$$P = \rho N^3 D^5 \phi \left[\frac{Q}{ND^3}, \frac{D}{B}, \frac{N^2 D^2}{gH}, \frac{\rho ND^2}{\mu} \right]$$

   Derive an expression for the specific speed of a hydraulic turbine.

   (b) A 1:20 scale model of a hydraulic turbine operates under a constant head of 10 m. The prototype will operate under a head of 150 m at a speed of 300 rev/min. When running at the corresponding speed, the model generates 1.2 kW at a discharge of 13.6 L/s. Determine the corresponding speed, power output and discharge of the prototype.

10. The wave action and forces on a proposed sea wall are to be studied on a 1:10 scale model. The design wave has a period of 9 seconds and a height from crest to trough of 5 m. The depth of water in front of the wall is 7 m.

    Assuming that the wave is a gravity wave in shallow water and that the celerity $c = \sqrt{gy}$ (where y is the water depth), determine the wave period, wavelength and wave height to be reproduced in the model. If a force of 4 kN due to wave breaking on a 0.5 m length of the model sea wall were recorded, what would be the corresponding force per unit length on the prototype?

Chapter 10
Ideal Fluid Flow and Curvilinear Flow

10.1 Ideal fluid flow

The analysis of ideal fluid flow is also referred to as **potential flow**. The concept of an ideal fluid is that of one which is inviscid and incompressible; the flow is also assumed to be irrotational. Since flow in boundary layers is rotational and strongly influenced by viscosity, the analytical techniques of ideal fluid flow cannot be applied in such circumstances. However, in many situations the flow of real fluids outside the boundary layer, where viscous effects are small, approximates closely to that of an ideal fluid.

The object of the study of ideal fluid flow is to obtain the flow pattern and pressure distribution in the fluid flow around prescribed boundaries. Examples are the flow over airfoils, through the passages of pump and turbine blades, over dam spillways and under control gates. The governing differential equations of ideal fluid flow have also been successfully applied to oscillatory wave motions, groundwater and seepage flows.

10.2 Streamlines, the stream function

A **streamline** is a continuous line drawn through the fluid such that it is tangential to the velocity vector at every point. In steady flow the streamlines are identical with the **pathlines** or tracks of discrete liquid elements. No flow can occur across a streamline, and the concept of a **stream tube** in two-dimensional flow emerges as the flow per unit depth between adjacent streamlines.

From Figure 10.1a,

$$\frac{dy}{dx} = \frac{v}{u}; \qquad u\,dy - v\,dx = 0 \qquad [10.1]$$

The continuity equation for two-dimensional steady ideal fluid flow is

$$\frac{\partial u}{\partial x} + \frac{\partial v}{\partial y} = 0 \qquad [10.2]$$

Nalluri & Featherstone's Civil Engineering Hydraulics: Essential Theory with Worked Examples, Sixth Edition. Martin Marriott.
© 2016 John Wiley & Sons, Ltd. Published 2016 by John Wiley & Sons, Ltd.
Companion Website: www.wiley.com/go/Marriott

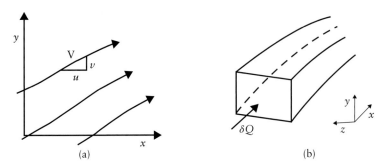

Figure 10.1 Two-dimensional ideal fluid flow: (a) streamlines and (b) streamtube.

Equation 10.2 is satisfied by the introduction of a stream function denoted by ψ such that

$$u = \frac{\partial \psi}{\partial y} \quad \text{and} \quad v = -\frac{\partial \psi}{\partial x}$$

whence Equation 10.2 becomes

$$\frac{\partial^2 \psi}{\partial y\, \partial x} - \frac{\partial^2 \psi}{\partial x\, \partial y} = 0$$

Substitution of u and v in Equation 10.1 yields

$$\frac{\partial \psi}{\partial y}\, dy + \frac{\partial \psi}{\partial x}\, dx = 0$$

Now mathematically

$$d\psi = \frac{\partial \psi}{\partial y}\, dy + \frac{\partial \psi}{\partial x}\, dx$$

whence $d\psi/ds = 0$, where s is the direction along a streamline. Thus, ψ is constant along a streamline and the pattern of streamlines is obtained by equating the stream function to a series of numerical constants.

Since $\delta Q = Vb$, where b is the spacing of adjacent streamlines and δQ the discharge per unit depth between the streamlines, the velocity vector V is universely proportional to the streamline spacing.

In polar coordinates the radial and tangential velocity components v_r and v_θ, respectively, are expressed by

$$v_r = \frac{1}{r} \frac{\partial \psi}{\partial \theta}; \qquad v_\theta = -\frac{\partial \psi}{\partial r} \qquad [10.3]$$

10.3 Relationship between discharge and stream function

Let δQ be the discharge per unit depth between adjacent streamlines (Figure 10.2).

$$\delta Q = u \sin \theta\, \delta s - v \cos \theta\, \delta s$$
$$= u\, \delta y - v\, \delta x$$
$$= \frac{\partial \psi}{\partial y}\, \delta y + \frac{\partial \psi}{\partial x}\, \delta x$$

Now $\delta \psi = \dfrac{\partial \psi}{\partial y}\, \delta y + \dfrac{\partial \psi}{\partial x}\, \delta x;$ whence $\delta Q = \delta \psi = \psi_2 - \psi_1$ [10.4]

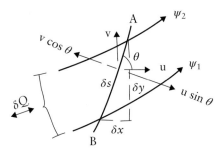

Figure 10.2   Adjacent streamlines.

10.4 Circulation and the velocity potential function

Circulation is the line integral of the tangential velocity around a closed contour expressed by

$$K = \int v_s\, ds \tag{10.5}$$

Velocity potential is the line integral along the s-direction between two points (see Figure 10.3).

$$\phi_A - \phi_B = \int_A^B V \sin\alpha\, ds \tag{10.6}$$

If $\alpha = 0$, $\phi_A - \phi_B = 0$; thus, the potential ϕ along AB is constant.
Note that

$$\frac{\partial\phi}{\partial s} = V\sin\alpha, \qquad \text{that is,} \qquad \frac{\partial\phi}{\partial s} = v_s \tag{10.7}$$

Lines of constant velocity potential are orthogonal to the streamlines, and the set of equipotential lines and the set of streamlines form a system of curvilinear squares described as a flow net (see Figure 10.4).

10.5 Stream functions for basic flow patterns

(a)   Uniform rectilinear flow in x-direction (Figure 10.5a):

$$u = \frac{\partial\psi}{\partial y}; \qquad \psi = uy + f(x) \tag{i}$$

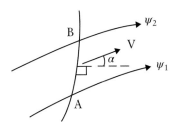

Figure 10.3   Velocity potential.

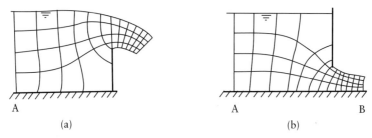

Figure 10.4   Flow nets: (a) flow over a sharp-crested area and (b) flow under a sluice gate.

$$v = \frac{-\partial \psi}{\partial x}; \qquad \psi = -vx + f(y) \qquad \text{(ii)}$$

Since $v = 0$ and Equations (i) and (ii) are identical,

$$f(x) = 0 \quad \text{and} \quad \psi = uy \qquad \text{(10.8)}$$

(b)   Uniform rectilinear flow in the y-direction (Figure 10.5b): Similarly since $u = 0$,

$$\psi = -vx \qquad \text{(10.9)}$$

(c)   Line source (Figure 10.5c): A line source provides an axi-symmetric radial flow. Using polar coordinates,

$$v_r = \frac{1}{r}\frac{\partial \psi}{\partial \theta}; \qquad \psi = rv_r\theta + f(r)$$

$$v_\theta = -\frac{\partial \psi}{\partial r}; \qquad \psi = -rv_\theta + f(\theta)$$

$$v_\theta = 0, \quad \text{whence } \psi = rv_r\theta$$

$$v_r = \frac{q}{2\pi r}$$

where q is the 'strength' of the source, which is equal to discharge per unit depth. Therefore,

$$\psi = \frac{q\theta}{2\pi} \qquad \text{(10.10)}$$

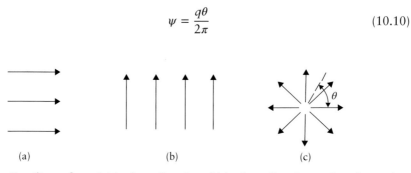

Figure 10.5   Rectilinear flows (a) in the x-direction, (b) in the y-direction and (c) from a line source.

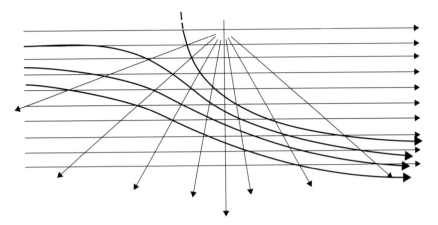

Figure 10.6   Combinations of basic flow patterns.

Similarly a line 'sink' which is a negative source is defined by

$$\psi = -\frac{q\theta}{2\pi} \qquad (10.11)$$

10.6 Combinations of basic flow patterns

10.6.1 Source in uniform flow (See Figure 10.6)

The stream function of the resultant flow pattern is obtained by addition of the component stream functions. Thus

$$\psi = uy + \frac{q\theta}{2\pi} \qquad [10.12]$$

The streamlines are obtained by solving Equation 10.12 for a number of values of ψ. Alternatively superimpose the streamlines for the individual flow patterns and algebraically add the values of the stream function where they intersect. Obtain the new streamlines by drawing lines through points having the same value of stream function.

10.7 Pressure at points in the flow field

The pressure p at any point (r, θ) in the flow field is obtained from application of Bernoulli's equation:

$$\frac{p_0}{\rho g} + \frac{V_0^2}{2g} = \frac{p}{\rho g} + \frac{V^2}{2g}$$

where p_0 and V_0 are the pressure and velocity vector in the undisturbed uniform flow; and V is the velocity vector at (r, θ). V is obtained from the orthogonal spacing of the streamlines at (r, θ) or analytically from $V = \sqrt{v_r^2 + v_\theta^2}$.

10.8 The use of flow nets and numerical methods

The analytical methods, as described in Section 10.6 and illustrated in Example 10.1, can be applied to other combinations of basic flow patterns to simulate, for example, the flow round a cylinder, flow round a corner and vortex flow. The reader is referred to texts on fluid mechanics such as Massey and Ward-Smith (2012) for a more detailed treatment of these applications.

In civil engineering hydraulics, however, the flows are generally constrained by non-continuous or complex boundaries. A typical example is the flow under a sluice gate (Figure 10.4b), and such cases are incapable of solution by analytical techniques.

10.8.1 The use of flow nets

One method of solution in such cases is the use of the flow net described in Section 10.4. Selecting a suitable number of **streamtubes**, streamlines 'ψ' are drawn starting from equally spaced points where uniform rectilinear flow exists such as section A (Figure 10.4a) and A and B (Figure 10.4b). A system of equipotential lines 'ϕ' is now added such that they intersect the 'ψ' lines orthogonally. If the streamlines have been drawn correctly to suit the boundary conditions, the resulting flow net will correctly form a system of curvilinear 'squares'. As a final test, circles drawn in each 'square' should be tangential to all sides. On the first trial the test will probably fail and successive adjustments are made to the 'ψ' and 'ϕ' lines until the correct pattern is produced.

Local velocities are obtained from the streamline spacings (or the spacing between the equipotential lines, since $\Delta\psi$ and $\Delta\phi$ are locally equal) in relation to the rectilinear flow velocities, and hence local pressures are calculated from Bernoulli's equation (Example 10.1). The technique has also been widely used in seepage flow problems under water-retaining structures usually covered in geotechnical texts.

10.8.2 Numerical methods

Where computer facilities are available, the streamline pattern can be obtained for complex boundary problems. With computer graphics a plot of the streamlines can be produced in addition. However, the problems can be solved using an electronic calculator. The method involves the solution of Laplace's equation using finite difference methods.

Theory: The assumption of irrotational flow when applied to a liquid element yields

$$\frac{\partial u}{\partial x} - \frac{\partial v}{\partial y} = 0 \qquad [10.13]$$

$$\text{that is, } \frac{\partial^2 \psi}{\partial x^2} + \frac{\partial^2 \psi}{\partial y^2} = 0 \quad \text{or} \quad \nabla^2 \psi = 0 \qquad [10.14]$$

Equation 10.14 is known as Laplace's equation.

Since we have seen that the stream function has numerical values at all points over the flow field, the streamline pattern can be produced if Equation 10.14 can be solved in ψ at discrete points in the field.

Superimpose a square or rectangular mesh of straight lines in the x- and y-directions to generally fit the boundaries of the physical system (see Figure 10.7).

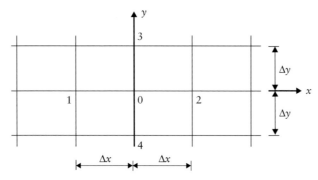

Figure 10.7   Rectangular mesh – numerical method.

Equation 10.14 is to be solved at points such as (x, y) where the grid lines intersect and must first be expressed in finite difference form.

Using the notation

$$f'(x) = \frac{\partial f(x)}{\partial x}; \qquad f''(x) = \frac{\partial^2 f(x)}{\partial x^2}; \qquad \text{etc.}$$

Taylor's theorem gives

$$f(x + \Delta x, \ y) = f(x, \ y) + \Delta x f'(x, \ y) + \frac{\Delta x^2}{2!} f''(x, \ y)$$

$$+ \frac{\Delta x^3}{3!} f'''(x, \ y) + \frac{\Delta x^4}{4!} f^{iv}(x, \ y) + \cdots$$

$$\text{and } f(x - \Delta x, \ y) = f(x, \ y) - \Delta x f'(x, \ y) + \frac{\Delta x^2}{2!} f''(x, \ y) - \frac{\Delta x^3}{3!} f'''(x, \ y)$$

$$+ \frac{\Delta x^4}{4!} f^{iv}(x, \ y) + \cdots$$

Adding gives

$$f(x + \Delta x, \ y) + f(x - \Delta x, \ y) = 2f(x, \ y) + \Delta x^2 f''(x, \ y)$$

neglecting Δx^4 and higher order terms.

Thus

$$f''(x, \ y) = \frac{[f(x + \Delta x, \ y) - 2f(x, \ y) + f(x - \Delta x, \ y)]}{\Delta x^2}$$

Since $\psi = f(x, y)$

$$\frac{\partial^2 \psi}{\partial x^2} = \frac{[\psi(x + \Delta x, \ y) - 2\psi(x, \ y) + \psi(x - \Delta x, \ y)]}{\Delta x^2}$$

Similarly

$$\frac{\partial^2 \psi}{\partial y^2} = \frac{[\psi(x, \ y + \Delta y) - 2\psi(x, \ y) + \psi(x, \ y - \Delta y)]}{\Delta y^2}$$

Chapter 10

Chapter 10

Using the grid notation of Figure 10.7 for simplicity, Equation 10.1 becomes

$$\frac{(\psi_1 - 2\psi_0 + \psi_2)}{\Delta x^2} + \frac{(\psi_3 - 2\psi_0 + \psi_4)}{\Delta y^2} = 0 \qquad [10.15]$$

where the point (x, y) is located at point 0, point $(x + \Delta x, y)$ is at point 2 and so on.
Or, if $\Delta x = \Delta y$,

$$\psi_1 + \psi_2 + \psi_3 + \psi_4 - 4\psi_0 = 0 \qquad [10.16]$$

10.8.2.1 Method of solution
The method of 'relaxation' originally devised by Southwell, for the numerical solution of elliptic partial differential equations such as Laplace's, Poisson's and biharmonic equations describing fluid flow and stress distributions in solid bodies, is not amenable to automatic computation and is not to be confused with the methods described here.

10.8.2.2 Boundary conditions
The solid boundaries and free water surfaces are streamlines and therefore have constant values of stream function. The allocation of values to the boundaries can be quite arbitrary: for example, in Figure 10.4 the bed can be allocated a value of $\psi = 0$ and the surface to a value of $\psi = 100$.

Where grid points do not coincide with a boundary the following form of Equation 10.17, for example, is obtained, assuming a linear variation of ψ along the grid lines (Figure 10.8).

$$\psi_0 = \frac{\psi_1 + (\psi_2/\lambda_2) + \psi_3 + \psi_4}{3 + (1/\lambda_2)} \qquad [10.17]$$

(i) *Matrix method.* Equations 10.15, 10.16 or 10.17 for the interior grid points together with the boundary conditions can be globally expressed in the form

$$[A][\psi] = [B]$$

Thus ψ_1, ψ_2 and so on can be found directly using Gaussian elimination on a computer.

(ii) *Method of successive corrections.* This method is also amenable to computer solution and is probably quicker than (i) above; it can also be executed using an electronic

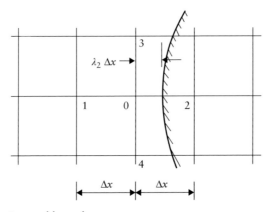

Figure 10.8   Grid points and boundary.

calculator. Values of ψ are allocated to the boundaries and, in relation to these, estimated values are given to the interior grid points. Considering each interior grid point in turn, the allocated value of ψ_0 is revised using Equations 10.15, 10.16 or 10.17 as appropriate. This procedure is repeated at all grid points as many times as necessary until the differences between the previous and revised values of ψ at every grid point are less than a prescribed limit. The discrete streamlines are then drawn by interpolation between the grid values of ψ.

10.9 Curvilinear flow of real fluids

The concepts of ideal fluid flow can be used to obtain the velocity and pressure distributions in curvilinear flow of real fluids in ducts and open channels. Curvilinear flow is also referred to as **vortex motion**.

Curvilinear flow is not to be confused with 'rotational' flow; 'rotation' relates to the net rotation of an element about its axis.

Theory: Consider an element of a fluid subjected to curvilinear motion (see Figure 10.9).

$$\text{Radial acceleration} = \frac{v_\theta^2}{r}$$

Equating radial forces,

$$dp\, r\, d\theta\, dy = \rho r\, d\theta\, dr\, dy \frac{v_\theta^2}{r}$$

whence

$$\frac{dp}{dr} = \frac{\rho v_\theta^2}{r}$$

or in terms of pressure head

$$\frac{dh}{dr} = \frac{v_\theta^2}{gr} \qquad\qquad [10.18]$$

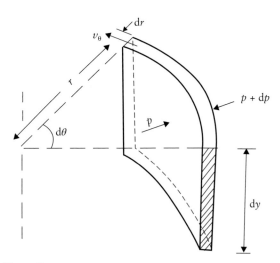

Figure 10.9 Curvilinear flow.

10.10 Free and forced vortices

(a) Free vortex motion occurs when there is no external addition of energy; examples occur in bends in ducts and channels, and in the 'sink' vortex.

   The energy at a point where the velocity is v_0 and pressure head is h is

$$E = h + \frac{v_\theta^2}{2g} \qquad [10.19]$$

where E is constant across the radius; hence

$$\frac{dh}{dr} + \frac{v_\theta}{g}\frac{d(v_\theta)}{dr} = 0$$

From Equation 10.19,

$$\frac{v_\theta^2}{gr} + \frac{v_\theta}{g}\frac{d(v_\theta)}{dr} = 0$$

whence

$$\frac{dv_\theta}{v_\theta} = \frac{dr}{r} \quad \text{or} \quad \ln(v_\theta r) = \text{Constant}$$

Hence

$$v_\theta r = \text{Constant} = K \quad \text{(circulation)} \qquad [10.20]$$

(b) Forced vortex motion is caused by rotating impellers or by rotating a vessel containing a liquid. The equilibrium state is equivalent to the rotation of a solid body where $v_\theta = r\omega$, where ω is the angular velocity (rad/s).

Worked examples

Example 10.1

A line source of strength 180 L/s is placed in a uniform flow of velocity 0.1 m/s.

(a) Plot the streamlines above the x-axis.
(b) Obtain the pressure distribution on the streamline denoted by $\psi = 0$.

Solution:

(a) It is convenient to consider the uniform flow to be in the x-direction. This results in a streamline having the value $\psi = 0$, which may be interpreted as the boundary of a solid body.

$$\psi = uy + \frac{q\theta}{2\pi} \quad \text{(Equation 10.12)}$$

or

$$\psi = ur\sin\theta + \frac{q\theta}{2\pi}$$

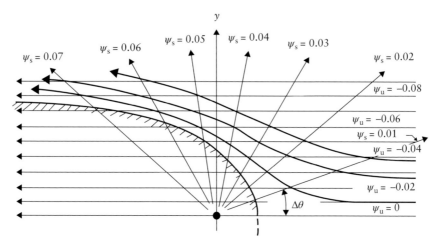

Figure 10.10   Flow net using polar coordinates.

Setting ψ successively to 0, −0.01, −0.02, ... and noting that $u = -0.1$, the coordinates (r, θ) along the streamlines can be obtained from Equation 10.12.

				Values of r (m)					
ψ	$\theta° = 10$	20	40	60	80	100	120	140	160
0	0.29	0.29	0.31	0.35	0.41	0.51	0.69	1.09	2.34
−0.01	0.86	0.58	0.47	0.46	0.51	0.61	0.81	1.24	2.63
−0.02	1.44	0.87	0.62	0.58	0.61	0.71	0.92	1.40	2.92
−0.03	2.02	1.17	0.78	0.69	0.71	0.81	1.04	1.56	3.21

The pairs of coordinates (θ, r) are plotted to give the streamline pattern. The graphical method gives identical results and is clearly quicker; however, the mathematical approach is appropriate in computer applications where computer graph-plotting facilities are available (see Figure 10.10).

The graphical construction proceeds as follows: the uniform flow is represented by a series of equidistant parallel straight lines (defined by ψ_u in Figure 10.10). Discharge between the streamlines was chosen to be 0.010 m³/s m in the analytical solution; thus, the spacing of the uniform flow streamlines in the diagram represents to scale a distance of 0.10 m. It is convenient to choose the streamlines of the source such that the discharge between them is also 0.010 m³/(s m); the angle $\Delta\theta$ is simply obtained since

$$\Delta\psi_s = \frac{q\Delta\theta}{2\pi}$$

that is, $0.010 = \dfrac{0.180 \times \Delta\theta}{2\pi}$

whence $\Delta\theta = 20°$

At the points of intersection of the uniform flow and source streamlines, the individual stream functions are added algebraically and the resultant flow pattern is obtained by sketching the streamlines through points of equal resultant stream function. The construction and final flow pattern are illustrated in Figure 10.10.

(b) The pressure distribution in the disturbed flow field at (r, θ) is obtained from

$$\frac{p_{r,\theta} - p_0}{\rho g} = \frac{V_0^2 - V_{r,\theta}^2}{2g} \tag{i}$$

where V_0 is the undisturbed velocity and $V_{r,\theta}$ the velocity vector at (r, θ).

$$V_{r,\theta} = \sqrt{v_r^2 + v_\theta^2}$$

$$v_r = \frac{1}{r}\frac{\partial \psi}{\partial \theta} \quad \text{(Equation 10.3)}$$

$$= \frac{ur\cos\theta}{r} + \frac{q}{2\pi r}$$

$$v_\theta = -\frac{\partial \psi}{\partial r} \quad \text{(Equation 10.3)}$$

$$= -u\sin\theta$$

$$\Rightarrow V_{r,\theta}^2 = u^2 + 2u\cos\theta\frac{q}{2\pi r} + \left(\frac{q}{2\pi r}\right)^2 \tag{ii}$$

Around the 'body' $(\psi = 0)$,

$$ur\sin\theta + \frac{q\theta}{2\pi} = 0$$

$$\text{i.e.} \quad \frac{q}{2\pi} = -\frac{ur\sin\theta}{\theta}$$

Thus

$$V_{r,\theta}^2 = u^2\left(1 - \frac{\sin 2\theta}{\theta} + \frac{\sin^2\theta}{\theta^2}\right)$$

whence, from Equation (i), since $V_0 = u$ in this case

$$\frac{p_{r,\theta} - p_0}{\rho g} = \frac{u^2}{2g\theta}\left(\sin 2\theta - \frac{\sin^2\theta}{\theta}\right) \tag{iii}$$

Substitution of $u = -0.1$ m/s into Equation (iii) yields the pressure distribution for a range of θ.

Example 10.2

A discharge of 7 m³/(s m) width flows in a rectangular channel. A vertical sluice gate situated in the channel has an opening of 1.5 m. $C_c = 0.62$; $C_v = 0.95$. Assuming that downstream conditions permit free flow under the gate, draw the streamline pattern.

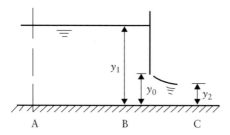

Figure 10.11 Flow under gate.

Solution:

See Figure 10.11.

$$y_2 = C_c \times y_0 = 0.62 \times 1.5 = 0.93 \text{ m}$$

$$y_1 + \frac{V_1^2}{2g} = y_2 + \frac{V_2^2}{2g} + h_L \tag{i}$$

$$\text{and } h_L = \left[\left(\frac{1}{C_v} \right)^2 - 1 \right] \frac{V_2^2}{2g}; \quad V_2 = \frac{7}{0.93} = 7.527 \text{ m/s}$$

$$h_L = 0.31 \text{ m}$$

whence from Equation (i), $y = 3.97$ m (say 4.0 m).

Divide the field into a 1.0 m square grid (see Figure 10.11) and allocate boundary values and interior grid values of ψ.

Numbering rows and columns from the top left-hand corner, note that at point (4, 7) two of the adjacent grid points do not coincide with the boundary. Using the notation of Figure 10.8, $\lambda_2 = 0.6$ and $\lambda_3 = 0.5$.

The results of successive corrections using Equation 10.17 in the form

$$\psi_0 = \frac{\psi_1 + \psi_2 + \psi_3 + \psi_4}{4}$$

at all interior grid points with the exception of point (4, 7) where Equation 10.17

$$\psi_0 = \frac{\psi_1 + (\psi_2/\lambda_2) + (\psi_3/\lambda_3) + \psi_4}{2 + (1/\lambda_2) + (1/\lambda_3)}$$

is used are shown in the following table:

First correction

			Column			
Row	2	3	4	5	6	7
2	80.00	85.00	89.50	93.40	97.00	100.00
3	60.00	70.00	79.40	88.19	93.82	100.00
4	35.00	43.75	52.03	57.55	61.59*	75.57

*Maximum correction = −28.41.

Second correction

Row	Column					
	2	3	4	5	6	7
2	80.00	84.88	89.40	93.67	96.87	100.00
3	58.75	66.69	74.08	79.78	84.56	100.00
4	31.87	37.65	42.32	45.92*	51.51	73.79

*Maximum correction = −11.63.

Values (all corrections < |1.0|) after 8 iterations

Row	Column					
	2	3	4	5	6	7
2	76.49	77.86	79.69	83.02	89.39	100.00
3	51.90	53.78	56.59	62.20	74.27	100.00
4	26.22	27.53	29.69	34.35	45.34	72.71

The above is probably sufficiently accurate for most purposes, but computations can be continued if required.

Values after 12 iterations

Row	Column					
	2	3	4	5	6	7
2	75.78	76.84	78.71	82.33	89.07	100.00
3	51.08	52.62*	55.50	61.44	73.91	100.00
4	25.76	26.88	29.09	33.92	45.13	72.67

*Maximum correction = |0.125|.

The discrete streamline can now be drawn (see Figure 10.12).

Example 10.3

Water flows under pressure round a bend of inner radius 600 mm in a rectangular duct 600 mm wide and 300 m deep. The discharge is 360 L/s. If the pressure head at the entry to the bend is 3.0 m, calculate the velocity and pressure head distributions across the duct at the bend.

Solution:

See Figure 10.13.

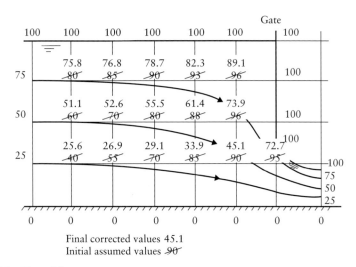

Figure 10.12   Plot of flow net.

Energy at entry to bend

$$E = h + \frac{V^2}{2g} \quad \text{(where } V = \text{approach velocity } = 2 \text{ m/s)}$$

$$= 3.0 + \frac{2^2}{2g} = 3.204 \text{ m}$$

$$E = \text{Constant} = h_A + \frac{v_{\theta,A}^2}{2g} = h_B + \frac{v_{\theta,B}^2}{2g}$$

$dh/dr = v_\theta^2/gr$ from Equation 10.18, and therefore the variation of v_θ with radius is required.

Now $v_\theta = K/r$, and to evaluate K express the discharge Q as

$$Q = w \int_{r_A}^{r_B} v_\theta \, dr$$

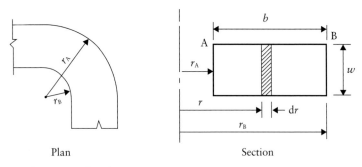

Plan   Section

Figure 10.13   Flow across bend.

that is, $Q = w \int_{r_A}^{r_B} \frac{K}{r} \, dr = Kw \ln \left(\frac{r_B}{r_A} \right)$

$$0.36 = K \times 0.3 \times \ln \left(\frac{1.2}{0.6} \right) = 0.208K$$

or

$$K = 1.73$$

$$v_{\theta,A} = \frac{1.73}{0.6} = 2.88 \text{ m/s}; \quad v_{\theta,C} = \frac{1.73}{0.9} = 1.92 \text{ m/s}$$

$$v_{\theta,B} = \frac{1.73}{1.2} = 1.44 \text{ m/s}$$

$$E(=3.204 \text{ m}) = h + \frac{v_\theta^2}{2g}; \quad h = 3.204 - \frac{v_\theta^2}{2g}$$

whence

$$h_A = 2.78 \text{ m}; \quad h_C = 3.02 \text{ m}; \quad h_B = 3.10 \text{ m}$$

Example 10.4

Water flows round a horizontal 90° bend in a square duct of side length 200 mm, the inner radius being 300 mm. The differential head between the inner and outer sides of the bend is 200 mm of water. Determine the discharge in the duct.

Solution:

See Figure 10.14.

$$\frac{dh}{dr} = \frac{V_\theta^2}{gr} \quad \text{(Equation 10.18)}$$

$$\Delta h = h_A - h_B = \int_A^B dh = \int_{r_A}^{r_B} \frac{v_{\theta,r}^2 \, dr}{gr}$$

where $v_{\theta,r}$ means the tangential velocity at r, and $v_{\theta,r} = K/r$ (Equation 10.20). Therefore

$$\Delta h = h_A - h_B = \frac{K^2}{g} \int_{r_A}^{r_B} \frac{dr}{r^3} = \frac{K^2}{2g} \left[-\frac{1}{r^2} \right]_{r_A}^{r_B} \tag{i}$$

$$Q = w \int_{r_A}^{r_B} v_\theta \, dr = w \int_{r_A}^{r_B} \frac{K}{r} \, dr$$

$$Q = wK \ln \left(\frac{r_B}{r_A} \right); \quad K = \frac{Q}{w \ln \left(r_B/r_A \right)}$$

Therefore, in Equation (i)

$$\Delta h = \frac{Q^2}{w^2 [\ln(r_B/r_A)]^2} \left(\frac{1}{r_A^2} - \frac{1}{r_B^2} \right) \frac{1}{2g}$$

$$\text{where } Q = \sqrt{2g\Delta h} \, w \ln \left(\frac{r_B}{r_A} \right) \sqrt{\frac{r_B^2 \times r_A^2}{r_B^2 - r_A^2}}$$

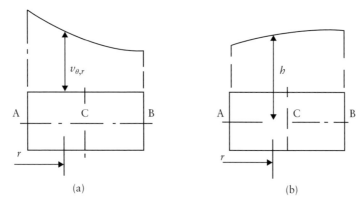

Figure 10.14   (a) Velocity and (b) pressure variations, across bend.

$$r_A = 0.3 \text{ m}; \quad r_B = 0.5 \text{ m}; \quad w = 0.2 \text{ m}; \quad \Delta h = 0.2 \text{ m}$$
$$\Rightarrow Q = 0.076 \text{ m}^3/\text{s}$$

Example 10.5

A cylindrical vessel is rotated at an angular velocity of ω. Show that the surface profile of the contained liquid under equilibrium conditions is parabolic.

Solution:

This is an example of forced vortex motion (see Figure 10.15).
   For all types of curvilinear flow, Equation 10.18 is applicable.

$$\frac{dh}{dr} = \frac{v_{\theta,r}^2}{gr} \tag{i}$$

$$\text{and } v_{\theta,r} = r\omega \quad \text{(for the forced vortex).} \tag{ii}$$

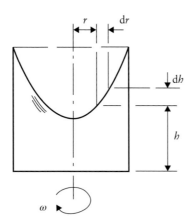

Figure 10.15   Forced vortex motion.

Therefore

$$h_2 - h_1 = \int_1^2 dh = \frac{1}{g} \int_{r_1}^{r_2} \frac{v_{\theta,r}^2}{r} \, dr$$

$$\text{i.e. } h_2 - h_1 = \frac{1}{g} \int_{r_1}^{r_2} r\omega^2 \, dr = \frac{\omega^2}{2g} \left(r_2^2 - r_1^2 \right)$$

Taking the origin ($r = 0, h = 0$) at 0,

$$h = \frac{\omega^2}{2g} r^2$$

which is a paraboloid of revolution.

Example 10.6

A siphon spillway of constant cross section 3.0 m wide by 2.0 m deep operates under a head of 6.0 m. The crest radius is 3.0 m and the siphon has a total length of 18.0 m, the length from inlet to crest being 5.0 m along the centre line.

(a) Assuming the inlet head loss to be $0.3V^2/2g$, the bend loss $0.5V^2/2g$, the Darcy friction factor 0.012 and $\alpha = 1.2$, calculate the discharge through the siphon.
(b) If the level in the reservoir is 0.5 m above the crest of the siphon, calculate the pressures at the crest and cowl and comment on the result.

Solution:

Notes: Siphon spillways are used on dams to discharge floodwater. They are particularly useful where the available crest length of a free overall spillway would be inadequate. Once a siphon spillway has 'primed', it operates like full-bore pipe flow under the head between the reservoir level and the outlet and has a high discharge capacity per unit area. Unlike a free overfall spillway which provides a gradual increase in discharge (see Example 11.9), the siphon discharge reaches a peak very quickly, which may cause a surge to propagate downstream. For this reason a number of siphons may have their crests set at different levels so that their priming times are not simultaneous.

Since siphons incorporate one vertical bend at a high level, the resulting vortex motion can produce very low pressures at the crest, which may result in air entrainment, cavitation and vibration. The design of the crest radius is therefore of utmost importance.

In this example while head losses in the direction of flow are taken into account (i.e. real fluid flow is considered), no energy losses across the flow occur and the curvilinear flow is treated as in ideal fluid flow (see Figure 10.16).

Using the principles of resistance to flow in non-circular ducts,

$$H = \text{entry loss} + \text{velocity head} + \text{bend loss} + \text{friction loss}$$

$$\text{that is, } H = \frac{0.3V^2}{2g} + \frac{\alpha V^2}{2g} + \frac{0.5V^2}{2g} + \frac{\lambda L V^2}{8gR} \tag{i}$$

$$R = \text{hydraulic radius} = 0.6 \text{ m}$$

$$6 = \frac{V^2}{2g} \left(0.3 + 1.2 + 0.5 \frac{0.012 \times 18}{4 \times 0.6} \right)$$

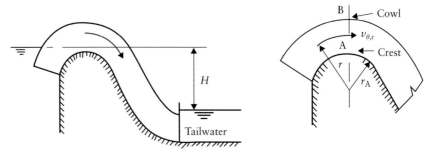

Figure 10.16 Siphon spillway.

$$6 = \frac{V^2}{2g} \times 2.09 \quad \text{whence } V = 6.37 \, \text{m/s}$$

$$\text{Discharge} = 6 \times 6.37 = 38.22 \, \text{m}^3/\text{s}$$

(Note that the discharge may be expressed as $Q = C_d A \sqrt{2gh}$, where C_d is the overall discharge coefficient.)

Net head at crest H_A of bend $= 0.5 - \text{losses}$

$$H_A = 0.5 - \frac{0.3V^2}{2g} - \frac{\lambda L_A V^2}{8gR} = 0.5 - 2.87 \times \left(0.3 + \frac{0.012 \times 5}{4 \times 0.6} \right) \quad \text{(relative to crest)}$$

$$= -0.433 \, \text{m}$$

At the crest A,

$$H_A = \frac{p_A}{\rho g} + \frac{v_{\theta,A}^2}{2g} \tag{ii}$$

whence

$$\frac{p_A}{\rho g} = H_A - \frac{v_{\theta,A}^2}{2g}$$

$$v_{\theta,A} = \frac{K}{r_A} \tag{iii}$$

The discharge across the section of the duct at the crest,

$$Q = \int_{r_A}^{r_B} v_\theta b \, dr$$

where b is the width of the duct.

$$Q = b \int_{r_A}^{r_B} \frac{K}{r} dr = bK \ln \left(\frac{r_B}{r_A} \right)$$

$$38.22 = 3 \times K \times \ln \left(\frac{5}{3} \right) = 1.532K$$

whence $K = 24.94$

$$\Rightarrow v_{\theta,A} = \frac{24.940}{3} = 8.313 \, \text{m/s}; \quad \frac{v_{\theta,A}^2}{2g} = 3.52 \, \text{m}$$

Therefore from Equation (iii),

$$\frac{p_A}{\rho g} = -0.433 - 3.52 = -3.953 \text{ m}$$

At B (the cowl),

$$H_B = \frac{p_B}{\rho g} + \frac{v_{\theta,B}^2}{2g} + 2.0 \quad \text{(relative to crest)}$$

Since $H_B = H_A$,

$$\frac{p_B}{\rho g} = H_A - \frac{v_{\theta,B}^2}{2g} - 2.0$$

$$v_{\theta,B} = \frac{K}{r_B} = \frac{24.940}{5} = 4.988; \quad \frac{v_{\theta,B}^2}{2g} = 1.268$$

$$\Rightarrow \frac{p_B}{\rho g} = -0.433 - 1.268 - 2.0 = -3.701 \text{ m}$$

Comment: Neither $p_A/\rho g$ nor $p_B/\rho g$ are particularly low, and there should be no danger of cavitation. The spillway could satisfactorily operate under a larger gross head, provided the crest pressure does not fall below −7 m, that is, a gauge pressure head of 7 m below atmospheric pressure.

Example 10.7

At a discharge of 10 m³/s the depth of uniform flow in a rectangular channel 3 m wide is 2.2 m. The water flows round a 90° bend of inner radius 5.0 m. Assuming no energy loss at the bend, calculate the depth of water at the inner and outer radii of the bend.

Solution:

See Figure 10.17.

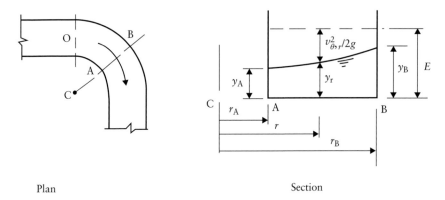

Plan   Section

Figure 10.17 Flow across a rectangular channel bend.

Since there is no energy loss between the entry to the bend and points within the bend, $E_0 = E_A = E_B = E$; that is,

$$y_0 + \frac{V_0^2}{2g} = y_A + \frac{v_{\theta,A}^2}{2g} = y_B + \frac{v_{\theta,B}^2}{2g} \tag{i}$$

Hence, if $v_{\theta,A}$ and $v_{\theta,B}$ can be determined, y_A and y_B can be calculated.

$$y_0 = 2.2 \text{ m}; \quad V_0 = \frac{10}{3 \times 2.2} = 1.515 \text{ m/s}; \quad \frac{V_0^2}{2g} = 0.117 \text{ m}$$

$$\Rightarrow E_0 = 2.2 + 0.117 = 2.317 \text{ m}$$

$$v_{\theta,r} = \frac{K}{r} \quad \text{and} \quad Q = \int_{r_A}^{r_B} v_{\theta,r} y_r \, dr \tag{ii}$$

$$y_r = E - \frac{v_{\theta,r}^2}{2g} = E - \frac{K^2}{2gr^2}$$

And substituting in Equation (ii),

$$\Rightarrow Q = K \int_{r_A}^{r_B} \frac{1}{r} \left(E - \frac{K^2}{2gr^2} \right) dr = K \int_{r_A}^{r_B} \left(\frac{E}{r} - \frac{K^2}{2gr^3} \right) dr$$

$$Q = K \left[E \ln \left(\frac{r_B}{r_A} \right) + \frac{K^2}{4g} \left(\frac{1}{r_B^2} - \frac{1}{r_A^2} \right) \right]$$

$$10 = K \left[2.317 \ln \left(\frac{8}{5} \right) + \frac{K^2}{4g} \left(\frac{1}{8^2} - \frac{1}{5^2} \right) \right]$$

By trial, $K = 9.71$.

$$\Rightarrow v_{\theta,A} = \frac{9.71}{5} = 1.942; \quad \frac{v_{\theta,A}^2}{2g} = 0.192 \text{ m}$$

$$v_{\theta,B} = \frac{9.71}{5} = 1.2137; \quad \frac{v_{\theta,B}^2}{2g} = 0.075 \text{ m}$$

$$\Rightarrow y_A = 2.317 - 0.192 = 2.125 \text{ m}$$
$$y_B = 2.317 - 0.075 = 2.242 \text{ m}$$

References and recommended reading

Massey, B. S. and Ward-Smith, J. (2012) *Mechanics of Fluids*, 9th edn, Taylor & Francis, Abingdon, UK.

United States Bureau of Reclamation (USBR) (1987) *Design of Small Dams*, 3rd edn, US Department of the Interior, Washington, DC.

Problems

1. Determine the stream function for a uniform rectilinear flow of velocity V inclined at α to the x-axis in (a) Cartesian and (b) polar forms.

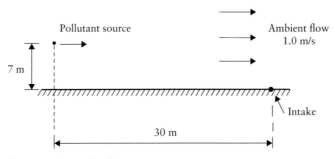

Figure 10.18   Source release of pollutant.

2. A stream function is defined by $\psi = xy$. Determine the flow pattern and the velocity potential function.

3. Draw the streamlines defining **streamtubes** conveying 1 m³/(s m) depth for a source of strength 12 m³/(s m) in a rectilinear uniform flow of velocity -1.0 m/s.

4. In the system described in Problem 3, a sink of strength 12 m³/(s m) is situated 5 m downstream from the origin of source in the direction of the x-axis. Draw the streamlines and determine the shape of the 'body' defined by the streamline $\psi = 0$.

5. A pollutant is released steadily from the vertical outlet of an outfall into a river (Figure 10.18) such that it rises vertically without radial flow (neglecting the effects of entrainment). The pollutant is carried downstream by a uniform current of 1.0 m/s. A water abstraction is situated 30 m downstream of the outfall; this may be considered as a line sink, the total inflow over the 2 m of the stream being 6 m³/s. Neglecting the effects of dispersion, investigate the possibility of the pollutant entering the intake (see Figure 10.18).

6. Water flows through a rectangular duct 4 m wide and 1 m deep at a rate of 20 m³/s. The flow passes through the side contraction shown in Figure 10.19. Assuming ideal fluid flow and using either a flow net construction or a numerical method of streamline plotting, determine the pressure distribution through the transition between AA and BB (a) along the centre line of the duct and (b) along the boundary if the pressure head at AA is 10 m of water (see Figure 10.19).

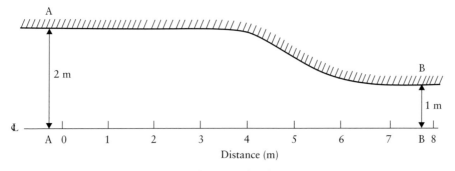

Figure 10.19   Flow through a contracted rectangular duct.

7. A discharge of 6 m³/s per unit width approaches a vertical sluice gate in a rectangular channel at a depth of 4 m. The vertical gate opening is 1.5 m and $C_d = 0.6$. Assuming that downstream conditions do not affect the natural flow through the gate opening, verify that the flow through the opening is supercritical, draw the streamline pattern and determine the pressure distribution and force per unit width on the gate. Compare the value of force obtained with that obtained using the momentum equation.

8. Water flows under pressure in a horizontal rectangular duct of width b and depth w. The pressure head difference across the duct, at a horizontal bend of radius R to the centre line, is h. Show that if $R = 1.5\,b$, the discharge is obtained from the expression $Q = 3.545\ w\ b\sqrt{h}$.

9. Water flows round a horizontal 90° bend in a square duct of side length 200 mm, the inner radius being 400 mm. The differential head between the inner and outer sides of the bend is 150 mm of water. Determine the discharge in the duct.

10. The depth of uniform flow of water in a rectangular channel 5 m wide conveying 40 m³/s is 2.5 m. The water flows round a 90° bend of inner radius 10 m. Assuming no energy loss through the bend, determine the depth of water at the inner and outer radii of the bend at the discharge of 150 m³/s.

11. A proposed siphon spillway is of uniform rectangular sections 5 m wide and 3 m deep. The crest radius is 3 m, and the length of the approach to the crest along the centre line is 7 m. The overall discharge coefficient is 0.65; the entry head loss $= 0.2\,V^2/2g$, where V is the mean velocity; and the Darcy friction factor is 0.015. Determine the discharge when the siphon operates under a head of 6 m with the upstream water level 0.5 m above the crest, and determine the pressure heads at the crest and cowl.

12. Water discharges from a tank through a circular orifice 25 mm in diameter in the base. The discharge coefficient under conditions of radial flow towards the orifice in the tank is 0.6. A free vortex forms when water is discharging under a head of 150 mm; at a horizontal distance of 10 mm from the centre line of the orifice, the water surface is 50 mm below the top water level. Determine the discharge through the orifice.

13. A cylindrical vessel 0.61 m in diameter and 0.97 m in depth, open at the top, is rotated about a vertical axis at 105 rev/min. If the vessel was originally full of water, how much water will remain under equilibrium conditions?

Chapter 11
Gradually Varied Unsteady Flow from Reservoirs

11.1 Discharge between reservoirs under varying head

Figure 11.1 shows two reservoirs, of constant area, interconnected by a pipeline through which water transfer occurs under gravity. Reservoir A receives an inflow I_1, while a discharge Q_2 is withdrawn from B; I_1 and Q_2 may be time variant.

In the general case the head h and hence the transfer discharge will vary with time, and the object is to obtain a differential equation describing the rate of variation of head with time, dh/dt. The corresponding rate of change of discharge is considered to be sufficiently small so that the steady-state discharge–head relationship for the pipeline flow may be applied at any instantaneous head; compressibility effects are also neglected. Such unsteady flow situations are therefore sometimes referred to as 'quasi-steady' flow.

Let h be the gross head at any instant, Δh_1 the change in level in A during a small time interval Δt and Δh_2 the change in level in B in time Δt; then

$$\text{change in total head} = \Delta h = \Delta h_1 - \Delta h_2 \qquad [11.1]$$

$$\text{Continuity equation for A:} \quad I - Q_1 = A_1 \frac{\Delta h_1}{t} \qquad [11.2]$$

$$\text{Continuity equation for B:} \quad Q_1 - Q_2 = A_2 \frac{\Delta h_2}{t} \qquad [11.3]$$

Note that Q_1 is the discharge in the pipeline and hence is the inflow rate into B.

From the steady-state head–discharge relationship for the pipeline,

$$h = \left(K_{\mathrm{m}} + \frac{\lambda L}{D} \right) \frac{Q_1^2}{2g A_{\mathrm{p}}^2}$$

where A_{p} is the area of pipe and K_{m} the minor loss coefficient. Whence

$$Q_1 = K h^{1/2} \qquad [11.4]$$

Nalluri & Featherstone's Civil Engineering Hydraulics: Essential Theory with Worked Examples, Sixth Edition. Martin Marriott.
© 2016 John Wiley & Sons, Ltd. Published 2016 by John Wiley & Sons, Ltd.
Companion Website: www.wiley.com/go/Marriott

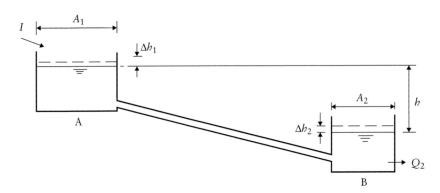

Figure 11.1 Flow through pipe with varying head.

in which

$$K = A_p \sqrt{\frac{2g}{K_m + (\lambda L / D)}}$$

From Equations 11.1, 11.2 and 11.3,

$$\Delta h = \left(\frac{I - Q_1}{A_1} - \frac{Q_1 - Q_2}{A_2} \right) \Delta t$$

Introducing Equation 11.4,

$$\Delta h = \left[\frac{1}{A_1} - Kh^{1/2} \left(\frac{1}{A_1} + \frac{1}{A_2} \right) + \frac{Q_2}{A_2} \right] \Delta t$$

and in the limit $\Delta t \to 0$,

$$dt = \frac{dh}{\left(\frac{I}{A_1} + \frac{Q_2}{A_2} \right) - Kh^{1/2} \left(\frac{1}{A_1} + \frac{1}{A_2} \right)} \qquad [11.5]$$

In a similar dynamic system where the upper reservoir discharges to atmosphere through a pipeline, orifice or valve (Figure 11.2), the term Δh_2 in Equations 11.1 and 11.3 disappears and a similar treatment, or alternatively removing the irrelevant terms A_2 and Q_2 from Equation 11.5, yields

$$dt = \frac{A_1 \, dh}{I - Kh^{1/2}} \qquad [11.6]$$

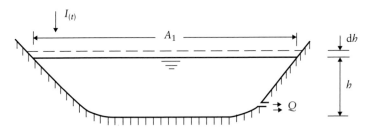

Figure 11.2 Flow through orifice under varying head.

or

$$I - Kh^{1/2} = A_1\frac{dh}{dt} \qquad [11.7]$$

where $Kh^{1/2}$ represents the appropriate steady-state discharge–head relationship for the outlet device.

11.2 Unsteady flow over a spillway

The computation of the time variation in reservoir elevation and spillway discharge during a flood inflow to the reservoir is essential in design of the spillway to ensure safety of the impounding structure (see Figure 11.3).

The continuity equation is

$$I_{(t)} - Q_{(t)} = \frac{dS}{dt} \qquad [11.8]$$

where dS/dt is the rate of change of storage, or volume, S. dS/dt may be expressed as $A(dh/dt)$, where A is the instantaneous plan area of the reservoir and dh/dt the instantaneous rate of change of depth.

Assuming that in the case of a fixed-crest free overfall spillway, the discharge rate Q may be expressed by the steady-state relationship

$$Q = \frac{2}{3}\sqrt{2g}C_dLh^{3/2} \qquad [11.9]$$

where L is the crest length and C_d the discharge coefficient,

$$\text{that is, } Q = Kh^{3/2} \qquad [11.10]$$

Equation 11.7 becomes

$$I_{(t)} - Kh^{3/2} = \frac{dS}{dt} = A\frac{dh}{dt} \qquad [11.11]$$

$I_{(t)}$ is the known time-variant inflow rate. Except in the rather special case where A does not vary with depth and $I_{(t)}$ is constant, Equation 11.11 is not directly integrable and in general must be evaluated numerically.

This is known as **reservoir routing**, also referred to as **level pond routing**, and the numerical solution of the flood routing equation may be carried out by either a semi-graphical or a finite difference approach. The storage is treated simply as a function of the head

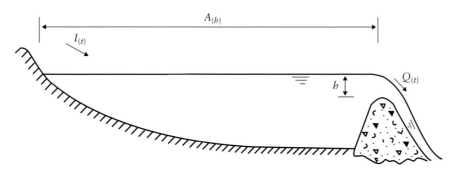

Figure 11.3   Reservoir routing.

Chapter 11

above the outlet crest, assuming negligible backwater effect through the reservoir. If that assumption does not apply, then a channel routing approach is required as described, for example, in Mansell (2003) or Chadwick *et al.* (2013).

11.3 Flow establishment

Figure 11.4 shows a constant-head tank discharging to atmosphere through a pipeline terminated by a control valve. If the valve, which is initially closed, is suddenly opened the discharge will not instantaneously attain its equilibrium value, and the object is to determine the time taken for this to be attained (i.e. the time of steady-flow establishment).

Initially the total head H is available to accelerate the flow but this decreases as the velocity increases due to the associated head losses. At any instant the available head is $H - KV^2$ (where K is the head loss coefficient)

$$= \left(\frac{\lambda L}{D} + K_m \right) \frac{1}{2g}$$

where K_m is the minor loss coefficient.

The equation of motion (force = mass × acceleration)

$$\rho g A_p (H - KV^2) = \rho A_p L \frac{dV}{dt} \qquad [11.12]$$

where A_p is the area of pipe cross section. Whence

$$T = \int_{t_1}^{t_2} dt = \frac{L}{g} \int_{V_1}^{V_2} \frac{dV}{H - KV^2} \qquad [11.13]$$

Assuming λ to be constant in the interval between velocities V_1 and V_2 and writing $a^2 = H$ and $b^2 = K$ (constant), Equation 11.13 becomes

$$T = \frac{L}{g} \int_{V_1}^{V_2} \frac{dV}{a^2 - b^2 V^2}$$

$$= \frac{L}{2ag} \int_{V_1}^{V_2} \left(\frac{1}{a + bV} + \frac{1}{a - bV} \right) dV$$

$$= \frac{L}{2gab} \left[\ln \left(\frac{a + bV}{a - bV} \right) \right]_{V_1}^{V_2}$$

$$= \frac{L}{2g\sqrt{KH}} \left[\ln \left(\frac{\sqrt{H} + \sqrt{KV}}{\sqrt{H} - \sqrt{KV}} \right) \right]_{V_1}^{V_2} \qquad [11.14]$$

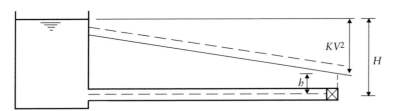

Figure 11.4   Flow establishment through pipeline.

Hence the time for flow establishment is the value of the integral (Equation 11.13) between $V = 0$ and $V = V_0$, where V_0 is the steady-state velocity in the pipeline under the head H. Therefore, since $H = KV_0^2$, Equation 11.14 becomes

$$T = \frac{LV_0}{2gH} \ln \left[\frac{V_0 + V}{V_0 - V} \right]_{V=0}^{V_0} \qquad [11.15]$$

If the variation of λ with discharge is taken into account during the accelerating period, the time of flow establishment must be evaluated by numerical or graphical integration of Equation 11.13.

Worked examples

Example 11.1

A circular orifice 20 mm in diameter with a discharge coefficient $C_d = 0.6$ is fitted to the base of a tank having a constant cross-sectional area of 1.5 m. Determine the time taken for the water level to fall from 3.0 m to 0.5 m above the orifice.

Solution:

Since there is no inflow ($I = 0$), Equation 11.7 becomes

$$-Kh^{1/2} = A_1 \frac{dh}{dt}$$

whence

$$dt = -A_1 \frac{dh}{Kh^{1/2}}$$

Integrating

$$\text{time, } T = \int_{t=0}^{t_1} dt = -\frac{A_1}{K} \int_{h_0}^{h_1} \frac{dh}{h^{1/2}}$$

$$T = -\frac{A_1}{K} \left[2h^{1/2} \right]_{h_0}^{h_1} = \frac{2A_1}{K} \left[h^{1/2} \right]_{h_1}^{h_0}$$

$$K = C_d A_0 \sqrt{2g} = 0.6 \times 3.142 \times 4.43 = 8.35 \times 10^{-4}$$

whence

$$T = \frac{2 \times 1.5}{8.35 \times 10^{-4}} \left[h^{1/2} \right]_{0.5}^{3.0} = 3682.4 \text{ s}$$

Example 11.2

If the tank in Example 11.1 has a variable area expressed by $A_1 = 0.0625(5 + h)^2$, calculate the time for the head to fall from 4 m to 1 m.

Solution:

$$dt = \frac{A_1 \, dh}{Kh^{1/2}}$$

whence

$$T = -\frac{0.0625}{0.000835} \int_{3.0}^{0.5} \frac{(5+h)^2}{h^{1/2}} \, dh$$

$$= 74.85 \int_{0.5}^{3.0} (25h^{-1/2} + 10h^{1/2} + h^{3/2}) \, dh$$

$$= 74.85 \left[50h^{1/2} + \frac{20}{3}h^{3/2} + \frac{2}{5}h^{5/2} \right]_{0.5}^{3.0}$$

$$= 6713.75 \text{ s}$$

Example 11.3

A pipeline 1000 m in length, 100 mm in diameter and with a roughness size of 0.03 mm discharges water to atmosphere from a tank having a cross-sectional area of 1000 m². Find the time taken for the water level to fall from 20 m to 15 m above the pipe outlet.

Solution:

$$\text{Continuity equation:} \quad I - Q = A_1 \frac{dh}{dt} \tag{i}$$

The pipeline discharge Q may be expressed by the Darcy–Colebrook–White equation:

$$Q = -2A_p \sqrt{2gD\frac{h_f}{L}} \log \left[\frac{k}{3.7D} + \frac{2.51\nu}{D\sqrt{2gDh_f/L}} \right] \tag{ii}$$

However, if this were substituted into Equation (i), the resulting equation would need to be evaluated using graphical or numerical integration methods. Such a procedure is fairly straightforward, but if constant value of λ over the range of discharges is adopted, a direct solution is obtained. In this latter case, Q is expressed by

$$Q = \frac{A_p \sqrt{2g}h^{1/2}}{\sqrt{K_m + (\lambda L/D)}} = Kh^{1/2} \quad \text{(Equation 11.4)}$$

Thus Equation (i) reduces to Equation 11.6: $dt = A_1 \, dh/(1 - Kh^{1/2})$.

To evaluate K calculate the pipe velocities at values of $h_f = 20$ m and 15 m using Equation (ii) (i.e. neglecting minor losses), and evaluate λ from the Darcy–Weisbach equation:

$$\lambda = \frac{2gD}{V^2}\frac{h_f}{L}$$

(a)   When $h_f = h = 20$ m, $V = 1.46$ m/s and $\lambda = 0.0184$.
(b)   When $h_f = h = 15$ m, $V = 1.25$ m/s and $\lambda = 0.0188$.

Adopting $\lambda = 0.0186$ and $K_m = 1.5$, $K = 0.00254$. Whence

$$T = \int_{h_1}^{h_2} \frac{A_1 \, dh}{Kh^{1/2}} \quad \text{(since } I = 0\text{)}$$

$$T = \frac{2 \times 1000}{0.00254}(20^{1/2} - 15^{1/2})$$

$$= 471\ 660 \text{ s}$$

$$= 131.02 \text{ h}$$

Example 11.4

If in Example 11.3 a constant inflow of 5 L/s enters the tank, determine the time for the head to fall from 20 m to 18 m.

Solution:

$$dt = \frac{A_1 \, dh}{1 - Kh^{1/2}} \tag{i}$$

$$T = A_1 \int_{h_0}^{h_1} \frac{dh}{I - Kh^{1/2}} \tag{ii}$$

Write $y = Kh^{1/2} - I$, whence

$$h = \frac{1}{K^2}(y^2 + 2Iy + I^2) \tag{iii}$$

$$dh = \frac{1}{K^2}(2y + 2I)dy$$

Equation (ii) becomes

$$T = -2\frac{A_1}{K^2} \int_{y_0}^{y_1} \frac{y + I}{y} \, dy$$

i.e. $T = -2\dfrac{A_1}{K^2} \displaystyle\int_{y_0}^{y_1} \left(1 + \dfrac{I}{y}\right) dy = -2\dfrac{A_1}{K^2}[y + I \ln(y)]_{y_0}^{y_1}$

Substituting for y from Equation (iii),

$$T = 2\frac{A_1}{K^2}\left[Kh^{1/2} - I + I \ln(Kh^{1/2} - I)\right]_{18}^{20}$$

Using $K = 0.00254$ (as in Example 11.3),

$$T = \frac{2 \times 1000}{0.00254^2} \left\{ 0.00254 \left(\sqrt{20} - \sqrt{18} \right) + 0.005 \right.$$
$$\left. \times \left[\ln \left(0.00254 \sqrt{20} - 0.005 \right) - \ln \left(0.0254 \sqrt{18} - 0.005 \right) \right] \right\}$$

$$= 329\,579.8 \text{ s}$$

$$= 91.55 \text{ h}$$

Example 11.5

Reservoir A with a constant surface area of 10 000 m² delivers water to Reservoir B with a constant area of 2500 m² through a 10 000 m long, 200 mm diameter pipeline of roughness 0.06 mm. Minor losses including entry and velocity head total $20V^2/2g$. A steady inflow of 10 L/s enters Reservoir A, and a steady flow of 20 L/s is drawn from Reservoir B. If the initial level difference is 100 m, determine the time taken for this to become 90 m.

Solution:

Refer to Section 11.2 and Figure 11.1.
   From Equation 11.5,

$$dt = \frac{dh}{\left(\frac{I}{A_1} + \frac{O_2}{A_2} \right) - Kh^{1/2} \left(\frac{1}{A_1} + \frac{1}{A_2} \right)} \tag{i}$$

Since in this case I, Q_2, A_1 and A_2 are constant, Equation (i) can be directly integrated. Let

$$W = \left(\frac{I}{A_1} + \frac{Q_2}{A_2} \right) \quad \text{and} \quad Z = K \left(\frac{1}{A_1} + \frac{1}{A_2} \right)$$

Then in Equation (i),

$$dt = \frac{dh}{W - Zh^{1/2}} \quad \text{and} \quad T = \int_{h_1}^{h_2} \frac{dh}{W - Zh^{1/2}} \tag{ii}$$

Using the mathematical technique of Example 11.4, this integral (Equation (ii)) becomes

$$T = \frac{2}{Z^2} \left[Z(h_1^{1/2} - h_2^{1/2}) - W \ln \left(\frac{Zh_1^{1/2} - W}{Zh_2^{1/2} - W} \right) \right] \tag{iii}$$

Now $K = \frac{A_p \sqrt{2g}}{\sqrt{K_m + (\lambda L/D)}}$ and λ is evaluated as in Example 11.4.
   Adopting $\lambda = 0.0173$, $K = 4.678 \times 10^{-3}$, $W = 9.0 \times 10^{-6}$ and $Z = 2.339 \times 10^{-6}$.
Thus from Equation (iii),

$$T = 725\,442 \text{ s}$$
$$= 201.51 \text{ h}$$

Example 11.6

For the system described in Example 11.5, find the time taken for the level in Reservoir A to fall by 1.0 m.

Solution:

Continuity equation for Reservoir A (from Equation 11.1):

$$I - Q_1 = A_1 \frac{dh_1}{dt} \tag{i}$$

Since $Q_1 = Kh^{1/2}$,

$$I - Kh^{1/2} = A_1 \frac{dh_1}{dt} \tag{ii}$$

Thus the rate of change of level in A is related to the instantaneous gross head h. A numerical or graphical integration method must be used to evaluate the time taken for the level in A to change by a specified amount since, in Equation (ii), the head h also varies with time.

Thus

$$\Delta h_1 = \frac{I - Kh^{1/2}}{A_1} \Delta t \tag{iii}$$

By taking a series of values of h, the variation of h with time is evaluated using Equation (iii) in Example 11.5.

Head (m)	Time (s)
99	69 786.6
98	140 148.3
97	211 100.0
96	282 656.0
95	354 827.0

Plotting head versus time (Figure 11.5a), values of h at discrete time intervals Δt, say 100 000 s, are obtained. Equation (iii) is then evaluated for Δh_1 in the time interval Δt.

The following values are obtained from Figure 11.5a and Equation (iii). $\bar{h}$ is the average head in the time interval. The negative sign for Δh_1 indicates a falling level in A.

Time (s $\times 10^5$)	0	1	2	3
h (m)	99	98.5	97.2	95.0
$h^{1/2}$ (m)		9.94	9.89	9.89
Δh_1 (m)		−0.355	−0.353	−0.349
$\sum \Delta h_1$ (m)		−0.355	−0.708	−1.057

From the graph of $\sum \Delta h_1$ versus t (Figure 11.5b), $t = 2.95 \times 10^5$ s when $\sum \Delta h_1 = -1.0$ m.

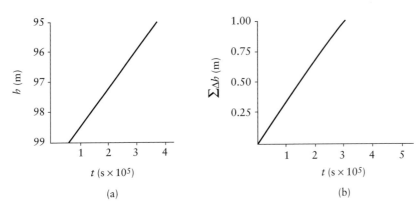

Figure 11.5   Variation of reservoir levels with time.

Example 11.7

An impounding reservoir is to be partially emptied from the level of the spillway crest through three 1.0 m diameter valves with $C_d = 0.95$, set at the same level with their axes 20 m below the spillway crest. The variation of reservoir surface area (A_1) with height (h) above the valve centre line is tabulated (see table below).

Assuming that a continuous inflow of 0.5 m³/s enters the reservoir, determine the time required to lower the level from the spillway crest to 1.0 m above the valves.

h (m)	0	5	10	15	20
A_1 (m² × 10⁶)	2.5	6.0	11.5	15.0	17.0

Solution:

$$I - Q = A_1 \frac{dh}{dt} \tag{i}$$

$$Q \text{ (the outflow)} = 3 \times C_d A_0 \sqrt{2g} h^{1/2} = K h^{1/2}$$

where $K = 3 \times 0.95 \times 0.7854\sqrt{2g} = 9.915$.

From Equation (i),

$$T = \int_{t_0}^{t_1} dt = \int_{h_0}^{h_1} \frac{A_1 \, dh}{I - k h^{1/2}} \tag{ii}$$

Since A_1 is not readily expressible as a continuous function of h, Equation (ii) is not directly integrable, but may be evaluated graphically or numerically.

Writing

$$X(h) = \frac{A_1}{I - K h^{1/2}}, \qquad T = \int_{h_0}^{h_1} X(h) \, dh$$

$X(h)$ is evaluated for a number of discrete values of h and the integral

$$\int_{h_0}^{h_1} X(h) \, dh$$

obtained from the area under the $X(h)$ versus h curve.

h (m)	20	15	10	5	1
$A_1(h)$ (m$^2 \times 10^6$)	17.0	15.0	11.5	6.0	3.0
$X(h)$ (s/m $\times 10^6$)	−0.38776	−0.39571	−0.37272	−0.27687	−0.318640

The negative sign for $X(h)$ indicates that h is decreasing with time. Plotting $X(h)$ versus A between $h = 20.0$ and $h = 1.0$ yields

$$T = 6.61 \times 10^6 \text{ s}$$
$$= 1836 \text{ h}$$

Example 11.8

Flood water discharges from an impounding reservoir over a fixed-crest spillway 100 m long; $C_d = 0.7$. The variation of surface area with head above the crest (h) and the inflow hydrograph are shown in the next three tables.

Calculate the outflow hydrograph and state the maximum water level and peak outflow, assuming that an outflow of 20 m^3/s exists at $t = 0$ h.

Solution:

See Figure 11.3.

The continuity equation (Equation 11.8) is

$$I_{(t)} - Q_{(t)} = \frac{dS}{dt} \tag{i}$$

where S is the volume of storage.

$$Q_{(t)} = \text{Outflow rate over spillway crest}$$
$$= \frac{2}{3}\sqrt{2g}C_d L h^{3/2}$$
$$= K h^{3/2} \tag{ii}$$

Thus Equation (i) becomes

$$I_{(t)} - Kh^{3/2} = \frac{dS}{dt} = \frac{A\,dh}{dt} \tag{iii}$$

Since $I_{(t)}$ is not a function of h and S is not a readily expressible function of h, Equation (iii) is best evaluated using a numerical method. Taking small time intervals Δt, Equation (iii) may be written as

$$\bar{I} - K\bar{h}^{3/2} = \bar{A}\,\frac{\Delta h}{\Delta t} \tag{iv}$$

where

$$\bar{I} = \frac{I_1 + I_2}{2}; \qquad \bar{h}^{3/2} = \frac{(h_1 + \Delta h)^{3/2} + h_1^{3/2}}{2}$$

$$\bar{A} = \frac{A_1 + A_2}{2}; \qquad A_1 = A_{(h_1)}; \qquad A_2 = A_{(h_1 + \Delta h)}$$

Chapter 11

Subscripts 1 and 2 indicate values at the beginning and end of each time interval, respectively.

Δh is estimated and adjusted until Equation (iv) is satisfied and the computation proceeds to the next time interval. A curve, or numerical relationship, relating A with h is required. This procedure is best carried out on a digital computer.

A simpler, explicit method is obtained by writing Equation (i) in finite difference form taking discrete time intervals Δt.

$$\frac{I_1 + I_2}{2} - \frac{(Q_1 + Q_2)}{2} = \frac{S_2 - S_1}{\Delta t} \tag{v}$$

$$\text{i.e. } I_1 + I_2 + \frac{2S_1}{\Delta t} - Q_1 = \frac{2S_2}{\Delta t} + Q_2 \tag{vi}$$

At each time step the values of S_1 and Q_1 are known; hence, the value of the left-hand side (LHS) is known and hence S_2 and Q_2 can be found from curves relating $\left(\frac{2S}{\Delta t} + Q\right)$ versus h and Q versus h.

Taking $S = \sum\left(A_{(h)} \times \Delta h\right)$ and $\Delta t = 1\ h = 3600\ s$.

Table of calculations of outflow rate and head

t (h)	I (m³/s)	$\dfrac{2S}{\Delta t} - Q$	Value of LHS (vi)	Q (m³/s)	h (m)
0	20	1160		20	0.21
1	40	1180	1220	20	0.21
2	70	1242	1290	24	0.23
3	100	1360	1412	26	0.25
4	128	1528	1588	30	0.27
5	150	1730	1806	38	0.32
6	155	1949	2035	43	0.35
7	140	2144	2244	50	0.39
8	112	2286	2396	55	0.41
9	73	2359	2471	56	0.42
10	46		2478	56	0.42

Note: At $t = 0$, $Q = 20$ m³/s, and hence $h = 0.21$ m, and $\frac{2S}{\Delta t} + Q = 1200$; $\frac{2S}{\Delta t} - Q = \frac{2S}{\Delta t} + Q - (2Q)$; hence, Column 3 is completed.

Variation of surface area

h (m)	0	0.1	0.2	0.3	0.4	0.5	0.6	0.7	0.8	0.9	1.0
A (m² × 10⁶)	10.00	10.10	10.20	10.34	10.46	10.60	10.75	10.92	11.10	11.30	11.50

Inflow hydrograph

t (h)	0	1	2	3	4	5	6	7	8	9	10
I (m³/s)	20	40	70	100	128	150	155	140	111	73	46

Table of calculations of $\left(\frac{2S}{\Delta t} + Q\right)$ *versus* h

$h(m)$	0.1	0.2	0.3	0.4	0.5	0.6	0.7	0.8	0.9	1.0
$Q = K(h)^{3/2}$	6.54	18.49	33.97	52.29	73.08	96.07	121.06	147.91	176.49	206.71
$A\,(h)\,(m^2\times 10^6)$	10.10	10.20	10.34	10.46	10.60	10.75	10.92	11.10	11.30	11.50
$\frac{2S}{\Delta t} + Q$ (m^3/s)	567	1146	1736	2335	2945	3565	4197	4840	5497	6166

At $t = 1$ h,

$$I_1 + I_2 + \frac{2S}{\Delta t} - Q = 20 + 40 + 1160 = 1220$$

$$\frac{2S}{\Delta t} + Q = 1220, \quad \text{whence } h = 0.21 \text{ m and } Q = 20 \text{ m}^3/\text{s}$$

Peak outflow $= 58$ m^3/s at $t = 9.5$ h; $\quad h_{max} = 0.42$ m

Plot the inflow and outflow hydrographs (see Figure 11.6), and note that Q_{max} coincides with the falling limb of the inflow hydrograph.

Example 11.9

A pipeline 5000 m long, 300 mm in diameter and with roughness size 0.03 mm discharges water from a reservoir to atmosphere through a terminal valve. The difference in level between the reservoir and the valve is 20 m, which may be assumed constant. If the valve, which is initially closed, is suddenly opened, determine the time for steady flow to become established, neglecting compressibility effects. Assume minor losses $= 5V^2/2g$ including velocity head and entry loss.

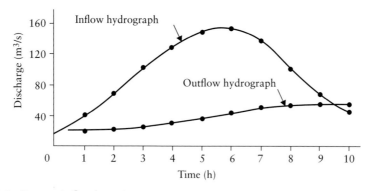

Figure 11.6   Reservoir flood routing.

Chapter 11

Solution:

From Equation 11.15 and Section 11.3, the theoretical time for flow establishment is

$$T = \frac{LV_0}{2gH} \ln \left[\frac{V_0 + V}{V_0 - V} \right]^{V_0}_{V=0} \tag{i}$$

Using the techniques of Chapter 4 the steady-state velocity V_0 under the head of 20 m is calculated to be 1.23 m/s. Theoretically, the time taken to attain steady flow is infinite (substituting $V = V_0$ in Equation (i)); therefore, adopt $V_0 = 0.99 \times 1.23 = 1.2$ m/s (say) whence in Equation (i),

$$T = \frac{5000 \times 1.23}{19.62 \times 20} \ln \left[\frac{2.43}{0.03} \right] = 68.9 \text{ s}$$

The above solution assumes a constant value of λ. The reader should evaluate T, taking account of the variation of λ with velocity, using a graphical or numerical integration of Equation 11.13.

References and recommended reading

Chadwick, A., Morfett, J. and Borthwick, M. (2013) *Hydraulics in Civil and Environmental Engineering*, 5th edn, Spon, Abingdon, UK.
Mansell, M. G. (2003) *Rural and Urban Hydrology*, Thomas Telford, London.

Problems

1. A reservoir with a constant plan area of 20 000 m² discharges water to atmosphere through a 2000 m long pipeline of 300 mm diameter and roughness 0.15 mm. The reservoir receives a steady inflow of 20 L/s. If the head between the reservoir surface and the pipe outlet is initially 40 m, determine (neglecting minor losses) the time taken for the head to fall to 35 m.

2. An impounding reservoir of constant area 100 000 m² discharges to a service reservoir through 20 km of 400 mm diameter pipeline of roughness 0.06 mm. Minor losses amount to $20V^2/2g$. The impounding reservoir of constant plan area 10 000 m² receives a steady inflow of 30 L/s, while a steady outflow of 10 L/s takes place from the service reservoir. If the initial level difference is 50 m, determine the time taken for the head to become 48 m and the time for the level in the upper reservoir to fall by 0.5 m.

3. An impounding reservoir delivers water to a hydroelectric plant through four pipelines, each 2.0 m in diameter and 1000 m in length with a roughness of 0.3 mm. The reservoir is to be drawn down using the four pipelines with bypasses around the turbines to discharge into the tailrace which has a constant level. Allowing $5V^2/2g$ for local losses including velocity head, entry and bypass losses, determine the time taken for the level in the reservoir to fall from 50 m above the tailrace to 20 m above the tailrace assuming a constant inflow of 1.0 m³/s. Reservoir surface area data are given in the table below.

Level above tailrace (m)	20	30	40	50
Surface area of reservoir (m² × 10⁶)	2.0	4.0	6.8	12.2

4. Using the data given for the reservoir in Example 11.8, obtain the outflow hydrograph resulting from the following inflow hydrograph if the spillway is 50 m long with $C_d = 0.7$. Assume the outflow rate at $t = 0$ to be 10 m³/s.

Time (h)	0	1	2	3	4	5	6	7	8	9	10	11	12
Inflow (m³/s)	10	30	80	130	216	250	228	176	120	80	52	44	20

Chapter 12
Mass Oscillations and Pressure Transients in Pipelines

12.1 Mass oscillation in pipe systems – surge chamber operation

When compressibility effects are not significant the unsteady flow in pipelines is called surge. A typical example of surge occurs in the operation of a medium- to high-head hydro-electric scheme (Figure 12.1).

If, while running under steady power conditions, the turbine is required to be closed down, values in the inlet to the turbine runner passages will be closed slowly. This will result in pressure transients, which involve compressibility effects, occurring in the penstocks between the turbine inlet valve and the surge chamber. (For details of pressure transients, see Section 12.5 onwards.) The pressure transients do not proceed beyond the surge chamber and hence high pressures in the tunnel are prevented, resulting in a reduced cost of construction.

Due to the presence of the surge chamber the momentum of the water in the tunnel is not destroyed quickly and water continues to flow, passing into the surge chamber at a level which stops rising when the pressure in the tunnel at the surge chamber inlet is balanced by the pressure created by the head in the chamber.

At this time the level in the chamber will be higher than that in the reservoir and reversed flow will occur, setting up a long-period oscillation between the two. Figure 12.2 shows a typical time variation of level in a surge chamber, the oscillations being eventually damped out by friction in the tunnel, and losses at the inlet to the chamber and in the chamber itself. Since there are many different types of surge chambers and types of inlets, the reader is referred to the recommended reading for such details.

The governing equations describing the mass oscillations in the reservoir–tunnel–surge chamber system are as follows:

(a) *The dynamic equation*:

$$\frac{L}{g}\frac{dV}{dt} + z + F_s V_s |V_s| + F_T V |V| = 0 \qquad [12.1]$$

Nalluri & Featherstone's Civil Engineering Hydraulics: Essential Theory with Worked Examples, Sixth Edition. Martin Marriott.
© 2016 John Wiley & Sons, Ltd. Published 2016 by John Wiley & Sons, Ltd.
Companion Website: www.wiley.com/go/Marriott

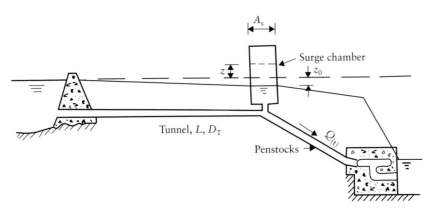

Figure 12.1   Medium-head hydropower scheme.

(b) *The continuity equation*:

$$VA_T = A_s \frac{dz}{dt} + Q \qquad [12.2]$$

where L is the length of tunnel, z elevation of water in the surge chamber above that in the reservoir, F_s head loss coefficient at the surge chamber inlet (throttle), V_s velocity in the surge chamber (=dz/dt), F_T head loss coefficient for tunnel friction head loss (=$\lambda L/2gD_T$), D_T diameter of the tunnel, V velocity in the tunnel, A_T area of cross section of the tunnel, A_s area of cross section of the surge chamber and Q discharge to turbines.

Equations 12.1 and 12.2 can be integrated directly only for cases of sudden load rejection; tunnel friction and throttle losses may be included.

12.2 Solution neglecting tunnel friction and throttle losses for sudden discharge stoppage

F_s, F_T and $Q = 0$ and Equations 12.1 and 12.2 become

$$\frac{L}{g} \frac{dV}{dt} + z = 0 \qquad [12.3]$$

$$VA_T = A_s \frac{dz}{dt} \qquad [12.4]$$

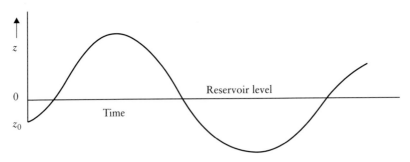

Figure 12.2   Surge chamber oscillations.

Differentiating Equation 12.4,

$$\frac{dV}{dt} = \frac{A_s}{A_T} \frac{d^2z}{dt^2} \qquad [12.5]$$

$$\text{Hence } \frac{L}{g} \frac{A_s}{A_T} \frac{d^2z}{dt^2} + z = 0 \qquad [12.6]$$

This is a linear, homogeneous, second-order differential equation with constant coefficient, the solution to which is

$$z = C_1 \cos \frac{2\pi t}{T} + C_2 \sin \frac{2\pi t}{T} \qquad [12.7]$$

where T is the period of oscillation. Since the tunnel is assumed to be frictionless, $z = 0$ at $t = 0$.

$$\text{Hence } z = C_2 \sin \frac{2\pi t}{T} \qquad [12.8]$$

$$T = 2\pi \sqrt{\frac{L}{g} \frac{A_s}{A_T}} \qquad [12.9]$$

$$\frac{dz}{dt} = C_2 \frac{2\pi}{T} \cos \frac{2\pi t}{T} \quad \text{(from Equation 12.8)} \qquad [12.10]$$

$$\text{and } \frac{dz}{dt} = V \frac{A_T}{A_s} \quad \text{(from Equation 12.4)} \qquad [12.11]$$

$$\text{Hence } V = \frac{A_s}{A_T} C_2 \frac{2\pi}{T} \cos \frac{2\pi t}{T}$$

$$\text{when } t = 0, V = V_0 \quad \text{and} \quad V_0 = \frac{A_s}{A_T} C_2 \frac{2\pi}{T} \qquad [12.12]$$

Substituting for T from Equation 12.9,

$$C_2 = V_0 \sqrt{\frac{L}{g} \frac{A_T}{A_s}} \qquad [12.13]$$

whence (from Equation 12.8),

$$z = V_0 \sqrt{\frac{L}{g} \frac{A_T}{A_s}} \sin \frac{2\pi t}{T} \qquad [12.14]$$

12.3 Solution including tunnel and surge chamber losses for sudden discharge stoppage

Since $Q = 0$, $V_s A_s = V_T A_T$ whence Equation 12.1 becomes

$$\frac{L}{g} \frac{dV}{dt} + z + \left[F_s \left(\frac{A_T}{A_s} \right)^2 + F_T \right] V|V| = 0$$

$$\text{or } \frac{L}{g} \frac{dV}{dt} + z + F_R V|V| = 0 \qquad [12.15]$$

From Equations 12.4, 12.5 and 12.15, the following relationships are obtainable:

$$V^2 = -\frac{z}{F_R} + \frac{LA_T}{2gA_sF_R^2} + C\exp\left(\frac{-2gA_sF_Rz}{LA_T}\right) \qquad [12.16]$$

when $t = 0$, $V = V_0$ and $z_0 = F_TV_0^2$ whence we obtain

$$\frac{V^2 + (z/F_R) - (LA_T/2gA_sF_R^2)}{V_0^2 + (F_TV_0^2/F_R) - (LA_T/2gA_sF_R^2)} = \exp\left(\frac{-2gA_sF_R(z + F_TV_0^2)}{LA_T}\right) \qquad [12.17]$$

z_{max} occurs when $V = 0$.

Note that since $dz/dt = V_s$, $dt = dz/V_s$, and the time corresponding with any value of z is

$$T_z = \int_0^t dt = \int_0^z \frac{dz}{V_s} = \int_0^z \frac{dz}{VA_T/A_s} \qquad [12.18]$$

Equation 12.18 can be evaluated by using a graphical integration or numerical integration method, in the latter case taking small intervals of z.

12.4 Finite difference methods in the solution of the surge chamber equations

Numerical methods of analysis using digital computers provide solutions to a wide range of operating conditions, and types and shapes of surge chambers.

Considering the general case of a surge chamber with a variable area and taking a finite interval Δt during which V changes by ΔV and z changes by Δz, Equations 12.1 and 12.2 become

$$\frac{L}{g}\frac{\Delta V}{\Delta t} + z_m + F_TV_m|V_m| + F_sV_s|V_s| = 0 \qquad [12.19]$$

$$V_mA_T = A_{s,m}\frac{\Delta z}{\Delta t} + Q_m \qquad [12.20]$$

where subscript m indicates the average value in the interval and $A_{s,m}$ is the average area of the surge chamber between z and $z + \Delta z$.

(a) Solution by successive estimates. In each time interval, estimate ΔV. Then, $V_m = V_i + (\Delta V/2)$, and from Equation 12.19 calculate z_m ($=z_i + (\Delta z/2)$) whence z is calculated, noting that $V_s = \Delta z/\Delta t$. Subscript i indicates values at the beginning of the time interval and which are therefore known. Q_m is known since the time variation of discharge to the turbines will be prescribed and substitution of Δz into Equation 12.20 yields V_m. If the two values of V_m agree, the estimated value of ΔV is correct; otherwise adjust ΔV and repeat until agreement is achieved and proceed to the next time interval.

Alternatively, estimate Δz and proceed in similar fashion; this is preferable if the chamber has a variable area. In both cases the time variation of z is obtained.

Such calculations are ideally carried out on a computer spreadsheet.

(b) Direct solution of Equations 12.19 and 12.20. From Equation 12.20,

$$\Delta z = \frac{\Delta t}{A_{s,m}} \left(V_i A_T + \frac{A_T}{2} \Delta V - Q_m \right)$$ [12.21]

$$\text{where } V_m = V_i + \frac{\Delta V}{2}$$

Also $z_m = z_i + (\Delta z / 2)$ and $V_s = \Delta z / \Delta t$; Equation 12.19 becomes

$$\frac{L}{g} \frac{\Delta V}{\Delta t} + z_i + \frac{\Delta t}{2 A_s} \left(V_i A_T + \frac{A_T}{2} \Delta V - Q_m \right) \pm F_T \left(V_i + \frac{\Delta V}{2} \right)^2$$

$$\pm \frac{F_s}{A_s^2} \left[A_T^2 \left(V_i^2 + V_i \Delta V + \frac{\Delta V^2}{4} \right) - 2 A_T \left(V_i + \frac{\Delta V}{2} \right) Q_m + Q_m^2 \right] = 0$$

Rearranging

$$\pm \frac{F_R}{4} \Delta V^2 + \left[\frac{L}{g \Delta t} + \frac{A_T}{4 A_{s,m}} \Delta t \pm \left(F_R V_i - \frac{F_s A_T Q_m}{A_s^2} \right) \right] \Delta V + z_i$$

$$+ \frac{A_T}{2 A_{s,m}} V_i \Delta t - \frac{Q_m}{2 A_{s,m}} \Delta t \pm \left[F_R V_i^2 + \frac{F_s}{A_s^2} Q_m \left(- 2 V_i A_T + Q_m \right) \right] = 0 \quad [12.22]$$

which is of the form

$$a \Delta V^2 + b \Delta V + c = 0$$ [12.23]

$$\text{whence } \Delta V = \frac{-b + \sqrt{b^2 - 4ac}}{2a}$$ [12.24]

ΔV is therefore determined explicitly in each successive time step Δt, and the corresponding change in z is obtained from Equation 12.21. Note that if V becomes negative (i.e. on the downswing) the negative value of F_R is used. As with most finite difference methods, in this case Δt should be small since the use of average values of the variables implies a linear time variation. A 10 s time interval usually gives a sufficiently accurate solution.

12.5 Pressure transients in pipelines (waterhammer)

Changes in the discharge in pipelines, caused by valve or pump operation, either closure or opening, result in pressure surges which are propagated along the pipeline from the source. If the changes in control are gradual, the time variation of pressures and discharge may be achieved by assuming the liquid to be incompressible and neglecting the elastic properties of the pipeline such as in the problems on surge analysis dealt with in Sections 12.1–12.4.

In the case of rapid valve closure or pump stoppage, the resulting deceleration of the liquid column causes pressure surges having large pressure differences across the wave front. The speed (celerity) of the pressure wave is dependent on the compressibility of the liquid and the elasticity of the pipeline, and these parameters are therefore incorporated in the analysis.

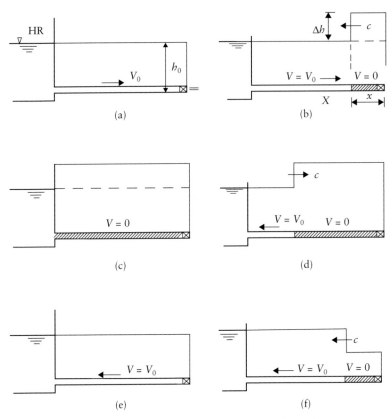

Figure 12.3   Pressure transients in uniform pipeline due to sudden valve closure.

The simplest case of waterhammer, that due to an instantaneous valve closure, can be used to illustrate the phenomenon (Figure 12.3).

At time Δt after closure, the pressure wave has reached a point $x = c\Delta t$, where c is the celerity of the wave. In front of the wave the velocity is V_0, and behind it the water has come to rest. The pressure within the region 0–x will have increased significantly, and the pipe diameter will have increased due to the increased stress. The density of the liquid will increase due to its compressibility. Note that it takes a time $t = L/c$ for the whole column to come to rest. At this time the wave has reached the reservoir where the energy is fixed at HR (Figure 12.3c). Thus the increased stored strain energy in the pipeline cannot be sustained and in its release water is forced to flow back into the reservoir in the direction of the pressure gradient. The wave front retreats to the valve at celerity c (Figure 12.3d) and arrives at $t = 2L/c (=T)$ (Figure 12.3e). Due to the subsequent arrest of the retreating column, a reduced pressure wave is propagated to the reservoir and the whole sequence repeated. In practice, friction eventually damps out the oscillations.

It will be shown later that in the case of an instantaneous stoppage the pressure head rise $\Delta h = cV_0/g$, where V_0 is the initial steady velocity. c may be of the order of 1300 m/s for steel or iron pipelines. For example, if $V_0 = 2$ m/s, $\Delta h = (1300 \times 2)/9.81 = 265$ m, thus giving some idea of the potentially damaging effects of waterhammer.

12.6 The basic differential equations of waterhammer

The continuity and dynamic equations applied to the element of flow δx (Figure 12.4) yield

$$\text{Continuity equation: } \frac{\partial h}{\partial t} + V\frac{\partial h}{\partial x} + \frac{c^2}{g}\frac{\partial V}{\partial x} = 0 \qquad [12.25]$$

$$\text{Dynamic equation: } \frac{\partial h}{\partial x} + \frac{V}{g}\frac{\partial V}{\partial x} + \frac{1}{g}\frac{\partial V}{\partial t} + \frac{\lambda}{2Dg}V|V| = 0 \qquad [12.26]$$

c is the speed of propagation of the pressure wave given by

$$c = \sqrt{\frac{1}{\rho\left(\frac{1}{K} + \frac{C_1 D}{TE}\right)}} \qquad [12.27]$$

where K is the bulk modulus of liquid, ρ density of liquid, E elastic modules of pipe material, T pipe wall thickness, D pipe diameter, and C_1 constant, depending on the method of pipeline anchoring.

For a thin-walled pipe fixed at the upper end, containing no expansion joints but free to move in the longitudinal direction, $C_1 = (5/4) - \eta$, where η is Poisson's ratio for the pipe wall material. For steel and iron, $\eta = 0.3$. $C_1 = 1 - \eta^2$ for a pipe without expansion joints and anchored throughout its length.

$C_1 = 1 - \eta/2$ for a pipe with expansion joints throughout its length. If longitudinal effects are ignored, C_1 is taken to equal 1.

Equation 12.27 can be expressed in the form

$$c = \sqrt{\frac{K^*}{\rho}}$$

where K^* is the effective bulk modulus of the fluid in the flexible pipeline. Since the speed of propagation of a pressure wave (or speed of sound) in an infinite fluid or in a rigid pipeline is $c = \sqrt{K/\rho}$, the effect of the term $C_1 D/TE$ is to reduce the speed of propagation.

$$\text{For water, } K = 2.1 \times 10^9 \text{ N/m}^2$$
$$\text{For steel, } E = 2.1 \times 10^{11} \text{ N/m}^2$$

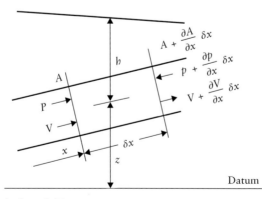

Figure 12.4   Unsteady flow field.

12.7 Solutions of the waterhammer equations

Equations 12.25 and 12.26 can be solved analytically only if certain simplifying assumptions are made, such as the neglect of certain terms, and for simple boundary conditions such as reservoirs and valves. Some methods neglect friction losses, which can lead to serious errors in the calculation of pressure transients.

Since the advent of the digital computer, Equations 12.25 and 12.26 when expressed in discretised form can be readily evaluated for a whole range of boundary conditions and pipe configurations including networks. The computations proceed in small time steps, and the pipeline is 'divided' into equal sections. At every 'station' between adjacent sections, the transient pressure head and velocity are calculated at each time step. In this way the complete history of the waterhammer in space and time is revealed. No simplifying assumptions to the basic equations need be made. Indeed, if interior boundary conditions such as low pressures causing cavitation arise, especially in sloping pipelines, these can be incorporated in the analysis.

The most commonly used numerical method is probably the method of 'characteristics' which reduces the partial differential equations to a pair of simultaneous ordinary differential equations. These when expressed in numerical (finite difference) form can be programmed for automatic evaluation on a digital computer. Such methods are, however, beyond the scope of this text.

In the present day, therefore, it hardly seems justifiable to use simplified methods. However, in the following sections some of these methods will be illustrated. These were developed before the advent of computers and represent examples of classic analytical techniques. In some cases friction can be included, and the results, for example for the pressure transients at closing valves, are very similar to those obtained using the 'characteristics' method. Friction losses are often simplified by assuming them to be localised, for example at an 'orifice' at the outlet from a reservoir. Such losses may also be discretely distributed along the pipeline but this makes the analysis more laborious.

The analytical methods which are included hereafter illustrate some aspects of the waterhammer phenomenon. The inclusion of additional features has been kept to a minimum in view of the superiority of numerical analysis for practical problems.

12.8 The Allievi equations

The differential Equations 12.25 and 12.26 cannot be solved analytically unless certain simplifications are carried out.

The term $V\partial h/\partial x$ is of the order of $[V/(V + c)]/(\partial h/\partial t)$ and may therefore be small. $\partial V/\partial x$ is small compared with $\partial V/\partial t$ since

$$\frac{\mathrm{d}V}{\mathrm{d}t} = -\frac{V}{\partial x}\frac{\partial V}{\partial x} + \frac{\partial V}{\partial t}$$
$$= \frac{\partial V}{\partial t}\left(1 - \frac{V}{\partial x}\frac{\partial V}{\partial V}\frac{\partial t}{\partial V}\right)$$
$$= \frac{\partial V}{\partial t}\left(1 - \frac{V}{\partial x/\partial t}\right)$$

in which the last term is small.

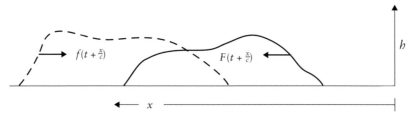

Figure 12.5 Propagation of pressure wave.

Neglecting the friction loss term in addition yields

$$\frac{\partial V}{\partial x} = -\frac{g}{c^2}\frac{\partial h}{\partial t} \quad \text{(continuity equation)} \qquad [12.28]$$

$$\frac{\partial V}{\partial t} = -g\frac{\partial h}{\partial x} \quad \text{(dynamic equation)} \qquad [12.29]$$

Riemann obtained a solution to such simultaneous differential equations which may be expressed in the form

$$h_{tx} = h_0 + F\left(t - \frac{x}{c}\right) + f\left(t + \frac{x}{c}\right) \qquad [12.30]$$

$$V_{tx} = V_0 - \frac{g}{c}\left[F\left(t - \frac{x}{c}\right) - f\left(t + \frac{x}{c}\right)\right] \qquad [12.31]$$

where F and f mean 'a function of' and $+x$ is measured as in Figure 12.5 so that the signs in Equations 12.28 and 12.29 become positive. The functions F and f have the dimension of head.

An observer travelling along the pipeline in the $+x$ direction will be at a position $X_1 = ct + X_0$ at time t, where X_0 is his position at $t = 0$.

For the observer, the function

$$F\left(t - \frac{x}{c}\right) = F\left(t - \frac{ct + X_0}{c}\right) = F\left(\frac{X_0}{c}\right) = \text{Constant}$$

Thus the function $F(t - x/c)$ is a pressure (head) wave which is propagated upstream at the wave speed c.

By similar argument the function $f(t + x/c)$ is a pressure wave propagated in the $-x$ direction at the wave speed c (Figure 12.5).

An $F(t + x/c)$ wave generated, for example, by valve operation at $x = 0$ will propagate upstream towards the reservoir at which it will be completely and negatively reflected as an f wave at time $T/2 = (L/c)$, where T is the waterhammer period $(2L/c)$.

Denoting i as the discrete time period with interval T, Equations 12.30 and 12.31 can be written for the downstream control, for example valve.

$$h_i = h_0 + F_i + f_i \qquad [12.32]$$

$$V_i = V_0 - \frac{g}{c}(F_i - f_i) \qquad [12.33]$$

At $i = 0, t = 0$, then $h_0 = h_0 + F_0 + f_0$

$$V_0 = V_0 - \frac{g}{c}(F_0 - f_0)$$

$$\text{At } i = 1, t = T, \text{ then } h_1 = h_0 + F_1 + f_1$$
$$V_1 = V_0 - \frac{g}{c}(F_1 - f_1)$$

and so on.

Since f is the reflected F wave, $f_0 = 0, f_1 = -F_0, f_2 = -F_1$ and so on. Thus,

$$h_1 = h_0 + F_1 - F_0$$
$$V_1 = V_0 - \frac{g}{c}(F_1 + F_0)$$

and

$$h_2 = h_0 + F_2 - F_1$$
$$V_2 = V_0 - \frac{g}{c}(F_2 + F_1)$$

Adding successive pairs of head equations and subtracting successive pairs of velocity equations,

$$h_1 + h_0 = 2h_0 + F_1$$
$$h_2 + h_1 = 2h_0 + F_2 - F_0$$
$$V_0 - V_1 = \frac{g}{c}(F_1 + F_0)$$
$$V_1 - V_2 = -\frac{g}{c}(F_0 - F_2)$$

whence, in general,

$$F_i - F_{i-2} = \frac{c}{g}(V_{i-1} - V_i)$$
$$h_i + h_{i-1} - 2h_0 = \frac{c}{g}(V_{i-1} - V_i) \qquad [12.34]$$

Boundary conditions may be set at the valve at the downstream end.

The discharge through a valve of area $A_{v,i}$ at time i when the pressure head behind it is h_i is given by

$$Q = C_{d,i} A_{v,i} \sqrt{2gh_i} \qquad [12.35]$$

and the velocity in the pipe immediately upstream is

$$V_i = \frac{C_{d,i} A_{v,i}}{A_p} \sqrt{2gh_i} \qquad [12.36]$$

where A_p is the area of the pipe and $C_{d,i}$ the discharge coefficient.

The ratio

$$\frac{V_i}{V_0} = \frac{C_{d,i} A_{v,i}}{C_{d,0} A_{v,0}} \frac{\sqrt{h_i}}{\sqrt{h_0}}$$

Denoting

$$\eta_i = \frac{C_{d,i} A_{v,i}}{C_{d,0} A_{v,0}} \quad \text{and} \quad \xi_i^2 = \frac{h_i}{h_0}$$
$$V_i = V_0 \eta_i \xi_i \qquad [12.37]$$
$$\text{and } h_i = h_0 \xi_i^2 \qquad [12.38]$$

Substituting into Equation 12.34 yields

$$h_0\xi_i^2 + h_0\xi_{i-1}^2 - 2h_0 = \frac{cV_0}{g}(\eta_{i-1}\xi_{i-1} - \eta_i\xi_i)$$ [12.39]

and denoting $cV_0/2gh_0$ by the symbol ρ (not fluid density), [12.39a]

$$\xi_i^2 + \xi_{i-1}^2 - 2 = 2\rho(\eta_{i-1}\xi_{i-1} - \eta_i\xi_i)$$ [12.40]

Equation 12.40 represents a series of equations which enable the heads, at time intervals (T) $i = 1, 2, 3, \ldots$, at the valve to be calculated for a prescribed closure pattern ($A_{v,i}$ vs. t). The equations are known as the **Allievi interlocking equations**.

If the valve closure is instantaneous, $\eta_{1=0}$ in Equation 12.40 and

$$\xi_i^2 + \xi_0^2 - 2 = 2\rho(\eta_0\xi_0 - 0)$$

Now $\eta_0 = 1$; $\xi_0 = 1$. Whence

$$\xi_1^2 - 1 = 2\rho = \frac{cV_0}{gh_0}$$

$$\Rightarrow h_0\left(\frac{h_1}{h_0} - 1\right) = \frac{cV_0}{g}$$

$$\text{or } \Delta h = h_1 - h_0 = \frac{cV_0}{g}$$

Note therefore that if the valve is closed in any time $t < 2L/c$, the result is the same as that for instantaneous closure since $\eta_1 = 0$.

12.9 Alternative formulation

For a downstream valve boundary condition at discrete time intervals T, from Equation 12.36

$$V_i = \frac{C_{d,i}A_{v,i}}{A_p}\sqrt{2gh_i}$$

$$\text{Then } V_i = B_i\sqrt{h_i}; \qquad h_i = \frac{V_i^2}{B_i^2}$$

$$\text{where } B_i = \frac{C_{d,i}A_{v,i}\sqrt{2g}}{A_p}$$

Substituting in Equation 12.32 and adding Equation 12.33 yield

$$\frac{V_i^2}{B_i^2} + \frac{cV_i}{g} - h_0 - 2f - \frac{cV_0}{g} = 0$$

$$\text{whence } V_i = -\frac{B_i^2c}{2g} + B_i\sqrt{\left(\frac{B_ic}{2g}\right)^2 + \frac{cV_0}{g} + h_0 + 2f}$$ [12.41]

Note that $f_i = -F(i-1)$ (see Section 12.8)

$$F_i = \frac{c}{g}(V_0 - V_i) + f_i \qquad\qquad [12.42]$$

$$\text{and } h_i - h_0 = F_i + f_i \qquad\qquad [12.43]$$

Worked examples

Example 12.1

A surge chamber 10 m in diameter is situated at the downstream end of a low-pressure tunnel 10 km long and 3 m in diameter. At a steady discharge of 36 m³/s the flow to the turbines is suddenly stopped by closure of the turbine inlet valves. Determine the maximum rise in level in the surge chamber and its time of occurrence.

Solution:

$$V_0 = \frac{36}{7.063} = 5.093 \text{ m/s}; \qquad A_T = 7.069 \text{ m}^2; \qquad A_s = 78.54 \text{ m}^2$$

From Section 12.2,

$$T = 2\pi\sqrt{\frac{10\,000}{9.81} \times \frac{78.54}{7.069}} = 668.67 \text{ s} \quad (\text{Equation 12.9})$$

From Equation 12.12,

$$C_2 = V_0 \frac{A_T}{A_s} \frac{T}{2\pi} = 5.093 \times \frac{7.069}{78.54} \times \frac{668.67}{2\pi} = 48.78$$

From Equation 12.8,

$$z = C_2 \sin\frac{2\pi t}{T} = 48.78 \sin\frac{2\pi t}{T}$$

$z = $ maximum, when $t = T/4$, $= 48.78$ m occurring after 167.17 s.

Example 12.2

A surge chamber 100 m² in area is situated at the end of a 10 000 m long, 5 m diameter tunnel, with $\lambda = 0.01$. A steady discharge of 60 m³/s to the turbines is suddenly stopped by the turbine inlet valve. Neglecting surge chamber losses, determine the maximum rise in level in the surge chamber and its time of occurrence.

Solution:

Substitution of a series of values of z into Equation 12.17 and use of Equation 12.16 enable the z versus V relationship to be plotted. When $V = 0$ (maximum upsurge), z is found to be about 37.1 m at a time of 123 s.

$$z_0 = -9.42 \text{ m}; \qquad V_0 = 3.056 \text{ m/s}$$

Example 12.3

A low-pressure tunnel, 8000 m long, 4 m in diameter, with $\lambda = 0.012$, delivers a steady discharge of 45 m³/s to hydraulic turbines. A surge chamber of constant area 100 m² is situated at the downstream end of the tunnel, $F_s = 1.0$. Calculate the time variation of tunnel velocity V and level in surge chamber using the finite difference form of the governing differential equations given by Equations 12.19 and 12.20 if the flow to the turbines is suddenly stopped.

Solution:

Individual steps in the iterative procedure are not given; the solution is shown in the 'Summary of computations using iterative method' table at the end of this solution. The reader should carry out a few calculations.

Note that $F_R = (\lambda L/2gD) + F_s(A_T/A_s)^2$.

Solving Example 12.3 by the direct method (Equation 12.24), take

$$\Delta t = 10 \text{ s}$$

$$A_T = 12.566 \text{ m}^2; \qquad A_s = 100 \text{ m}^2 \text{ (constant)}$$

$$F_R = F_s \left(\frac{A_T}{A_s}\right)^2 + \frac{\lambda L}{2gD_T}$$

$$= 1.0 \left(\frac{12.566}{100}\right)^2 + \frac{0.012 \times 8000}{19.62 \times 4} = 1.239$$

$$V_0 = \frac{Q_0}{A_T} = \frac{45}{12.566} = 3.581 \text{ m/s}$$

$$z_0 = F_T V_0^2 = -15.686 \text{ m}$$

In Equation 12.24,

$$a = \frac{F_R}{4} = 0.3097$$

$$b = \frac{L}{g\Delta t} + \frac{A_T}{4A_{s,m}} \Delta t + F_R V_i$$

V_i being the velocity at the beginning of the time step (= 3.581 m/s).

$$\Rightarrow \quad b = \frac{8000}{9.81 \times 10} + \frac{12.566 \times 10}{4 \times 100} + 1.239 \times 3.581$$

$$= 86.90 \text{ (s}^{-1})$$

$$c = z_i + \frac{A_T}{2A_{s,m}} V_i \Delta t - \frac{Q_{m,\Delta t}}{2A_{s,m}} + F_R V_i^2$$

$$= -15.686 + \frac{12.566 \times 3.581 \times 10}{2 \times 100} + 1.239 \times 3.581^2 \quad \text{(since } Q_m = 0\text{)}$$

$$= 2.4525 \text{ m}$$

Hence, $\Delta V = \dfrac{-86.3 + \sqrt{86.3^2 - 4 \times 0.3097 \times 2.4525}}{2 \times 0.3097}$ \quad (Equation 12.24)

$$= -0.0284 \text{ m/s}$$

$$\Rightarrow V_{i+1} = 3.552 \text{ m/s}$$

From Equation 12.21,

$$\Delta z = \frac{\Delta t}{A_{s,m}} \left(V_i A_{\mathrm{T}} + \frac{A_{\mathrm{T}}}{2} \Delta V - Q_m \right)$$

$$= \frac{10}{100} \left(3.581 \times 12.566 - \frac{12.566}{2} \times 0.0284 - 0 \right)$$

$$= 4.482 \text{ m}$$

that is, $z_{i+1} = (z_i + \Delta z) = -15.686 + 4.48$

$$= -11.204 \text{ m}$$

The values V_{i+1} and z_{i+1} become V_i and z_i for the next time step and computations proceed in the same manner, as shown in the 'Summary' table.

Summary of computations using iterative method

t (s)	0	10	20	30	40	50	60	70
z (m)	-15.686	-11.204	-6.788	-2.497	1.620	5.519	9.161	12.515
V (m/s)	3.581	3.552	3.475	3.355	3.197	3.007	2.790	2.550

t (s)	80	90	100	110	120	130	140
z (m)	15.554	18.254	20.597	22.568	24.153	25.344	26.134
V (m/s)	2.288	2.010	1.720	1.417	1.106	0.789	0.468

t (s)	150	160	170	180
z (m)	26.52	26.49	26.06	25.23
V (m/s)	0.143	-0.182	-0.50	-0.82

Example 12.4

Calculate the speed of propagation of a pressure wave in a steel pipeline 200 mm in diameter with a wall thickness of 15 mm,

(a) assuming the pipe to be rigid
(b) assuming the pipe to be anchored at the reservoir, with no expansion joints, and free to move longitudinally
(c) assuming the pipe to be provided with expansion joints.

In each case determine the pressure head rise due to sudden valve closure when the initial steady velocity of flow is 1.5 m/s.

Solution:

$$K = 2.1 \times 10^9 \text{ N/m}^2; \qquad E = 2.1 \times 10^{11} \text{ N/m}^2; \qquad \eta = 0.3$$

(a)

$$c = \sqrt{\frac{K}{\rho}} = \sqrt{\frac{2.1 \times 10^9}{1000}} = 1450 \text{ m/s}$$

$$\Delta h = \frac{cV_0}{g} = \frac{1450 \times 1.5}{9.81} = 221.7 \text{ m}$$

(b)

$$C_1 = \frac{5}{4} - \eta = 0.95$$

$$c = \sqrt{\frac{1}{\rho\left(\frac{1}{K} + \frac{C_1 D}{TE}\right)}} = \sqrt{\frac{1}{1000\left(\frac{1}{2.1 \times 10^9} + \frac{0.95 \times 0.2}{0.015 \times 2.1 \times 10^{11}}\right)}}$$

$$c = 1365.2 \text{ m/s}; \qquad \Delta h = 208.7 \text{ m}$$

(c)

$$C_1 = 1 - \frac{\eta}{2} = 1 - 0.15 = 0.85$$

$$c = 1373.4 \text{ m/s}; \qquad \Delta h = 210.0 \text{ m}$$

Example 12.5

A steel pipeline 1500 m in length and 300 mm in diameter discharges water from a reservoir to atmosphere through a valve at the downstream end. The speed of the pressure wave is 1200 m/s. The valve is closed gradually in 20s, and the area of gate opening varies as shown in the 'A_v vs. t for Example 12.5' table. Neglecting friction, calculate the variation of pressure head at the valve during closure if the initial head at the valve is 10 m, using (a) the Allievi method and (b) the method of Section 12.9.

Solution:

$$\text{Waterhammer period, } T = \frac{2L}{c} = \frac{3000}{1200} = 2.5 \text{ s}$$

Working in time intervals of 2.5 s, the corresponding valve areas by interpolation are shown in the 'Valve area at discrete time intervals' table.

(a) *Allievi method*: From Equation 12.40,

$$\xi_i^2 + \xi_{i-1}^2 - 2 = 2\rho\left(\eta_{i-1}\xi_{i-1} - \eta_i \xi_i\right) \tag{i}$$

$$\text{where } \xi_i^2 = \frac{h_i}{h_0}; \qquad \eta_i = \frac{C_{d,i} A_{v,i}}{C_{d,0} A_{v,0}}$$

A_v vs. t for Example 12.5

Time (s)	0	2	4	6	8	10	12	14	16	18	20
A_v (m²)	0.03	0.025	0.02	0.015	0.010	0.008	0.006	0.005	0.004	0.002	0

Valve area at discrete time intervals; $\Delta t = \dfrac{2L}{c}$

Time (s)	0	2.5	5.0	7.5	10.0	12.5	15.0	17.5	20.0
A_v (m²)	0.03	0.0234	0.0165	0.0104	0.008	0.0058	0.0044	0.0022	0.0

where i indicates the discrete time intervals separated by $T = 2L/c$ (s) and $V_i = V_0 \eta_i \xi_i$ (Equation 12.37); $C_d = 0.6$ (constant); $V_0 = (C_d A_{v,0}/A_p)\sqrt{2gh_0} = 3.567$ m/s; and $\rho = cV_0/2gh_0$ (Equation 12.39a).

Table of calculations

(1)	(2)	(3)	(4)	(5)	(6)
T_i	t (s)	η	ξ	h (m)	v (m/s)
0	0	1	1	10	3.567
1	2.5	0.780	1.2644	15.988	3.518
2	5.0	0.550	1.6908	28.588	3.317
3	7.5	0.347	2.2816	52.057	2.821
4	10.0	0.260	2.2952	52.678	2.128
5	12.5	0.193	2.1508	46.260	1.483
6	15.0	0.147	1.8753	35.166	0.981
7	17.5	0.073	2.0117	40.470	0.526
8	20.0	0	2.0952	43.897	0

Notes: At $T = 1$,

$$\xi_1^2 + \xi_0^2 - 2 = 2\rho(\eta_0 \xi_0 - \eta_1 \xi_1)$$
$$\xi_1^2 + 1 - 2 = 2\rho(1 - 0.78\xi_1)$$
$$\xi_1^2 - 1 = 43.632(1 - 0.78\xi_1)$$

whence $\xi_1 = 1.2644$; $h_1 = \xi_1^2 \times h_0 = 15.988$ m
$V_1 = V_0 \eta \xi_i = 3.567 \times 0.78 \times 1.2644 = 3.518$ m/s

At $T = 2$,

$$\xi_2^2 + \xi_1^2 - 2 = 2\rho(\eta_1 \xi_1 - \eta_2 \xi_2)$$
$$\xi_2^2 + (1.2644)^2 - 2 = 43.632(0.78 \times 1.2644 - 0.55 \times \xi_2)$$

whence $\xi_2 = 1.6908$; $h_2 = 28.588$ m
$V_2 = 3.567 \times 0.55 \times 1.6908 = 3.317$ m/s

(b) Alternative algebraic method:

Table of calculations

(1)	(2)	(3)	(4)	(5)	(6)	(7)	(8)
T_i	t (s)	B	V (m/s)	F (m)	f (m)	Δh (m)	H (m)
0	0	1.128	3.567	0	0	0	10.0
1	2.5	0.880	3.518	5.988	0	5.988	15.988
2	5.0	0.620	3.317	24.577	−5.988	18.588	28.588
3	7.5	0.391	2.821	66.634	−24.577	42.057	52.057
4	10.0	0.293	2.128	109.313	−66.634	42.678	52.678
5	12.5	0.218	1.483	145.573	−109.313	36.260	46.260
6	15.0	0.165	0.991	170.739	−145.573	25.166	35.166
7	17.5	0.083	0.526	201.209	−170.739	30.470	40.470
8	20.0	0	0	235.107	−201.209	33.898	43.898

Chapter 12

Notes: i indicates time step; h_0, initial head at valve; and V_0, initial velocity in pipe.

$$\text{Column 3: } B_i = \frac{C_d A_{vi} \sqrt{2g}}{A_p}$$

$$\text{Column 4: } V_i = -\frac{B_i^2 c}{2g} + B_i \sqrt{\left(\frac{B_i c}{2g}\right)^2 + \frac{cV_0}{g} + h_0 + 2f_i}$$

$$\text{Column 5: } F_i = \frac{c}{g}(V_0 - V_i + f)$$

$$\text{Column 6: } f_i = -F_{i-1}$$

$$\text{Column 7: } \Delta h_i = F_i + f_i$$

$$\text{Column 8: } h_i = h_0 + \Delta h_i$$

References and recommended reading

Ellis, J. (2008) *Pressure Transients in Water Engineering: A Guide to Analysis and Interpretation of Behaviour*, Thomas Telford, London.

Fox, J. A. (1989) *Transient Flow in Pipes, Open Channels and Sewers*, Ellis Horwood, Hemel Hempstead.

Streeter, V. L., Wylie, E. B. and Bedford, K. W. (1997) *Fluid Mechanics*, 9th edn, McGraw-Hill, New York.

Thorley, A. R. D. (2004) *Fluid Transients in Pipeline Systems*, 2nd edn, John Wiley & Sons, London.

Problems

1. A low-pressure tunnel 4 m in diameter and 8000 m in length, having a Darcy friction factor of 0.012, delivers 45 m³/s to a hydraulic turbine. A surge chamber 8 m in diameter is situated at the downstream end of the tunnel. Taking $F_s = 1.0$ and using the method of Section 12.3, plot the variation of water level in the surge chamber relative to the reservoir level when the flow to the turbines is suddenly stopped.

2. (a) Repeat Problem 1 using a numerical method.
    (b) If the discharge to the turbines were to be reduced linearly to zero in 90 s, calculate the time variation of water level in the surge chamber and state the maximum upswing and time of occurrence.

3. A steel pipeline 2000 m in length and 300 mm in diameter discharges water from a reservoir to atmosphere through a control valve, the discharge coefficient of which is 0.6. The valve is closed so that its area decreases linearly from 0.065 m² to zero in (a) 15 s and (b) 30 s. If the initial head at the valve is 3.0 m and the wave speed is 1333.3 m/s, calculate, neglecting friction, the pressure head and velocity at the valve at the discrete waterhammer periods.

Chapter 13
Unsteady Flow in Channels

13.1 Introduction

River flood propagation, estuarial flows and surges resulting from gate operation or dam failure are practical examples of unsteady channel flows. Natural flood flows in rivers and the propagation of tides in estuaries are examples of gradually varied unsteady flow since the vertical component of acceleration is small. Surges are examples of rapidly varied unsteady flow.

Consider the two-dimensional propagation of a low wave which has a small height in relation to its wavelength (see Figure 13.1). The celerity or speed of propagation relative to the water is given by $\sqrt{gy}$, where y is the water depth. Therefore the velocity of the wave relative to a stationary observer is

$$c = \sqrt{gy} \pm V \qquad [13.1]$$

Note that the Froude number Fr expressed by $V/\sqrt{gy}$ is the ratio of water velocity to wave celerity. If the Froude number is greater than unity, which corresponds with supercritical flow, a small gravity wave cannot be propagated upstream. Waves of finite height are dealt with in Sections 13.3 and following sections.

13.2 Gradually varied unsteady flow

Examples of gradually varied unsteady flow are floodwaves and estuarial flows; in such waves the rate of change of depth is gradual.

In one-dimensional form (i.e. depth and width integrated) it can be shown that the two governing continuity and dynamic partial differential equations are

$$\frac{\partial Q}{\partial x} + \frac{\partial A}{\partial t} = 0 \qquad [13.2]$$

$$\frac{\partial y}{\partial x} + \frac{V}{g}\frac{\partial V}{\partial x} + \frac{1}{g}\frac{\partial V}{\partial t} = S_0 - S_f \qquad [13.3]$$

Nalluri & Featherstone's Civil Engineering Hydraulics: Essential Theory with Worked Examples, Sixth Edition. Martin Marriott.
© 2016 John Wiley & Sons, Ltd. Published 2016 by John Wiley & Sons, Ltd.
Companion Website: www.wiley.com/go/Marriott

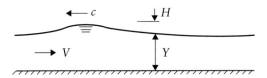

Figure 13.1 Propagation of low wave in channel.

where Q is the discharge at section located at x, with cross-sectional area A at time t; y is the depth; V is Q/A; S_0 is the bed slope; and S_f is the energy gradient.

These equations were first published by Saint-Venant. However analytical solutions of these equations are impossible unless, for example, the dynamic equation 13.3 is reduced to a 'kinematic wave' approximation by omitting the dynamic terms. In this form the equations are often applied to flood routing and overland flow computations. In general, the equations have to be evaluated at discrete space and time intervals using numerical methods such as finite difference methods. The availability of computers has enabled the governing equations to be applied to a wide range of practical problems. Such methods are, however, outside the scope of this text and the reader is referred to the recommended reading for more specialist literature.

However, the case of rapidly varied unsteady flow is, with certain simplifying assumptions, amenable to direct solution.

13.3 Surges in open channels

A surge is produced by a rapid change in the rate of flow, for example by the rapid opening or closure of a control gate in a channel. The former causes a positive surge wave to move downstream (Figure 13.2a); the latter produces a positive surge wave to moves upstream (Figure 13.2b).

A stationary observer therefore sees an increase in depth as the wave front of a positive surge wave passes. A negative surge wave, on the other hand, leaves a shallower depth as the wave front passes.

Negative waves are produced by an increase in the downstream flow, for example by the increased demand from a hydropower plant (Figure 13.3a) or downstream from a gate which is being closed (Figure 13.3b).

Figures 13.2 and 13.3 demonstrate that each type of surge can move either upstream or downstream.

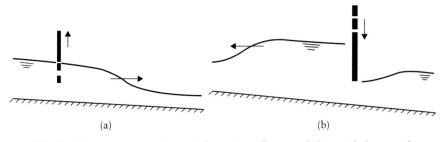

(a)　　　　　　　　　　　　　　　　　　(b)

Figure 13.2 Positive surge waves: (a) rapid opening of gate and (b) rapid closure of gate.

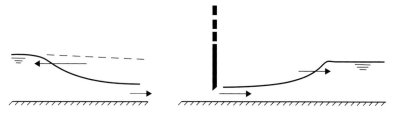

Figure 13.3   Negative surge waves.

13.4 The upstream positive surge

Consider the propagation of a positive wave upstream in a frictionless channel resulting from gate closure (Figure 13.4).

The front of the surge wave is propagated upstream at a celerity c relative to a stationary observer. To the observer, the flow situation is unsteady as the wave front passes; to an observer travelling at a speed c with the wave, the flow appears steady although non-uniform. Figure 13.5 shows the surge reduced to steady state.

The continuity equation is

$$A_1(V_1 + c) = A_2(V_2 + c) \qquad [13.4]$$

$$\text{or } V_2 = \frac{A_1 V_1 - c(A_2 - A_1)}{A_2} \qquad [13.5]$$

The momentum equation is

$$gA_1\bar{y}_1 - gA_2\bar{y}_2 + A_1(V_1 + c)(V_1 - V_2) = 0 \qquad [13.6]$$

where $\bar{y}_1$ and $\bar{y}_2$ are the respective depths of the centres of area.

Substituting for V_2 from Equation 13.5, Equation 13.4 yields

$$g(A_2\bar{y}_2 - A_1\bar{y}_1)\frac{A_2}{A_1(A_2 - A_1)} = (V_1 + c)^2$$

$$\text{whence } c = \left[gA_2\frac{(A_2\bar{y}_2 - A_1\bar{y}_1)}{A_1(A_2 - A_1)}\right]^{1/2} - V_1 \qquad [13.7]$$

In the special case of a rectangular channel,

$$A = by, \qquad \bar{y} = \frac{y}{2}$$

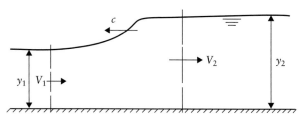

Figure 13.4   Upstream positive surge.

Chapter 13

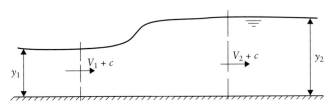

Figure 13.5   Upstream positive surge.

From Equation 13.7,

$$c = \left[\frac{gy_2}{2} \frac{(y_2^2 - y_1^2)}{y_1(y_2 - y_1)} \right]^{1/2} - V_1$$

$$\text{whence } c = \left[\frac{gy_2}{2} \frac{(y_2 + y_1)}{y_1} \right]^{1/2} - V_1 \qquad [13.8]$$

The hydraulic jump can be shown to be a stationary surge.
    Putting $c = 0$ in Equation 13.8,

$$V_1^2 = \frac{gy_2}{2} \frac{(y_2 + y_1)}{y_1}$$

$$\frac{2V_1^2 y_1}{g} = y_2^2 + y_2 y_1$$

Now

$$Fr_1^2 = \frac{V_1^2}{gy_1}$$

$$\Rightarrow y_2^2 + y_2 y_1 - 2Fr_1^2 y_1^2 = 0$$

$$\text{whence } y_2 = \frac{y_1}{2} \left(\sqrt{1 + 8Fr_1^2} - 1 \right)$$

which is identical with Equation 8.17 with $\beta = 1.0$.
    In the case of a low wave where y_2 approaches y_1, Equation 13.8 becomes

$$c = \sqrt{gy} - V_1 \qquad [13.9]$$

and in still water ($V_1 = 0$),

$$c = \sqrt{gy} \qquad [13.10]$$

13.5 The downstream positive surge

This type of wave may occur in the channel downstream from a sluice gate at which the
opening is rapidly increased (see Figure 13.6).
    Reducing the flow to steady state:

$$\text{Continuity: } (c - V_1)A_1 = (c - V_2)A_2 \qquad [13.11]$$

$$\text{that is, } V_1 = \frac{cA_1 - cA_2 + V_2 A_2}{A_1} \qquad [13.12]$$

$$\text{Momentum: } gA_1\bar{y}_1 - gA_2\bar{y}_2 + (c - V_2)A_2(V_2 - V_1) = 0$$

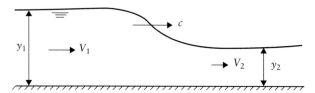

Figure 13.6 Downstream positive surge.

Substituting for V_1 yields

$$c = \left[\frac{g(A_1 \bar{y}_1 - A_2 \bar{y}_2)}{(A_1 - A_2)} \frac{A_1}{A_2} \right]^{1/2} + V_2 \qquad [13.13]$$

In the case of a rectangular channel,

$$c = \left[\frac{g y_1}{2 y_2} (y_1 + y_2) \right]^{1/2} + V_2 \qquad [13.14]$$

13.6 Negative surge waves

The negative surge appears to a stationary observer as a lowering of the liquid surface. Such waves occur in the channel downstream from a control gate the opening of which is rapidly reduced or in the upstream channel as the gate is opened. The wave front can be considered to be composed of a series of small waves superimposed on each other. Since the uppermost wave has the greatest depth it travels faster than those beneath; the retreating wave front therefore becomes flatter (Figure 13.7).

Figure 13.8 shows a small disturbance in a rectangular channel caused by a reduction in downstream discharge; the wave propagates upstream.

Reducing the flow to steady state, the continuity equation becomes

$$(V + c)y = (V - \delta V + c)(y - \delta y)$$

Neglecting the product of small quantities,

$$\delta y = -\frac{y \delta V}{(V + c)} \qquad [13.15]$$

The momentum equation is

$$\frac{\rho g}{2}[y^2 - (y - \delta y)^2] + \rho y(V + c)\{V + c - (V - \delta V + c)\} = 0$$

$$\text{whence } \frac{\delta y}{\delta V} = -\frac{(V + c)}{g}$$

$$\text{or } \delta y = -\frac{\delta V(V + c)}{g} \qquad [13.16]$$

Figure 13.7 Propagation of negative surge.

Chapter 13

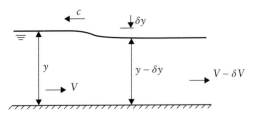

Figure 13.8   Negative surge propagation (upstream).

Equating 13.15 and 13.16,

$$\frac{y\delta V}{(V+c)} = \frac{(V+c)\delta V}{g}$$

$$\text{whence } c = \sqrt{gy} - V \qquad [13.17]$$

Substituting for $(V+c)$ from Equation 13.16 into Equation 13.17 yields

$$\delta y = -\frac{\delta V}{g}\sqrt{gy}$$

and in the limit as $\delta y \to 0$

$$\frac{dy}{\sqrt{y}} = -\frac{dV}{\sqrt{g}} \qquad [13.18]$$

For a wave of finite height (Figure 13.9), integration of Equation 13.18 yields

$$V = -2\sqrt{gy} + \text{Constant}$$

When $y = y_1$, $V = V_1$. Whence Constant $= V_1 + 2\sqrt{gy_1}$,

$$V = V_1 + 2\sqrt{gy_1} - 2\sqrt{gy} \qquad [13.19]$$

From Equation 13.17, $c = \sqrt{gy} - V$ and substituting in Equation 13.19 yields

$$c = 3\sqrt{gy} - 2\sqrt{gy_1} - V_1 \qquad [13.20]$$

The wave speed at the crest is therefore

$$c_1 = \sqrt{gy_1} - V_1 \qquad [13.21]$$

and at the trough

$$c_2 = 3\sqrt{gy_2} - 2\sqrt{gy_1} - V_1 \qquad [13.22]$$

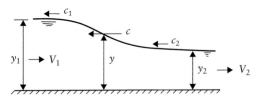

Figure 13.9   Negative surge of finite height.

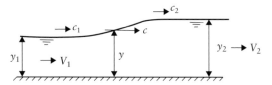

Figure 13.10   Downstream negative surge.

In the case of a downstream negative surge in a frictionless channel (Figure 13.10), a similar approach yields

$$c = \sqrt{gy} + V \qquad\qquad [13.23]$$

$$V = 2\sqrt{gy} - 2\sqrt{gy_2} + V_2 \qquad\qquad [13.24]$$

$$c = 3\sqrt{gy} - 2\sqrt{gy_2} + V_2 \qquad\qquad [13.25]$$

$$c_1 = 3\sqrt{gy_1} - 2\sqrt{gy_2} + V_2 \qquad\qquad [13.26]$$

$$c_2 = \sqrt{gy_2} + V_2 \qquad\qquad [13.27]$$

13.7 The dam break

The dam, or gate, holding water upstream at depth y_1 and zero velocity is suddenly removed (see Figure 13.11).
From Equation 13.20,

$$c = 3\sqrt{gy} - 2\sqrt{gy_1}$$

The equation to the surface profile is therefore

$$x = (ct) = (3\sqrt{gy} - 2\sqrt{gy_1})t$$

If $x = 0, y = 4y_1/9$ and remains constant with time. The velocity at $x = 0$ is

$$V = V_1 + 2\sqrt{gy_1} - 2\sqrt{gy} \quad \text{(from Equation 13.19)}$$

$$\text{that is, } V = \frac{2}{3}\sqrt{gy_1} \quad (\text{since } V_1 = 0)$$

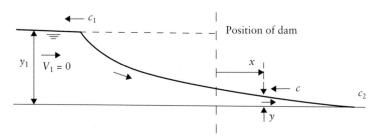

Figure 13.11   Sudden gate or dam bursting.

Worked examples

Example 13.1

A rectangular channel 4 m wide conveys a discharge of 25 m³/s at a depth of 3 m. The downstream discharge is suddenly reduced to 12 m³/s by partial closure of a gate. Determine the initial depth and celerity of the positive surge wave.

Solution:

Referring to Figure 13.4,

$$y_1 = 3 \text{ m}; \quad V_1 = 2.083 \text{ m/s}; \quad V_2 = \frac{12}{4 \times y_2} = \frac{3}{y_2}$$

From Equation 13.4,

$$(V_1 + c)y_1 = (V_2 + c)y_2 \qquad \text{(i)}$$

$$\text{that is, } (2.083 + c) \times 3 = \left(\frac{3}{y_2} + c\right) y_2$$

$$\text{whence } c = \frac{3.249}{(y_2 - 3)}$$

Substituting in Equation 13.8 yields

$$\frac{3.249}{(y_2 - 3)} = \left[\frac{gy_2}{2} \frac{(y_2 + 3)}{3}\right]^{1/2} - 2.083$$

By trial, $y_2 = 3.75$ m; $c = 4.35$ m/s.

Example 13.2

A tidal channel, which may be assumed to be rectangular, 40 m wide, bed slope 0.0003 and the Manning roughness coefficient $n = 0.022$, conveys a steady freshwater discharge of 60 m³/s. A tidal bore is observed to propagate upstream with a celerity of 5 m/s. Determine the depth of flow and the discharge immediately after the bore has passed, neglecting the density difference between the freshwater and saline water.

Solution:

The depth of uniform flow using the Manning equation $= 1.52$ m $(=y_1)$

$$V_1 = \frac{60}{40 \times 1.52} = 0.987 \text{ m/s}$$

y_2 can be determined using Equation 13.8; in other words,

$$c = \left[\frac{gy_2}{2} \frac{y_2 + y_1}{y_1}\right]^{1/2} - V_1$$

$$5.0 = \left[\frac{9.81 y_2}{2} \frac{(y_2 + 1.52)}{1.52}\right]^{1/2} - 0.987$$

By trial, $y_2 = 2.66$ m.

Using the continuity equation,

$$(V_1 + c)y_1 = (V_2 + c)y_2$$

$$\text{or} \quad V_2 = \frac{(V_1 + c)y_1}{y_2} - c$$

$$= \frac{(0.987 + 5.0) \times 1.52}{2.66} - 5.0$$

$$V_2 = -1.578 \text{ m/s}$$

$$\text{and } Q_2 = V_2 \, by_2 = -168.2 \text{ m}^2/\text{s (upstream)}$$

Example 13.3

A rectangular tailrace channel, 15 m wide, bed slope 0.0002 and Manning's roughness co-efficient 0.017, conveys a steady discharge of 45 m^3/s from a hydropower installation. A power increase results in a sudden increase in flow to the turbines to 100 m^3/s. Determine the depth and celerity of the resulting surge wave in the channel.

Solution:

Using the Manning equation the depth of uniform flow under initial conditions at a discharge of 45 m^3/s = 2.42 m.
  Using Equation 13.11,

$$(c - V_1)y_1 = (c - V_2)y_2$$

$$c = \frac{V_1 y_1 - V_2 y_2}{(y_1 - y_2)}$$

$$c = \frac{Q_1 y_1 / by_1 - V_2 y_2}{(y_1 - y_2)}$$

$$V_2 = \frac{Q_2}{by_2} = \frac{45}{15 \times 2.42} = 1.24 \text{ m/s}$$

$$Q_1 = 100 \text{ m}^3/\text{s}$$

$$\text{whence } c = \frac{6.67 - 3}{(y_1 - 2.42)} = \frac{3.67}{(y_1 - 2.42)}$$

Substituting in Equation 13.13 yields

$$\frac{3.67}{(y_1 - 2.42)} = \left[\frac{gy_1}{2 \times 2.42}(y_1 + 2.42) \right]^{1/2} + 1.24$$

By trial, $y_1 = 2.95$ m;

$$c = \frac{3.67}{(2.95 - 2.42)} = 6.92 \text{ m/s}$$

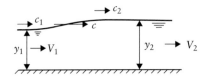

Figure 13.12 Wave front propagation.

Example 13.4

A steady discharge of 25 m³/s enters a long rectangular channel 10 m wide, bed slope 0.0001, the Manning roughness coefficient 0.017, regulated by a gate. The gate is rapidly partially closed resulting in a reduction of the discharge to 12 m³/s. Determine the depth and mean velocity at the trough of the wave, the surface profile and the time taken for the wave front to reach a point 1 km downstream neglecting friction (see Figure 13.12).

Solution:

$$y_2 = 2.86 \text{ m} \quad \text{(using the Manning equation)}$$

$$V_2 = \frac{25}{10 \times 2.86} = 0.874 \text{ m/s}$$

$$V_1 = \frac{12}{10y_1} = \frac{1.2}{y_1}$$

From Equation 13.24,

$$V = V_2 - 2\sqrt{g}(\sqrt{y_2} - \sqrt{y})$$

$$\text{whence } V_1 = V_2 - 2\sqrt{g}(\sqrt{y_s} - \sqrt{y_1})$$

$$\text{that is, } \frac{1.2}{y_1} = 0.874 - 6.26(1.691 - \sqrt{y_1})$$

Solving by trial,

$$y_1 = 2.637 \text{ m}$$
$$\text{Then } V_1 = 0.455 \text{ m/s}$$
$$c_2 = \sqrt{gy_2} + V_2 = 6.17 \text{ m/s}$$

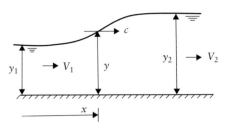

Figure 13.13 Wave front propagation.

Time taken to travel 1 km = 2.7 min.

$$\text{Surface profile: } x = ct = (3\sqrt{gy} - 2\sqrt{gy_2} + V_2)t$$
$$\text{or } x = (9.4\sqrt{y} - 10.6 + 0.874)t$$
$$x = (9.4\sqrt{y} - 9.726)t.$$

References and recommended reading

Abbot, M. B. and Basco, D. R. (1989) *Computational Fluid Dynamics*, Longman, Harlow.

Chadwick, A., Morfett, J. and Borthwick, M. (2013) *Hydraulics in Civil and Environmental Engineering*, 5th edn, Spon, Abingdon, UK.

Cunge, J. A., Holly, F. M. and Verwey, A. (1980) *Practical Aspects of Computational River Hydraulics*, Pitman, London.

French, R. H. (1994) *Open Channel Hydraulics*, 2nd edn, McGraw-Hill, New York.

French, R. H. (2007) *Open Channel Hydraulics*, Water Resources Publications, Highlands Ranch, CO.

Henderson, F. M. (1966) *Open Channel Flow*, Macmillan, New York.

Pickford, J. (1969) *Analysis of Surge*, Macmillan, London.

Problems

1. A rectangular channel 4 m wide conveys a discharge of 18 m³/s at a depth of 2.25 m. Determine the depth and celerity of the positive surge wave resulting from (a) sudden partial gate closure which reduces the downstream discharge to 10 m³/s and (b) sudden total gate closure.

2. At low tide the steady freshwater flow in an estuarial channel, 20 m wide, bed slope 0.0005 and the Manning roughness coefficient 0.02, is 20 m³/s. A tidal bore forms on the flood tide and is observed to propagate upstream at a celerity of 4 m/s. Neglecting the density difference between the freshwater and the saline water, determine the depth and discharge immediately after the bore has passed.

3. A rectangular channel, 10 m wide, bed slope 0.0001 and the Manning roughness coefficient 0.015, receives inflow from a reservoir with a gated inlet. When a steady discharge of 30 m³/s is being conveyed, the gate is suddenly opened to release a discharge of 70 m³/s. Calculate the initial celerity and depth of the surge wave.

4. A steady discharge of 30 m³/s is conveyed in a rectangular channel of bed width 9 m at a depth of 3.0 m. A control gate at the inlet is suddenly partially closed reducing the inflow to 10 m³/s. Assuming the channel to be frictionless, determine the depth behind the surge wave and the time taken for the trough of the wave to pass a point 500 m downstream.

5. A rectangular channel 30 m wide discharges 60 m³/s at a uniform flow depth of 2.5 m into a reservoir. The levels of water in the channel and reservoir at the reservoir inlet are initially equal. Water in the reservoir is released rapidly so that the level falls at the rate of 1 m/h. Neglecting friction and the channel slope, determine the time taken for the level in the channel to fall 0.5 m at a section 1 km upstream from the reservoir.

Chapter 14
Uniform Flow in Loose-Boundary Channels

14.1 Introduction

The loose boundary (consisting of movable material) of a channel deforms under the action of flowing water and the deformed bed with its changing roughness (bed forms) interacts with the flow. A dynamic equilibrium state of the boundary may be expected when a steady and uniform flow has developed.

   The resulting movement of the bed material (sediment) in the direction of flow is called **sediment transport** and a certain critical bed shear stress (τ_c) must be exceeded to start the particle movement. Such a critical shear stress is referred to as the incipient (threshold) motion condition, below which the particles will be at rest and the flow is similar to that on a rigid boundary.

14.2 Flow regimes

Shear stresses above the threshold condition disturb the initial plane boundary of the channel, and the bed and water surface assume various forms depending on the sediment and fluid flow characteristics. Two distinct regimes of flow may be identified with the increasing flows with the following bed forms:

(a)   Lower regime: ripples (for smaller sediment size <0.6 mm and low Froude number ≪1), dunes and ripples, dunes with increasing shear (τ_0) and Froude number Fr; further increases in τ_0 introduce transition to dunes/plane bed ($Fr \simeq 1$)
(b)   Upper regime: flat bed, antidunes, chutes and pools with large shear and Froude numbers (>1).

14.3 Incipient (threshold) motion

Shields (1936) introduced the concept of the dimensionless entrainment function ($\tau_0/\rho g \Delta d$) as a function of shear Reynolds number, $Re_* (=U_* d/\nu)$, where ρ is density of

Nalluri & Featherstone's Civil Engineering Hydraulics: Essential Theory with Worked Examples,
Sixth Edition. Martin Marriott.
© 2016 John Wiley & Sons, Ltd. Published 2016 by John Wiley & Sons, Ltd.
Companion Website: www.wiley.com/go/Marriott

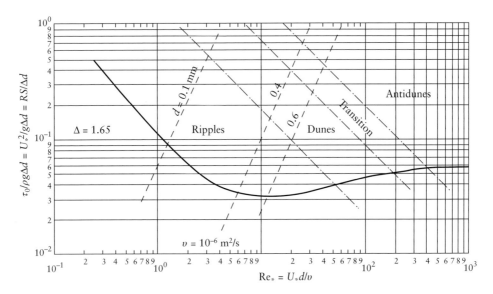

Figure 14.1 Shields diagram.

the fluid, Δ the relative density of sediment submerged in the fluid, d the diameter of sediment, g the acceleration due to gravity, U_* the shear velocity $(=\sqrt{\tau_0/\rho})$ and v the kinematic viscosity of the fluid.

Note that the submerged relative density Δ is equal to $(\rho_s - \rho)/\rho$, which equals $(s-1)$ where s is the relative density of the sediment as defined in Section 1.4.

The resulting Shields diagram is shown in Figure 14.1, with a curve defining the threshold condition.

When the flow is fully turbulent around the bed material $(Re_* > 400$ and $d \simeq 6$ mm$)$, the Shields criterion can be written as

$$\frac{\tau_0}{\rho g \Delta d} = 0.056 \qquad [14.1]$$

Combining Equation 14.1 with the uniform boundary shear equation

$$\tau_0 = \rho g R S \qquad [14.2]$$

gives the limiting particle size (with $\Delta = 1.65$) for incipient motion

$$d = 11RS \qquad [14.3]$$

where R is the hydraulic radius and S the friction gradient.

Combining Equation 14.3 and the Manning equation for mean velocity

$$V = \frac{1}{n}R^{2/3}S^{1/2} \qquad [14.4]$$

with

$$n = \frac{d^{1/6}}{26} \quad \text{(Strickler's equation)} \qquad [14.5]$$

gives

$$\frac{V_c}{\sqrt{gd}} \approx 1.9\sqrt{\Delta}\left(\frac{d}{R}\right)^{-1/6} \qquad [14.6]$$

where V_c is the critical velocity for the incipient motion of sediment particles.

Some recommended values of critical tractive forces and maximum permissible mean velocities for different sizes of bed material are listed in Table 8.3.

14.4 Resistance to flow in alluvial (loose-bed) channels

The resistance of an alluvial channel varies considerably with flow velocity once threshold has been passed. The bed introduces additional form drag due to bed formations, and the overall friction factor λ rises rapidly to three to four times its original value. Several attempts have been made to describe a relationship between the mean velocity V, the depth y_0 or hydraulic radius R, slope S and sediment size d, which can be broadly divided into two categories.

14.4.1 Total resistance approach

14.4.1.1 Regime channel equation
This was one of the earliest resistance relationships for alluvial channel flow proposed by Lacey (1930) in the form

$$V = 10.8R^{2/3}S^{1/3} \qquad [14.7]$$

in SI units based on the regime canal data from India. Its applicability in channels or rivers with different sediment sizes and flow depths is questionable.

14.4.1.2 Japanese equation
Sugio proposed the following equation using river data from Japan (Novak and Nalluri, 1984):

$$V = KR^{0.54}S^{0.27} \qquad [14.8]$$

in SI units where $K = 6.51$ for ripples, 9.64 for dunes and 11.28 for the transition regime.

14.4.1.3 Garde–Ranga Raju's formula
Garde and Ranga Raju (1966) analysed data from flumes, canals and natural streams, and a graphical relationship (Figure 14.2) between the parameters $K_1 V / \sqrt{(\Delta g R)}$ versus $K_2 (R/d)^{1/3} S / \Delta$, where K_1 and K_2 are functions of sediment size (see Figure 14.3), was proposed. Figures 14.2 and 14.3 facilitate the calculations of discharge in alluvial channels.

14.4.2 Grain and form resistance approach

This approach splits either the overall resistance into grain resistance λ' and form resistance λ'' (Alam and Kennedy, 1969) or U_* into U'_* and U''_* corresponding to grain and form resistances, respectively (Einstein and Barbarossa, 1952). Introducing the concept of bed hydraulic radius (R_b), charts and graphs have been produced to predict the resistance equations in alluvial channels. The proposed methods are out of scope of the present book.

In the case of rectangular channels (bed width B) with smooth sides, Einstein suggested the following equation for the hydraulic radius of the bed:

$$R_b = \left[1 + 2\left(\frac{y_0}{B}\right)\right]R - 2\frac{y_0}{B}R_w \qquad [14.9]$$

Chapter 14

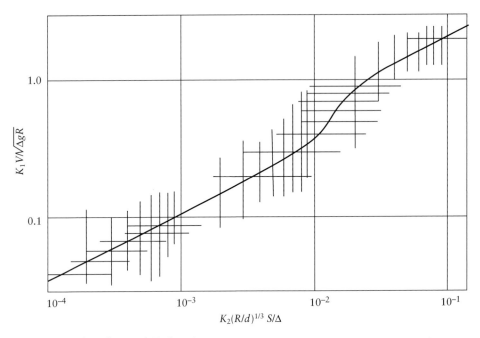

Figure 14.2 Plot of K_1 and K_2 functions.

where the hydraulic radius corresponding to the walls R_w is computed from the Manning equation, assuming it is applicable to the side walls and the bed independently.

Vanoni and Brooks (1957) proposed for rough channels with smooth sides that

$$R_b = \frac{\lambda_b V^2}{8gS} \qquad [14.10]$$

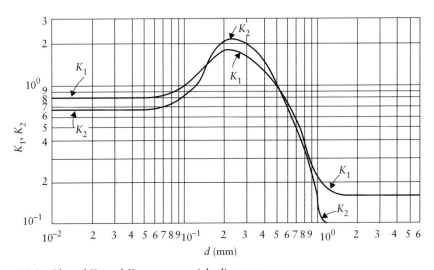

Figure 14.3 Plot of K_1 and K_2 versus particle diameter.

The bed friction factor λ_b can be found from

$$P\lambda = P_b\lambda_b + P_w\lambda_w \qquad [14.11]$$

where P is the wetted perimeter with suffixes 'b' and 'w' for bed and walls, respectively.

Resistance equations of the type Colebrook–White (or appropriate resistance plots; e.g. see Figure 14.10) can be used to predict λ_w in Equations 14.9 and 14.11, which is a function of Re/λ, where Re is the Reynolds number and λ the overall friction factor of the channel.

In the channels where roughness of the walls is different from the bed, the bed hydraulic radius may be used in place of the total hydraulic radius to determine the regimes, mean velocities and so on.

14.5 Velocity distributions in loose-boundary channels

Einstein's equation in the form

$$\frac{u}{U'_*} = 5.75 \log \left(\frac{30.2\, yx}{k_s} \right) \qquad [14.12]$$

where u is the temporal mean velocity at a distance y from the boundary is applicable universally for smooth, transition and rough beds. The correction factor x is a function of k_s/δ' (δ' sub-layer thickness given by $11.6v/U_*$) given in Table 14.1.

Equation 14.12 gives the mean velocity V as

$$V = 5.75 U_* \log \left(\frac{12.27 Rx}{k_s} \right) \qquad [14.13]$$

For $k_s/\delta' > 6.0$, the boundary is fully rough and the Manning–Strickler equation could conveniently be used to calculate the mean velocity.

14.6 Sediment transport

When flow characteristics (velocity, average shear stress, etc.) in an alluvial channel exceed the threshold condition for the bed material, the particles move in different modes along the flow direction. The mode of transport of the material depends on the sediment characteristics such as its size and shape, density ρ_s and movability parameter U_*/W_s, where W_s is the fall velocity of the sediment particle.

Fall velocities are equally of importance in reservoir sedimentation and settling processes and may be expressed as $W_s = f$ (shape and density of sediment, number of particles falling, and particle Reynolds number).

The fall velocity of a single spherical particle can be written as

$$W_s = \sqrt{\frac{4}{3}\frac{g\Delta d}{C_D}} \qquad [14.14]$$

Table 14.1 Correction factor x.

k_s/δ'	0.2	0.3	0.5	0.7	1.0	2.0	4.0	6.0	10.0
x	0.70	1.00	1.38	1.56	1.61	1.38	1.10	1.03	1.00

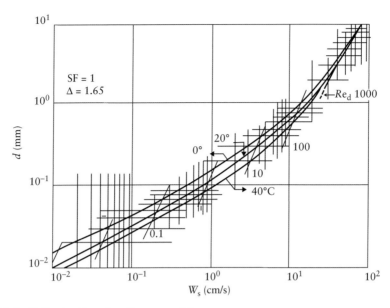

Figure 14.4   Fall velocities of sediment particles.

where C_D is the drag coefficient. The drag coefficient is a function of the particle Reynolds number ($Re_d = W_s d/\nu$).

For $Re_d < 1$,

$$C_D = \frac{24}{Re_d} \qquad [14.15]$$

For particles with a shape factor (SF) of 0.7 (natural sands) C_D is nearly equal to 1 when $Re_d > \simeq 200$, whereas it is around 0.4 in the case of spherical particles (SF = 1) for $Re_d > 2000$.

Figure 14.4 may be used to establish fall velocities of sediment particles of different sizes with SF = 1.

Some sediment particles roll or slide along the bed intermittently and some others saltate (hopping or bouncing along the bed). The material transported in one or both of these modes is called **bed load**. Finer particles (with low fall velocities) are entrained in suspension by the fluid turbulence and transported along the channel in suspension. This mode of transport is called **suspended load**. The combined transport derived from the bed material is called **total bed material load**. Sometimes finer particles from upland catchment (sizes which are not present in the bed material), called **wash load**, are also transported in suspension. The combined bed material and wash load is called **total load**.

14.7 Bed load transport

Several empirical equations from laboratory flume data have been proposed by many investigators with the basic assumptions that the sediment is homogeneous and noncohesive. The results differ appreciably and it is dangerous to transfer the information to outside the limits of the experiments. However, one can discern general trends of the transport rate by using several formulae (with some theoretical background). The following are the most commonly used equations:

1. The Shields equation

Shields used the concept of excess shear responsible for the transport and presented a dimensionally homogeneous equation

$$\frac{q_b \Delta}{qS} = \frac{10(\tau_0 - \tau_c)}{\rho g \Delta d} \qquad [14.16]$$

where q_b is the bed load per unit width and q the unit discharge in the channel. Equation 14.16 is based on the ranges of $0.06 < \Delta < 3.2$ and $1.56\,mm < d < 2.47\,mm$.

2. Schoklitsch's equation

The bed load g_b in kilograms per metre second is given by

$$g_b = 2500 S^{3/2}(q - q_{cr}) \qquad [14.17]$$

where q_{cr} is the unit discharge at threshold condition given by

$$q_{cr} = \frac{0.20(\Delta)^{5/3}\, d^{3/2}}{S^{7/6}} \qquad [14.18]$$

It must be noted that Equation 14.17 is not dimensionally homogeneous and is valid only for q and q_{cr} in metres cubed per metre second.

3. Kalinske's equation

For Shields function > 0.09, this can be written as

$$\frac{q_b}{U_* d} = 10 \left[\frac{U_*^2}{\Delta g d} \right]^2 \qquad [14.19]$$

Equation 14.19 is dimensionally homogeneous and may not be good for high transport rates.

4. Meyer–Peter and Muller's formula

The energy slope S is split into two parts, and only one part (μS) is considered to be responsible for transport (grain drag; the other is expended in the form drag). The factor μ is dependent on the bed form (ripple factor) and is expressed as

$$\mu = \left(\frac{C_{channel}}{C_{grain}} \right)^{3/2} \qquad [14.20]$$

where C is Chezy's coefficient given by

$$C = 18 \log \left(\frac{12R}{k} \right) \qquad [14.21]$$

in which $k = d$ for C_{grain} and k is a function of bed form ($\simeq$ dune height) for $C_{channel}$.

The ripple factor varies between 0.5 and 1.0 for dune to flat-bed condition. The bed load q_b is given by

$$q_b = 8\sqrt{\Delta g d^3} \left(\frac{\mu RS}{\Delta d} - 0.047 \right)^{3/2} \qquad [14.22]$$

Equation 14.22 is dimensionally homogeneous, covers a wide range of particle sizes and is widely used.

5. Einstein's equation

Introducing probability concepts of sediment movement, Einstein (1950) developed an empirical relationship

$$\phi = f(\psi) \qquad [14.23]$$

where

$$\text{shear intensity or flow parameter, } \psi = \frac{\Delta d}{\mu RS} \qquad [14.24]$$

$$\text{transport parameter, } \phi = \frac{q_b}{\sqrt{g \Delta d^3}} \qquad [14.25]$$

(*Note*: μR in Equation 14.24 may be treated as grain (bed) hydraulic radius R'.)

Figure 14.5 shows the functional relationship (Equation 14.23). For small values of ψ (<10) (ψ is around 20 for threshold conditions), the relationship between ϕ and ψ can be expressed as

$$\phi = 40 \left(\frac{1}{\psi} \right)^3 \qquad [14.26]$$

Rearranging Meyer–Peter and Muller's equation in terms of ϕ and ψ parameters results in

$$\phi = \left(\frac{4}{\psi} - 0.188 \right)^{3/2} \qquad [14.27]$$

which agrees well with Einstein's curve in Figure 14.5.

Einstein's relationship covers a very wide range of experimental data ($0.785\,\text{mm} < d < 28.65\,\text{mm}$; $0.052 < \Delta < 1.68$).

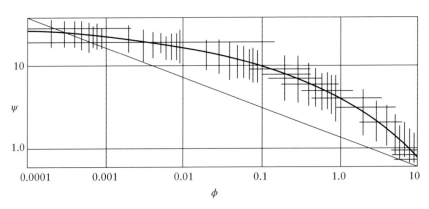

Figure 14.5 Plot of ϕ versus ψ functions.

14.8 Suspended load transport

1. Rouse's (1937) distribution equation

The vertical (suspended) mass balance equation in a two-dimensional flow was first expressed by O'Brien (1933) as

$$cW_s + \varepsilon_s \frac{dc}{dy} = 0 \qquad [14.28]$$

where c is the volumetric concentration of the sediment and ε_s the kinematic eddy viscosity (turbulence diffusion coefficient) in the presence of sediment, equal to $\beta\varepsilon$, ε being the eddy viscosity for clear water. β is of the order of unity in the presence of fine sediment and decreases with increasing particle size. Combining Equation 14.28 with the turbulent mixing theory (log law distribution of velocity) gives the solution for sediment concentration c at a height y in a channel as

$$\frac{c}{c_a} = \left[\frac{a(y_0 - y)}{y(y_0 - a)}\right]^{W_s/\beta\chi U} \qquad [14.29]$$

where c_a is the reference concentration at a height a from the bed and χ Karman's constant.

The theoretical distributions of the concentration (Equation 14.29) are shown in Figure 14.6 for different values of $W_s/\beta\chi U_*$ with $\beta = 1$ and $\chi = 0.4$ (i.e. clear water conditions).

Table 14.2 shows the state of suspension under different values of the movability parameter U_*/W_s.

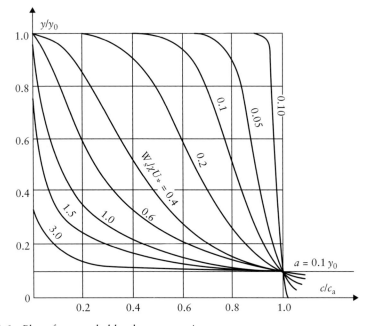

Figure 14.6 Plot of suspended load concentrations.

Chapter 14

Table 14.2   States of suspension.

States of suspension	Movability parameter U_*/W_s
Intensive saltation	0.25
Lower half in suspension	1
Particles reach surface	3
Well-developed suspension	20
Homogeneous suspension	200

The reference level a may be assumed to be around $2d$ (d being the diameter of suspended particles), and c_a as bed load corresponding to this diameter. Equation 14.29 must be used with care as the parameter $W_s/\beta_\chi U_*$ is not accurately computable.

The suspended load transport q_s can be obtained by summation as

$$q_s = \int_a^{y_0} cu\, dy \qquad [14.30]$$

where u is given by an appropriate velocity distribution. Equation 14.30 may be solved either numerically or graphically. Suspended load can generally be measured easily and accurately, and good field measurements of both c and u predict suspended load with reasonable accuracies.

2. Lane and Kalinske's (1939) approximate method

The suspended load q_s in wide channels is given by

$$q_s \approx qc_a P_e^{15(aW_s/y_0 U_*)} \qquad [14.31]$$

in which P is a function of the movability parameter, U_*/W_s and $n/y_0^{1/6}$, n being Manning's coefficient. Figure 14.7 shows the plot of P in SI units.

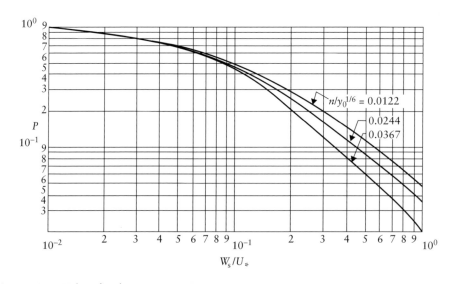

Figure 14.7   Values for the parameter P.

3. Empirical equations

Several practising engineers have reported several formulae of the type

$$q_s \propto q^b \tag{14.32}$$

the exponent b varying between 1.9 and 3.

Engelund (1970) alternatively proposed that

$$q_s = 0.5q \left(\frac{U_*}{W_s} \right)^4 \tag{14.33}$$

14.9 Total load transport

Total load includes both bed material load and wash load. Wash load is usually caused by land erosion, and a useful criterion for its existence may be taken as the particle Froude number ($=V/\sqrt{gd}$) around 20. Due to its small-size fractions, wash load moves in suspension and thus can be estimated from the total suspended load provided the suspended bed material load is known. The following approaches describe some of the available direct methods of estimating the total bed material load.

1. Laursen's (1958) approach

The cross-sectional mean concentration by volume C_v of the bed material load for quartz material was suggested

$$C_v = \frac{160 q_b}{q} \tag{14.34}$$

applicable to flume data with sand of $d < 0.2$ mm.

2. Garde's (1968) equation

Field and flume data with the sediment size range of 0.011 mm $< d <$ 0.93 mm gave

$$q_t \approx 10 \left(\frac{U_*}{d^3} \right) \left(\frac{RS}{\Delta} \right)^4 \tag{14.35}$$

3. Graf's approach

Similar to Einstein's bed load equation, Graf (1984) used several flume and stream data including closed conduit data to establish the equation

$$\phi_A = 10.39 (\psi_A)^{-2.52} \tag{14.36}$$

over a range of $10^{-2} < \phi < 10^3$, where

$$\phi_A = \frac{C_v VR}{\sqrt{g \Delta d^3}} \tag{14.37}$$

$$\text{and } \psi_A = \frac{\Delta d}{RS} \tag{14.38}$$

Table 14.3   Coefficients in Equation 14.39.

Coefficient	Fine and transitional	Coarse
	$1.0 < d_{gr} < 60$	$d_{gr} > 60$
n	$n = 1.00 - 0.56 \log d_{gr}$	0.00
A_{gr}	$A_{gr} = 0.14 + 0.23/\sqrt{d_{gr}}$	0.17
m	$m = 1.67 + 6.83/d_{gr}$	
C	$\log C = 2.79 \log d_{gr} - 0.98(\log d_{gr})^2 - 3.46$	0.025

4. Ackers–White's formula

This method introduced by Ackers and White (1973) was updated by Ackers (1993) using additional data. The total sediment transport load is predicted from the equation:

$$\frac{q_t y_0}{qd}\left(\frac{U_*}{V}\right)^n = C\left(\frac{F_{gr}}{A_{gr}} - 1\right)^m \qquad [14.39]$$

where

$$F_{gr} = \frac{U_*^n}{\sqrt{g\Delta d}}\left[\frac{V}{\sqrt{32}\log(10 y_0/d)}\right]^{1-n} \qquad [14.40]$$

With the dimensionless grain diameter d_{gr} defined by

$$d_{gr} = d\left[\frac{g\Delta}{v^2}\right]^{1/3} \qquad [14.41]$$

the coefficients in Equation 14.39 are shown in Table 14.3.

14.10 Regime channel design

1. Kennedy's approach

Regime equations were developed using data from stable channels in the Indian subcontinent, carrying moderate sediment loads of less than 500 ppm by weight. These equations do not consider sediment load variable and have limitations due to the fact that they are applicable to boundary characteristics similar to those found in the Indian subcontinent.

Kennedy's equation for nonsilting and nonscouring velocities V is given by

$$V = 0.55 m y_0^{0.64} \qquad [14.42]$$

where y_0 is the flow depth in metres and V is in metres per second; m is the critical velocity ratio ($= V/V_0$), a function of the sand size ($m = 1$ for the standard size, $d \approx 0.323$ mm). Table 14.4 shows m values for other sand sizes.

Equation 14.42 combined with the Manning equation (Equation 8.4b) gives two equations (Ranga Raju, 1993):

$$y_0 = \left[\frac{1.818Q}{(p + 0.5)m}\right]^{0.378} \qquad [14.43]$$

Table 14.4 m values as a function of sand size.

Type of sand	m	Remark
Fine silt	0.7	
Light sand silt	1.0	Standard size (data)
Coarse sand silt	1.1	
Sand loamy silt	1.2	
Coarser silt	1.3	

where $p = b/y_0$, b being the bed width of a trapezoidal channel with side slopes of 0.5 (horizontal):1 (vertical) (the final shape of the regime channel is not truly trapezoidal, and the final side slopes are much steeper due to silt deposition on banks), and

$$\frac{SQ^{0.02}}{n^2 m^2} = 0.3 \left[\frac{(p + 2.236)^{4/3}}{(p + 0.5)^{1.31}} \right] \qquad [14.44]$$

Equations 14.43 and 14.44 give any number of solutions for the three unknowns, b, y_0 and the slope S for given values of Q, m and Manning's n. Usually, the bed slope is assumed to be a reasonable value (based on past experience and surrounding terrain slope), and p and y_0, and hence b, are computed. Table 14.5 alternatively suggests the recommended values of p for stable channels as a function of Q.

2. Lacey's approach

Lacey proposed the following equations (Kennedy's equation does not specify the channel width, and experience suggests that this is an important parameter) for regime channel design:

$$P = 4.75 \sqrt{Q} \qquad [14.45]$$

$$R = 0.47 \left(\frac{Q}{f} \right)^{1/3} \qquad [14.46]$$

$$S = 3 \times 10^{-4} \frac{f^{5/3}}{Q^{1/6}} \qquad [14.47]$$

in which the silt factor f is given by

$$f = 1.76 \sqrt{d} \qquad [14.48]$$

where d is in millimetres, P and R are in metres and Q is in metres cubed per second. The silt factor f is a function of its size, as indicated in Table 14.6. Combining Equations 14.45–14.47, Lacey suggested the resistance equation

$$V = 10.8 R^{2/3} S^{1/3} \qquad [14.49]$$

Table 14.5 Recommended b/y_0 values.

Q (m³/s)	5.0	10.0	15.0	50.0	100.0	200.0	300.0
p $(=b/y_0)$	4.5	5.0	6.5	9.0	12.0	15.0	18.0

Chapter 14

Table 14.6 Silt factor f.

Type of sand	f	d (mm)
Very fine silt	0.5	0.081
Fine silt	0.6	0.120
Medium silt	0.85	0.233
Standard silt	1.0	0.323
Medium sand	1.25	0.505
Coarse sand	1.50	0.725

Equation 14.49 is commonly used in the Indian subcontinent practice in designing stable (regime) channels. Normally an additional margin of flow depth (freeboard) is provided in the design to allow any water level fluctuations. The recommended freeboards as functions of discharges are shown in Table 14.7.

3. Blench's approach

Blench developed more rational formulae (using flume and Indian subcontinent data), taking into account the effects of bank cohesiveness on channel geometry and sediment load. In a channel of mean width b and mean depth y_0, the discharge Q is written as

$$Q = Vby_0 \qquad [14.50]$$

Blench introduced bed and side factors as $f_b\ (=V^2/y_0)$ and $f_s\ (=V^3/b)$, respectively, and wrote

$$b = \sqrt{\frac{f_b Q}{f_s}} \qquad [14.51]$$

$$y_0 = \left(\frac{f_s Q}{f_b^2}\right)^{1/3} \qquad [14.52]$$

$$\frac{V^2}{gy_0 S} = 3.63\left(\frac{Vb}{v}\right)^{1/4} \qquad [14.53]$$

$$S = \frac{f_b^{5/6} f_s^{1/12} v^{1/4}}{11.91 g Q^{1/6}\,[1 + (c/2330)]} \qquad [14.54]$$

where c is the sediment concentration in parts per millimetre by weight and v the kinematic viscosity of water. He suggested $f_s = 0.1$–0.3 for slight to high cohesivity and $f_b = 1.9\sqrt{d}(1 + 0.012c)$, d being in millimetres.

Table 14.7 Recommended freeboards for canals.

Q (m³/s)	<0.75	0.75–1.50	1.50–85.0	>85.0
Freeboard (m)	0.45	0.60	0.75	0.90

Table 14.8 Regime equations of Simons and Albertson (Garde and Ranga Raju, 1991).

Type of channel		Sand bed and banks	Sand bed and cohesive banks	Cohesive bed and banks	Coarse noncohesive boundary
P	m	6.33	4.74	4.63	3.44
	n	0.512	0.512	0.512	0.512
A	m	2.57	2.25	2.25	0.939
	n	0.873	0.873	0.873	0.873
R	m	0.403	0.475	0.557	0.273
	n	0.361	0.361	0.361	0.361

4. Simons–Albertson's method

Regime channel data from the United States, Punjab and Sind (Indian subcontinent) were analysed by Simons and Albertson; their modified regime equations have a wider applicability. The channels are classified according to the nature of the bed and bank material (see Table 14.8) and the following equations were suggested:

$$b = 0.92B - 0.60 \tag{14.55}$$

where b is the average width and B the water surface width (in metres), and

$$P, A \text{ and } R = mQ^n \tag{14.56}$$

where the coefficients m and n are given in Table 14.8.
   The following resistance equations were also proposed by Simons and Albertson:

$$\text{Sand bed and banks:} \quad V = 9.33(R^2 S)^{1/3} \tag{14.57}$$

$$\frac{V^2}{g y_0 S} = 0.885 \left(\frac{Vb}{\nu} \right)^{0.37} \tag{14.58}$$

$$\text{Sand bed and cohesive banks:} \quad V = 10.8(R^2 S)^{1/3} \tag{14.59}$$

$$\frac{V^2}{g y_0 S} = 0.525 \left(\frac{Vb}{\nu} \right)^{0.37} \tag{14.60}$$

$$\text{Coarse noncohesive material:} \quad V = 4.75(R^2 S)^{0.286} \tag{14.61}$$

$$\frac{V^2}{g y_0 S} = 0.324 \left(\frac{Vb}{\nu} \right)^{0.37} \tag{14.62}$$

The slope equations (Equations 14.58, 14.60 and 14.62) are recommended for $Vb/\nu <$ 2×10^7, and Equations 14.57, 14.59 and 14.61 may be preferred to Equation 14.54 for the determination of the slope when $Vb/\nu > 2 \times 10^7$. It has been suggested that the flow Froude number be kept less than 0.30 for stability considerations.

Chapter 14

5. Non-scouring erodible boundary channel design

This method approaches the criterion that the bed material (coarse) does not move when the channel carries either clear water or water with fine silt in suspension (not depositing). The principle of design is to achieve a cross section in which the boundary material is on the verge of motion (initiation criterion). The method utilises the information on boundary shear distribution and Shields' initiation criterion (on both bed and banks) and establishes either permissible depth or slope (given one or the other). The Manning resistance equation with the appropriate n value $(=d^{1/6}/26)$ further establishes the bed width required to transport the design discharge. (See Example 14.6 for the detailed design procedures.)

In the above approach it is to be noted that not all the boundary particles are on the verge of motion (side slopes are less sustainable) and such a section is not economical/efficient. The most desirable section (bed and bank material at the incipient motion) is of the following profile (Glover and Florey, 1951):

$$y = y_0 \cos \left(\frac{0.8x}{y_0} \tan \phi \right) \qquad [14.63]$$

The design procedures using Equation 14.63 are illustrated in Example 14.7.

6. Design of stable erodible boundary channel

The most important physical processes in the formation of stable channels are now well documented, and White et al. (1980) proposed a solution using the Ackers–White sediment transport model coupled with the resistance equation for alluvial channels.

For a defined channel boundary material (i.e. known d and ρ_s) and water viscosity, there are six channel/sediment parameters: Q, Q_s, V, B, y_0 and S. Three equations (continuity, the Ackers–White transport equation and the resistance equation) and a fourth one based on the variational principle – minimum stream power (i.e. maximised transport with least energy expenditure) – would then facilitate the design procedures if two of the six parameters are stipulated. For further detailed information and design tables based on this method, see White et al. (1981), reviewed with proposed modifications by Valentine and Haidera (2005).

14.11 Rigid-bed channels with sediment transport

Rigid-bed channels are the conveyances with no boundary erosion and the sediment is fed from external source (e.g. lined irrigation canals carrying silt, sewers and outfalls). The channel is designed for no deposition criteria. The mode of transport and design criteria largely depend on the sediment and channel characteristics. A great deal of research into the areas of sediment initiation, transport, cohesivity aspects of sediments and so on has taken place (e.g. Novak and Nalluri, 1975; Butler et al., 2003).

Studies of noncohesive suspended silt reveal (Garde and Ranga Raju, 1991) that the limiting (for no deposition) concentration (C_v) is a function of sediment size (d), density (ρ_s), fall velocity (W_s), water discharge (Q), flow depth (y_0), water surface width (B), channel slope (S) and bed friction factor (λ_b). Figure 14.8 shows the proposed relationship valid for circular, rectangular and trapezoidal channels.

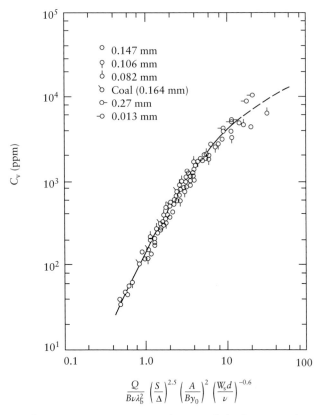

Figure 14.8 Limit-deposition concentration of suspended silt in circular, rectangular and trapezoidal channel.

Novak and Nalluri (1984) suggested the following equation for initiation of noncohesive coarser sediment (bed load):

$$\frac{V_c}{\sqrt{gd_{50}\Delta}} = 0.50 \left(\frac{d_{50}}{R}\right)^{-0.4}$$ [14.64]

The limit-deposition criterion in the case of rectangular channels is given by (Mayerle et al., 1991)

$$\frac{V_s}{\sqrt{g\Delta d_{50}}} = 11.59 D_{gr}^{-0.4} C_v^{0.15} \left(\frac{d_{50}}{R_b}\right)^{-0.43} \lambda_s^{-0.18}$$ [14.65]

where C_v is the limiting sediment concentration by volume that can be transported with a velocity V_s (self-cleansing). The overall friction factor λ_s (Nalluri and Kithsiri 1992) is given by

$$\lambda_s = 0.851 \lambda_c^{0.86} C_v^{0.04} D_{gr}^{0.03}$$ [14.66]

where λ_c is the channel's clear water friction factor given by Colebrook–White's equation (Equation 4.15). Equations 14.65, 14.66 and 4.15 will give (by iterative solution) the

Chapter 14

design velocity for the self-cleansing criterion in rigid-boundary rectangular channels with bed load transportation.

In the case of clean pipe channels, the limit-deposition criterion may be written as (Nalluri *et al.*, 1994)

$$\frac{V_s}{\sqrt{g\Delta d_{50}}} = 3.08 D_{gr}^{-0.09} C_v^{0.21} \left(\frac{d_{50}}{R}\right)^{-0.53} \lambda_s^{-0.21} \qquad [14.67]$$

with the friction factor λ_s given by

$$\lambda_s = 1.13 \lambda_c^{0.98} C_v^{0.02} D_{gr}^{0.01} \qquad [14.68]$$

where λ_c is the clear water friction factor given by Equation 4.15.

The limit-deposition criterion in the case of pipe channels with deposited flat beds of width b is given by

$$\frac{V_s}{\sqrt{g\Delta d_{50}}} = 1.94 C_v^{0.165} \left(\frac{b}{y_0}\right)^{-0.4} \left(\frac{d_{50}}{D}\right)^{-0.57} \lambda_{sb}^{0.10} \qquad [14.69]$$

The bed friction factor with transport λ_{sb} is given by

$$\lambda_{sb} = 6.6 \lambda_s^{1.45} \qquad [14.70]$$

where λ_s is given by

$$\lambda_s = 0.88 C_v^{0.01} \left(\frac{b}{y_0}\right)^{0.03} \lambda_c^{0.94} \qquad [14.71]$$

where λ_c is the clear water friction factor given by Equation 4.15.

It must be stressed that the proposed Equations 14.64–14.71 are developed by analysing experimental data and are valid within the experimental ranges of the data used.

Worked examples

Example 14.1

(a) Starting from first principles, show that the fall (sedimentation) velocity W_s of a particle size d in a fluid is given by

$$W_s = A\sqrt{g\Delta d}$$

where A is a function of the drag coefficient C_D and Δ the relative density of the particle in water $[=(\rho_s - \rho)/\rho]$.

Assuming that for particles of SF = 1,

$$C_D = \text{Constant} \ (=2) \text{ for large diameters}$$

$$\text{and } C_D = \frac{24}{Re_d} \text{ for very fine particles}$$

give the full equation for W_s in each of these cases.

(b) Examine the stability of the bed material (ρ_s = 2650 kg/m³, mean diameter = 1 mm) of a wide stream having a slope of 10^{-3} and carrying a flow at a depth of 0.3 m.

(c) What type(s) of transport and bed form, if any, do you expect in this stream?

Solution:

(a) Equating gravity force (weight of the particle) to drag force

$$\frac{(\rho_s - \rho)g\pi d^3}{6} = \frac{1}{2}C_D\rho\left(\frac{\pi d^2}{4}\right)W_s^2$$

or

$$W_s^2 = \frac{4\Delta gd}{3C_D}$$

or

$$W_s = \sqrt{\left(\frac{4}{3C_D}\right)(g\Delta d)}$$

Comparing this with the given equation,

$$A = \sqrt{\frac{4}{3C_D}} = f(Re_d)$$

since the drag coefficient C_D is a function of Re_d:

Coarse sediment: $C_D = 2$ (given)

$$\Rightarrow W_s = \sqrt{\left(\frac{2}{3}\right)g\Delta d}$$

Fine sediment: $C_D = \dfrac{24}{Re_d} = \dfrac{24v}{W_s d}$

$$\Rightarrow A = \sqrt{\frac{W_s d}{18v}}$$

Hence $W_s = g\Delta\dfrac{d^2}{18v}$

(b) Threshold condition: Shields criterion
Wide channel $\Rightarrow R \approx y_0 = 0.3$ m
By constructing a d (=1 mm) line on Figure 14.1 (Shields diagram) we obtain

$$\frac{\tau_c}{\rho g\Delta d} = 0.035$$

or $\tau_c = \rho U_*^2 = 0.035 \times \rho g \times 1.65 \times d$

Available boundary shear stress

$$\tau = \rho gRS = 2.943 \text{ N/m}^2$$

Thus minimum d for stability = 5.2 mm > 1 mm. Hence the bed material is not stable.

(c) Available $\tau/\rho g\Delta d = U_*^2/g\Delta d = RS/(1.65 \times 0.001) = 0.181$, and from Figure 14.1 this relates to bed dunes to high regime plane bed transition.

Chapter 14

Example 14.2

It is intended to stabilise a river bed section with the following data by depositing a layer of gravel or stone pitching:

$$\text{Channel width} = 20 \text{ m}$$
$$\text{Bed slope} = 0.0045$$
$$\text{Maximum discharge} = 500 \text{ m}^3/\text{s}$$
$$\text{Chezy's } C = 18 \log(12R/d) \quad (d \text{ is the mean diameter of the material})$$
$$\text{Submerged relative density of the bed material } \Delta = 1.65$$

Determine the depth of flow assuming the section to be rectangular and the minimum size of stone required for stability. Use Shields criterion for stability: $\tau/\rho g \Delta d = 0.05$.

Solution:

From Shields criterion,

$$d = 12RS$$
$$\Rightarrow \text{Chezy's } C = 18 \log\left(\frac{12R}{12RS}\right)$$
$$= 18 \log\left(\frac{1}{S}\right)$$
$$= 42.24 \text{ m}^{1/2}/\text{s}$$

$$\text{Mean velocity, } V = \frac{Q}{A} = \frac{500}{20 \times y_0} = 42.24\sqrt{RS} \quad \text{(i)}$$

$$\text{Hydraulic radius, } R = \frac{20y_0}{20 + 2y_0} \quad \text{(ii)}$$

Equations (i) and (ii) give

$$y_0 \approx 4.9 \text{ m} \quad \text{(by iteration)}$$
$$\Rightarrow R = 3.29 \text{ m} \quad \text{giving } d \, (=12RS) = 147 \text{ mm}$$

Hence provide an armour layer with 150 mm size stones.

Example 14.3

The following data relate to a wide stream:

$$\text{Slope} = 0.0001$$
$$\text{Bed material: size, } d = 0.4 \text{ mm}$$
$$\text{density, } \rho_s = 2650 \text{ kg/m}^3$$

(a) Find the limiting depth of flow at which the bed material just begins to move.
(b) Find the corresponding mean velocity in the stream.

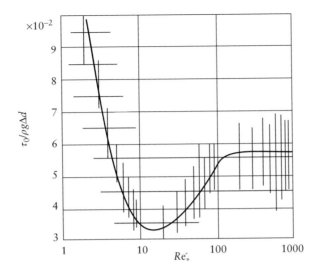

Figure 14.9 Modified Shields curve.

Solution:

(a) Problems of this kind may explicitly be solved by using the modified Shields diagram (Figure 14.9):
Modified Re_* can be written as

$$Re'_* = \frac{(Re_*)^{2/3}}{(\tau/\rho g \Delta d)^{1/3}}$$

$$= \frac{(\Delta g)^{1/3} d}{v^{2/3}} \quad (=d_{gr})$$

$$Re'_* = \frac{(1.65 \times 9.81)^{1/3} \times 0.0004}{(10^{-6})^{2/3}}$$

$$= 10.12$$

From Figure 14.9,

$$\frac{\tau_c}{\rho g \Delta d} = 0.035$$

$$\text{giving } \tau_c = 0.227 \text{ N/m}^2$$

Boundary shear in wide channel, $\tau_c = \rho g y_0 S$

Hence for critical condition, $\rho g y_0 S = 0.227$

$\Rightarrow$ The limiting flow depth, $y_0 = 0.231$ m

(b) At the threshold condition the bed is plane with roughness, $k = d$

$$\text{Chezy's } C \text{ (in transition)} = 18 \log \left[\frac{12R}{(k + 2\delta'/7)} \right] \quad \text{(Equation 8.41)}$$

Chapter 14

where δ' is the sub-layer thickness given by $\delta' = 11.6v/U_*$.

$$k = d = 0.0004 \, \text{m}, \quad \delta' = \frac{11.6 \times 10^{-6}}{\sqrt{9.81 \times 0.231 \times 0.0001}}$$
$$= 7.7 \times 10^{-5} \, \text{m}$$

$$\Rightarrow k + \frac{2\delta'}{7} = 0.0004 + 0.000022 = 0.000422 \, \text{m}$$

$$\Rightarrow C = 18 \log \left(\frac{12 \times 0.231}{0.000422} \right) = 54.7 \, \text{m}^{1/2}/\text{s}$$

Hence mean velocity, $V = C\sqrt{RS}$

$$= 54.7 \times \sqrt{0.231 \times 0.0001}$$
$$= 0.263 \, \text{m/s}$$

Example 14.4

A wide alluvial stream carries water with a mean depth of 1 m and slope of 0.0005. The mean diameter of the bed material is 0.5 mm with a relative density of 2.65. Examine the bed stability and bed form, if any. Also calculate the sediment transport rate that may exist in the channel.

Solution:

Wide channel $\Rightarrow R \approx y_0 = 1.0 \, \text{m}; S = 0.0005; d = 0.0005 \, \text{m}$
$$\Rightarrow d_{gr} = 12.5$$
From Figure 14.9,

$$\frac{\tau}{\rho g \Delta d} = 0.32$$
$$\Rightarrow \tau_c = 0.032 \times 1000 \times 9.81 \times 1.65 \times 0.0005$$
$$= 0.234 \, \text{N/m}^2$$

Channel boundary shear stress, $\tau_0 = \rho g R S = 4.9 \, \text{N/m}^2$
Since $\tau_0 > \tau_c$, sediment transport exists.

Type of transport:
Fall velocity of sediment particle from Figure 14.4

$$W_s = 0.075 \, \text{m/s}$$
$$\Rightarrow \text{Movability parameter,} \quad \frac{U_*}{W_s} = 0.93$$

Referring to Table 14.2, most part of the transport may be treated as bed load which may be computed using Shields' equation (Equation 14.16).

Discharge computations:

Referring to Garde and Ranga Raju's plots (Figures 14.2 and 14.3), mean velocity in the channel may be determined.

$$\text{For } d = 0.5 \text{ mm}; \ K_1 = K_2 = 0.95 \quad \text{(Figure 14.3)}$$

$$\Rightarrow K_2 \left(\frac{R}{d}\right)^{1/3} \frac{S}{\Delta} = 3.63 \times 10^{-3}$$

From Figure 14.2,

$$\frac{K_1 V}{\sqrt{g \Delta R}} = 0.20$$

Hence $V = 0.847$ m/s, giving $q = V y_0 = 0.847 \, \text{m}^3/(\text{m s})$.

From Shields' equation 14.19, the bed load transport

$$q_b = \frac{10(\tau_0 - \tau_c) \, qS}{\rho g \Delta^2 d}$$

$$= \frac{10 \times (4.9 - 0.292) \times 0.847 \times 0.0005}{1000 \times 9.81 \times 1.65^2 \times 0.0005}$$

$$= 1.46 \times 10^{-3} \, \text{m}^3/(\text{m s})$$

$$\approx 40 \, \text{N/(m s) or 4 kg/(m s)}$$

Bed form:

$$\frac{\tau_0}{\rho g \Delta d} = \frac{4.9}{1000 \times 9.81 \times 1.65 \times 0.0005} = 0.61$$

From Shields' curve (Figure 14.1), the bed form may be in high regime transition to antidunes; however, the flow Froude number ($=V/\sqrt{g y_0}$) is less than 1 and the bed may be in dune to high regime transition form.

Equivalent bed roughness (k) or dune height:

Using Chezy's equation, $V = C\sqrt{(RS)}$

$$\text{Chezy's } C = 0.847/\sqrt{(1 \times 0.0005)} = 37.88 \, \text{m}^{1/2}$$

$$\text{Hence from } C = 18 \log{(12R/k)} \text{ (assuming fully rough bed)}$$

$$k = 9.5 \times 10^{-2} \, \text{m or 95 mm}$$

Example 14.5

A laboratory rectangular flume with smooth sides and rough alluvial bed ($d = 6.5$ mm) of the following data carries 0.1 m³/s of water:

$$\text{Bed width} \quad = 0.5 \, \text{m}$$
$$\text{Depth of flow} = 0.25 \, \text{m}$$
$$\text{Slope} \quad = 3 \times 10^{-3}$$

Find the bed hydraulic radius using the resistance curve for smooth sides (Figure 14.9), bed shear stress and Manning's coefficient. Also examine the stability of the bed.

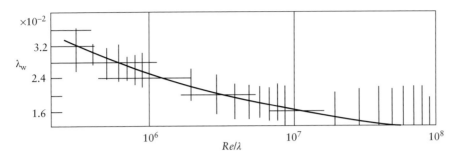

Figure 14.10 Resistance curve for smooth walls.

Solution:

$$\text{Overall hydraulic radius, } R = \frac{A}{P} = \frac{0.5 \times 0.25}{0.5 + 2 \times 0.25}$$
$$= 0.125 \, \text{m}$$

$$\text{Velocity, } V = \frac{Q}{A} = \frac{0.1}{0.5 \times 0.25} = 0.8 \, \text{m/s}$$

$$\Rightarrow \text{Reynolds number, } Re = \frac{4VR}{\nu} = 4 \times 10^5$$

$$\text{Overall friction factor, } \lambda = \frac{8gRS}{V^2} = 0.046$$

$$\Rightarrow \frac{Re}{\lambda} = 8.7 \times 10^6$$

and, from Figure 14.10, wall friction factor $\lambda_w = 0.017$.
From Equation 14.11,

$$\Rightarrow P\lambda = P_b\lambda_b + P_w\lambda_w$$
$$\text{Bed friction factor, } \lambda_b = 0.075$$

$$\Rightarrow \text{Bed hydraulic radius, } R_b = \frac{\lambda_b V^2}{8gS} \quad \text{(Equation 14.10)}$$
$$= 0.204 \, \text{m}$$

$$\text{Bed shear stress, } \tau_b = \rho g R_b S = 6 \, \text{N/m}^2$$
$$\text{From the Manning equation } V = \frac{1}{n_b}(R_b)^{2/3}S^{1/2}$$

$$n_b = 0.0237$$

Bed stability:

$$d_{gr} = 167.5$$

Hence $\tau_c/\rho g \Delta d = 0.056$ (Shields criterion) (see Figure 14.9) or critical shear stress for stability, $\tau_c = 5.9 \, \text{N/m}^2$.

As $\tau_b \simeq \tau_c$, the bed is just stable.

Note: Overall Manning's $n = R^{2/3}S^{1/2}/V = 0.017$, which may be compared with Strickler's equation giving $n = d^{1/6}/26 = 0.0166$ with $k = d$.

Example 14.6

Design a stable alluvial channel of trapezoidal cross section with the following data:

$$\text{Discharge} = 50 \text{ m}^3/\text{s}$$
$$\text{Bed material size} = 4 \text{ mm}$$
$$\text{Angle of repose, } \phi = 30°$$
$$\text{Bed slope} = 10^{-4}$$
$$\text{Channel side slopes} = 2 \text{ (horizontal)} : 1 \text{ (vertical)} \left(\tan\theta = \frac{1}{2}\right)$$

Solution:

$$Re'_* = \frac{(\Delta g)^{1/3}d}{v^{2/3}} = 100.8$$

From Figure 14.9,

$$Fr_d^2 = \frac{\tau}{\rho g \Delta d} = 0.054$$

$\Rightarrow$ Critical bed shear stress, $\tau_{bc} = 0.054 \times 1000 \times 9.81 \times 1.65 \times 0.004$
$$= 3.496 \text{ N/m}^2$$

Hence critical shear on slopes, $\tau_{sc} = K\tau_{bc}$

$$\text{where } K = \left(1 - \frac{\sin^2\theta}{\sin^2\phi}\right)^{1/2}$$
$$= \left(1 - \frac{0.2}{0.25}\right)^{1/2}$$
$$= 0.447$$
$$\Rightarrow \tau_{sc} = 0.447 \times 3.496 = 1.563 \text{ N/m}^2$$

Mean boundary shear distribution in a 2 (horizontal) : 1 (vertical) trapezoidal channel (see Figure 8.5):

$$\text{For } \frac{B}{y_0} > 10, \frac{\tau_{bm}}{\rho g y_0 S} \approx 0.985$$
$$\frac{\tau_{sm}}{\rho g y_0 S} \approx 0.78$$

Using critical shear on bed as the criterion for stability,

$$\tau_{bc} = \tau_{bm} = 0.985 \times \rho g y_0 S$$

Limiting flow depth, $y_0 = 3.62$ m

Using critical shear on sides as the criterion,

$$\tau_{sc} = \tau_{sm} = 0.78 \times \rho g y_0 S$$

Limiting flow depth, $y_0 = 2.04$ m

Choose $y_0 = 2$ m (smaller of the two criteria).

The channel boundary is in threshold condition; thus, the bed is plane, with roughness $k = d$.

$$\text{Manning's } n = \frac{d^{1/6}}{26} \quad \text{(Strickler's formula)}$$
$$= 0.0153$$
$$\text{Mean velocity, } V = \frac{1}{n} R^{2/3} S^{1/2} \quad \text{(the Manning formula)}$$

Select various B/y_0 values and calculate Q (see Table 14.9).

Adopt a trapezoidal section of bed width $= 23$ m

Flow depth $= 2$ m

Example 14.7

The most economical section (i.e. in which the particles are at threshold all over its perimeter) of an alluvial channel is given by (see Figure 14.11)

$$y = y_0 \cos\left(\frac{0.8x \tan \phi}{y_0}\right) \qquad [14.63]$$

where ϕ is the angle of repose of the bed material.

Calculate the maximum possible discharge that such a channel in an alluvium ($d = 4$ mm, $\phi = 30°$) with a bed slope of 10^{-3} can carry.

Design the most economical section to carry a discharge of 1 m³/s.

Table 14.9 Design of stable channel.

B/y_0	y_0 (m)	B (m)	A (m²)	P (m)	R (m)	V (m/s)	Discharge Q (m³/s)
10	2	20	48	28.94	1.659	0.916	43.97
11.5	2	23	54	31.04	1.691	0.924	50.09
12	2	24	56	32.94	1.700	0.931	52.14

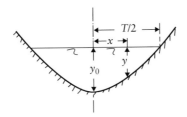

Figure 14.11 Cosine profile – most economical section.

Solution:

The parameters like area (A), perimeter (P), hydraulic radius (R) and water surface width (T) are all functions of flow depth y_0 and ϕ (see Table 14.10).

(*Note:* In order to accommodate lift forces on particles, the working value of ϕ may be taken as $\tan^{-1}(0.8 \tan \phi)$.)

Limiting bed shear for $d = 4\,\text{mm}$,

$$\tau_{bc} = 3.496\,\text{N/m}^2 \quad (\text{Shields})$$

As the shear distribution is uniform and equal to $\rho g y_0 S$,

$$\rho g y_0 S = 3.496$$
$$\text{flow depth, } y_0 = 0.356\,\text{m}$$

The boundary cross section of the channel is given by Equation 14.63:

$$y = 0.356 \cos (1.3x)$$

From Table 14.10 we can get for $\phi = \tan^{-1}(0.8 \tan 30°) = 24.8°$, $A = 0.54\,\text{m}^2$, $R = 0.214\,\text{m}$ and $T = 2.43\,\text{m}$.

$$\text{Manning's } n = \frac{d^{1/6}}{26} = 0.0153$$

$$\text{Mean velocity, } V = \frac{(0.214)^{2/3}(0.001)^{1/2}}{0.0153}$$
$$= 0.739\,\text{m/s}$$

$$\text{Maximum safe discharge, } Q_0 = AV = 0.4\,\text{m}^3/\text{s}$$
$$\text{Design discharge, } Q = 1\,\text{m}^3/\text{s}$$

as $Q > Q_0$ provides an additional central rectangular section (Figure 14.12) to accomodate the excess flow.

Table 14.10 Economic alluvial section as a function of ϕ.

ϕ	15°	20°	25°	30°	35°	40°	45°
A/y_0^2	7.5	5.4	4.21	3.46	2.79	2.31	2.00
P/y_0	12	8.8	7.0	5.8	4.9	4.2	3.85
R/y_0	0.625	0.615	0.602	0.588	0.570	0.550	0.520
T/y_0	11.72	8.63	6.74	5.44	4.49	3.74	3.14

Chapter 14

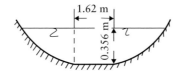

Figure 14.12 Cosine profile with flat bed for large discharges.

Velocity in the central section using the Manning equation

$$V = \frac{(0.356)^{2/3}(0.001)^{1/2}}{0.0153} = 1.038 \text{ m/s}$$

$$\text{Additional bed width, } b = \frac{(1.0 - 0.4)}{(1.038 \times 0.356)} = 1.62 \text{ m}$$

(*Note*: For discharges $Q < Q_0$, the section may be shortened by reducing T to $(T - 2x)$ as per Table 14.11.)

Example 14.8

The following data refer to an alluvial canal:

$$\text{Average water surface slope} = 5 \times 10^{-4}$$
$$\text{Average water depth} = 4.82 \text{ m}$$
$$\text{Width} = 52.5 \text{ m}$$
$$\text{Average velocity} = 2.43 \text{ m/s}$$
$$\text{Grain size distribution: } d_{90} = 50 \text{ mm}$$
$$d_{50} = 20 \text{ mm}$$

Estimate the bed load transport in the canal.

Solution:

$$\text{Hydraulic radius, } R = \frac{52.5 \times 4.82}{52.5 + 2 \times 4.82} = 4.07 \text{ m}$$

$$C_{\text{channel}} = \frac{V}{\sqrt{RS}} = 53.87 \text{ m}^{1/2}/\text{s}$$

$$C_{\text{grain}} = 18 \log \left(\frac{12R}{d_{90}} \right)$$

Table 14.11 Design table for $Q < Q_0$.

$2x/T$	0	0.2	0.4	0.6	0.8	1.0
Q/Q_0	1	0.615	0.310	0.110	0.015	0

(*Note*: Grain roughness exposed to flow is equivalent to d_{90} as d_{50} is eroded and kept in motion as sediment load.)

$$C_{grain} = 53.8\, m^{1/2}/s$$

As $C_{channel} = C_{grain}$,

Ripple factor, $\mu = 1 \Rightarrow$ flat bed

Discharge in canal $= 4.82 \times 2.43 = 11.7\, m^3/(m\,s)$

Bed shear, $\tau_0 = \rho gRS = 19.96\, N/m^2$

Critical shear stress for d_{50},

$$\tau_c = 0.056\, \rho g\Delta d = 18.1\, N/m^2$$

Bed load transport:

(i) Meyer–Peter and Muller's equation (Equation 14.22):

$$q_b = 8(1.65 \times 9.81)^{1/2}(0.02)^{3/2}\left(\frac{\mu RS}{\Delta d} - 0.047\right)^{3/2}$$

$$\frac{\mu RS}{\Delta d} = 0.062$$

$$\Rightarrow q_b = 1.67 \times 10^{-4}\, m^3/(m\,s)$$

or total bed load $\approx 23\, kg/s$

(ii) Einstein's curve (Figure 14.5):

$$\psi = \frac{\Delta d}{\mu RS} = 16.12 \quad \text{giving } \phi = 0.015$$

$$\Rightarrow q_b = 0.015(g\Delta d^3)^{1/2} = 1.71 \times 10^{-4}\, m^3/(m\,s)$$

or total bed load $\simeq 24\, kg/s$

(iii) Schoklitsch's equation (Equation 14.17):

Bed load in kilogram per metre second, $g_s = 2500S^{3/2}(q - q_{cr})$

From Equation 14.18,

$$q_{cr} = 9.27\, m^3/(m\,s)$$

$$\Rightarrow g_s = 6.79 \times 10^{-2}\, kg/(m\,s)$$

or total load $= 3.6\, kg/s$

(iv) Shields' equation (Equation 14.16):

$$q_b = 10qS\left(\frac{\tau_0 - \tau_c}{\rho g\Delta^2 d}\right)$$

$$= 5.2 \times 10^{-4}\, m^3/ms$$

or total load $= 72\, kg/s$

Shields' equation is applicable for the sediment range of $1.56 < d\, (mm) < 2.47$ and is not reliable in this case where $d = 20\, mm$. Also Schoklitsch's equation is not dimensionally homogeneous and does not include coarse material. Hence the probable total bed load in the canal is around 30 kg/s.

Example 14.9

The following data refer to a wide river:

$$\text{Depth} = 3\,\text{m}$$
$$\text{Mean velocity} = 1\,\text{m/s}$$
$$\text{Chezy's coefficient} = 50\,\text{m}^{1/2}/\text{s}$$
$$\text{Density of sediment} = 2650\,\text{kg/m}^3$$

A sample of wash load (mean diameter of 0.02 mm) taken at half depth of flow showed a concentration of 200 mg/L (200 ppm).

(i) Establish the wash load concentration distribution as a function of depth, and calculate the rate of suspended load assuming a homogeneous distribution.
(ii) Determine the transport rate using Lane and Kalinske's approximate method, and compare with the above result.

Solution:

From Chezy's formula, $(RS)^{1/2} = V/C = 0.02$,

$$\text{shear velocity, } U_* = (gRS)^{1/2} = 0.063\ \text{m/s}$$
$$\text{fall velocity, } W_s = 0.00035\ \text{m/s} \quad \text{(Figure 14.4)}$$
$$\text{movability parameter, } \frac{U_*}{W_s} = \frac{0.063}{0.00035} = 180$$

Referring to Table 14.2, the sediment is almost in homogeneous suspension.

$$\text{Total suspended load} = \frac{200 \times 3 \times 1 \times 1000}{10^6} = 0.6\ \text{kg/(m\,s)}$$

Distribution of sediment concentration:
From Equation 14.29,

$$\Rightarrow a = 1.5\ \text{m}; \quad c_a = 200\ \text{mg/L}; \quad y_0 = 3\ \text{m}$$
$$W_s = 0.00035\ \text{m/s} \quad \text{(Figure 14.4.)}$$
$$\beta = 1; \quad \chi = 0.4$$
$$c(y) = 200 \left[\frac{1.5(3-y)}{1.5y}\right]^{0.00035/(0.4\times0.063)}$$
$$= 200 \left(\frac{3-y}{y}\right)^{0.0139} \quad \text{mg/L}$$

Equation 14.31 (Lane and Kalinske):

$$\Rightarrow \frac{q_s}{q} = 200 P_e^{15(1/2)0.00555}$$

P from Figure 14.7:

$$\text{Manning's } n = \frac{R^{2/3}S^{1/2}}{V}$$

$$S^{1/2} = \frac{0.02}{3^{1/2}} = 0.01155$$

$$\Rightarrow n = 3^{2/3} \times \frac{0.01155}{1} = 0.024$$

$$\Rightarrow \frac{n}{y_0^{1/6}} = 0.02$$

and hence $P = 1$ at $W_s/U_* = 0.000555$.

$$\Rightarrow \frac{q_s}{q} = 200 \times 1 \times e^{0.0416} = 200 \times 1.04$$

Hence $q_s = 208 \times 1 \times 3 \times (1000/10^6) = 0.624$ kg/(m s).

Example 14.10

An alluvial river with the following data discharges water into a downstream reservoir of 10×10^6 m^3 capacity.

$$\text{Width} = 12 \text{ m}$$
$$\text{Depth} = 4 \text{ m}$$
$$\text{Slope} = 3 \times 10^{-4}$$
$$\text{Discharge} = 75 \text{ m}^3/\text{s}$$
$$\text{Bed material size, } d_{50} = 0.5 \text{ mm}$$
$$\text{Density of bed material} = 2650 \text{ kg/m}^3$$

Determine the total sediment transport rate by Ackers–White's formula in the river, and compare the result with that of Graf's formula. What is the life expectancy of the reservoir fed by this river?

Solution:

$$R = \frac{A}{P} = \frac{48}{20} = 2.4 \text{ m}$$

Graf's formula (Equation 14.36):

$$\Rightarrow \psi_A = \frac{\Delta d}{RS} = 1.146$$

$$\phi_A = 10.39 \times (1.146)^{-2.52} = 7.37$$

Hence

$$C_v \frac{VR}{\sqrt{g\Delta d^3}} = 7.37$$

$$\text{or} \quad \frac{1.56 \times 2.4 C_v}{4.5 \times 10^{-5}} = 7.37$$

$$C_v = 8.86 \times 10^{-5}$$

$$\text{Total transport rate} = 8.86 \times 10^{-5} \times 75 = 6.6 \times 10^{-3} \ \text{m}^3/\text{s}$$

Ackers–White's approach:
Using kinematic viscosity for water at 15°,

$$d_{gr} = 0.0005 \left[\frac{9.81 \times 1.65}{(1.14 \times 10^{-6})^2} \right]^{1/3} = 11.6 \quad \text{(Equation 14.41)}$$

For d_{gr} in transition $(1 < d_{gr} < 60)$,

$$n = 1 - 0.56 \log d_{gr} = 0.404$$

$$A_{gr} = 0.14 + \frac{0.23}{\sqrt{d_{gr}}} = 0.208$$

$$m = 1.67 + \frac{6.83}{d_{gr}} = 2.26$$

$$\log C = 2.79 \log d_{gr} - 0.98 (\log d_{gr})^2 - 3.46 = -1.60$$

$$\Rightarrow C = 0.0251$$

$$U_* = (9.81 \times 2.4 \times 0.0003)^{1/2} = 0.084 \ \text{m/s}$$

$$U_*^n = 0.368; \quad (g\Delta d)^{1/2} = 0.0900$$

$$V = \frac{Q}{A} = \frac{75}{48} = 1.56 \ \text{m/s}$$

$$\sqrt{32} \log \left(\frac{10 y_0}{d} \right) = 27.7$$

$$\Rightarrow F_{gr} = \left(\frac{0.368}{0.0900} \right) \left(\frac{1.56}{27.7} \right)^{1-0.404} = 0.736 \quad \text{(Equation 14.40)}$$

$$\frac{F_{gr}}{A_{gr}} = \frac{0.736}{0.208} = 3.54$$

$$\Rightarrow C \left(\frac{F_{gr}}{A_{gr}} - 1 \right)^{2.26} = 0.206$$

$$\frac{q_t}{q} = \left(\frac{0.206 \times 0.0005}{4} \right) \left(\frac{1.56}{0.084} \right)^{0.404} = 8.38 \times 10^{-5} \quad \text{(Equation 14.39)}$$

$$\Rightarrow Q_t = 8.38 \times 10^{-5} \times 75 = 6.3 \times 10^{-3} \ \text{m}^3/\text{s}$$

Reservoir life expectancy:

$$\text{Ackers–White} \Rightarrow \text{Annual deposit} = 6.3 \times 10^{-3} \times 365 \times 24 \times 60 \times 60$$

$$= 2.0 \times 10^5 \ \text{m}^3/\text{year}$$

$$\Rightarrow \text{Reservoir life} = \frac{10 \times 10^6}{2.0 \times 10^5} \approx 50 \ \text{years}$$

A similar estimate results in this case from Graf's approach.

Note that porosity and consolidation of sediments will affect the filling rate.

Chapter 14

References and recommended reading

Ackers, P. (1993) Sediment transport in open channels: Ackers and White update. *Proceedings of the Institution of Civil Engineers: Water Maritime and Energy*, 101, December, 247–249.

Ackers, P. and White, W. R. (1973) Sediment transport: new approach and analysis. *Journal of Hydraulic Engineering, Proc. ASCE*, 99, HY-11, 2041–2060.

Alam, Z. U. and Kennedy, J. F. (1969) Friction factors for flow in sand-bed channels. *Journal of Hydraulic Engineering, Proceedings of ASCE*, 95, 1973–1992, HY-6.

Butler, D., May, R. W. P. and Ackers, J. (2003) Self-cleansing sewer design based on sediment transport principles. *ASCE Journal of Hydraulic Engineering*, 129, 4, 276–282.

Einstein, H. A. (1950) *The Bed Load Function for Sediment Transportation in Open Channel Flows*, T. B. no. 1026 U.S. Department of Agriculture's Soil Conservation Service, Washington, DC.

Einstein, H. A. and Barbarossa, N. L. (1952) River channel roughness. *Transactions of ASCE*, 117, 1121–1146.

Engelund, F. (1970) Instability of erodible beds. *Journal of Fluid Mechanics*, 42, 2, 225–244.

Garde, R. J. (1968) Analysis of distorted river models with movable beds. *Journal of Irrigation and Power, India*, 25, 4.

Garde, R. J. and Ranga Raju, K. G. (1966) Resistance relationships for alluvial channel flow. *Journal of Hydraulic Division, Proceedings of ASCE*, 92, HY-4.

Garde, R. J. and Ranga Raju, K. G. (1991) *Mechanics of Sediment Transportation and Alluvial Stream Problems*, 2nd edn, Wiley Eastern Ltd., New Delhi, India.

Graf, W. H. (1984) *Hydraulics of Sediment Transport*, Water Resources Publications, Highland Ranch, CO.

Glover, R. E. and Florey, Q. L. (1951) *Stable Channel Profiles*, USBR Hyd. Lab. Report No. 325, US Bureau of Reclamation, Washington, DC.

Lacey, G. (1930) Stable channels in alluvium. *Proceeding of ICE*, 229, 259–384.

Lane, E. W. and Kalinske, A. A. (1939) The relation of suspended load to bed material in rivers. *Transactions American Geophysical Union*, 20, 637–641.

Laursen, E. M. (1958) Total sediment load of streams. *Journal of Hydraulic Division, Proceeding of ASCE*, 84, 1–36, HY-1.

Mayerle, R., Nalluri, C. and Novak, P. (1991) Sediment transport in rigid bed conveyances. *Journal of Hydraulic Reserch*, 29, 4, 475–495.

Nalluri, C., Ab Ghani, A. and El-Zaemey, A. K. S. (1994) Sediment transport over deposited beds in sewers. *Water Science Technology*, 29, 1, 125–133.

Nalluri, C. and Kithsiri, M. M. A. U. (1992) Extended data on sediment transport in rigid bed rectangular channels. *Journal of Hydraulic Research*, 30, 6, 851–856.

Novak, P. and Nalluri, C. (1975) Sediment transport in smooth fixed bed channels. *Journal of Hydraulic Division, Proceeding of ASCE*, 101, 1139–1154, HY-9.

Novak, P. and Nalluri, C. (1984) Incipient motion of sediment particles over fixed beds. *Journal of Hydraulic Research*, 22, 3, 181–197.

O'Brien, M. P. (1933) Review of the theory of turbulent flow and its relation to sediment transportation. *Transactions American Geophysical Union*, 14, 487–491.

Ota, J. J. (1999) Effect of particle size and gradation on sediment transport in storm sewers, PhD thesis, University of Newcastle, UK.

Ranga Raju, K. G. (1993) *Flow through Open Channels*, 2nd edn, Tata-McGraw Hill, New Delhi, India.

Raudkivi, A. J. (1990) *Loose Boundary Hydraulics*, 3rd edn, Pergamon, Oxford.

Rouse, H. (1937) Modern conceptions of mechanics of fluid turbulence. *Transactions of ASCE*, 102, 463–543.

Chapter 14

Shields, A. (1936) Anwendung der Aehnlichkeitsmechanik und der Turbulenzforschung auf die Geschiebebewegung. *Mitteilungen der Preussischen Versuchsanstalt für Wasserbau und Schiffbau*, 26, Berlin.

Valentine, E. M. and Haidera, M. A. (2005) A modification to the Wallingford rational regime theory. *Proceedings of the Institution of Civil Engineers: Water Management*, 158 WM2, pp. 71–80.

Vanoni, V. A. and Brooks, N. H. (1957) Laboratory studies of the roughness and suspended load of alluvial streams, Caltech Report No. E–68, California Institute of Technology, Pasadena, CA.

White, W. R., Paris, E. and Bettess, R. (1980) The frictional characteristics of alluvial streams: a new approach. *Proceedings of ICE*, Part 2 Research and Theory, 69, 737–750.

White, W. R., Paris, E. and Bettess, R. (1981) *Tables for the Design of Stable Alluvial Channels*, Report No. IT 208, Hydraulics Research Ltd. Wallingford, Wallingford.

Yalin, M. S. (1992) *River Mechanics*, Pergamon, Oxford.

Problems

1. (a) Using Shields' threshold criterion $\tau/\rho g \Delta d = 0.056$ and Strickler's equation for Manning's $n = d^{1/6}/26$, show, for wide channels,
   - (i) $d = 11RS$ (Equation 14.3)
   - (ii) $V_c/\sqrt{(gd)} = 1.9\sqrt{\lambda}(d/R)^{1/6}$ (Equation 14.6)
   - (iii) $q_{cr} = 0.2(\lambda)^{5/3}d^{3/2}/S^{7/6}$ (Equation 14.18)

   (b) A flood plain river bank in fine silty sand is experiencing extensive erosion. The bank-full discharge of the river is 60 m³/s, and the section is approximately 10 m wide and 2 m deep at this discharge. The flood plain is protected by a cover of 30 mm size stone rip-rap whose friction coefficient $\lambda = 0.022$. Examine the stability of the rip-rap cover using Shields' criterion.

2. A river bed of the following data is stabilised by the deposition of a gravel layer:

$$\text{Channel width} = 12 \, \text{m}$$
$$\text{Bed slope} = 5 \times 10^{-3}$$
$$\text{Maximum discharge} = 15 \, \text{m}^3/\text{s}$$

   Determine the limiting depth of flow, assuming the section to be rectangular if the gravel size is 30 mm.

3. A straight canal with side slopes of $2\frac{1}{2}$ (horizontal) : 1 (vertical) is carrying water with a mean depth of 1 m and mean velocity of 0.87 m/s. Chezy's coefficient is around 25 m$^{1/2}$/s. Determine the minimum size of broken gravel that can be used as a protective layer around the periphery of the canal. The angle of internal friction for the stone whose density is 2600 kg/m³ may be assumed as 35°.

4. A channel of trapezoidal cross section (with side slopes of 2 (horizontal) : 1 (vertical), bed width of 8 m and flow depth of 2 m) is excavated in gravel (mean diameter of 4 mm and ϕ of 30°). Determine the limiting bed slope of the channel.

5. A channel bed is protected with 40 kg stones ($\rho_s = 2800$ kg/m³). If the flow depth in the channel is 4 m, calculate the critical velocity at which the stability of the protective layer is in danger. If the velocity in the channel exceeds this critical velocity by 20%, determine the size of the stones that will be needed for its protection.

6. A long and wide laboratory flume is to be prepared to carry out experiments to check Shields' diagram. It is proposed to cover the bed of the flume with a layer of homogeneous noncohesive material.

   Determine the unit discharge rate and slope of the bed required for a flow depth of 2 m to investigate the studies using (i) sand with $d = 0.125$ mm and (ii) gravel with $d = 4$ mm. The density of both the materials may be assumed as 2650 kg/m³ with a water temperature of 12°C.

7. The following data refer to a wide river:

   Flow depth = 2 m
   Mean velocity = 0.71 m/s
   Slope = 1/12 000
   Grain size = 1 mm
   Density of grains = 2000 kg/m³
   Settling velocity = 0.10 m/s
   Kinematic viscosity of water = 10^{-6} m²/s

   (a) (i) Calculate the rate of sediment transport in newtons per day using Meyer–Peter and Muller's formula: $\phi = (4/\psi - 0.188)^{3/2}$.
       (ii) Determine the k value of the bed, identify the possible bed formation and explain whether the bed is hydraulically smooth or rough.
       (iii) Check whether there will be suspended load or not.
   (b) If this river is discharging into a lake of constant water level and the sediment transport is interrupted at 10 km upstream of the lake, discuss, with the help of neat sketches, the consquences of the river regime along this 10 km stretch of the river.

8. In a wide stream of 3 m depth, the shear stress on the bed is estimated to be 2.4 N/m². The concentration of suspended sediment is found to be 25.8 kg/m³ at a point 0.03 m from its bed. The settling velocity of the sediment is 9.14 mm/s in still water.
   (i) Plot the profile of the sediment concentration through the depth with Karman's constant, $\chi = 0.4$.
   (ii) Assuming the velocity profile as

   $$\frac{u}{u_*} = 5.8 + 2.5 \ln \left(\frac{u_* y}{v} \right)$$

   estimate the suspended load transport per unit width of the channel.

9. Suspended particles 50% by weight with $W_s = 9.1$ mm/s and 50% with $W_s = 15.2$ mm/s are admitted to a sedimentation tank with a mean velocity of 0.152 m/s and a depth of 1.52 m. Find the fraction of removal (F) of the total load if a tank length of 30 m with a bed slope of 10^{-4} is used. Use Sumer's equation: $x/y_0 = 12V \log (1 - F)/(U_* - 10W_s)$.

10. A mountainous creek is monitored to measure the suspended load by sampling the concentration of silt at its mid-depth. For a flow depth of 1 m, the mean concentration is found to be 21 N/m³ (dry weight) with $d = 30$ μm and $\rho_s = 2650$ kg/m³. The average width of the creek at the sampling section is 22 m, and from the topographic map of the area its slope was determined to be 3.6 m/km. Stage-discharge measurements of the creek indicated that its bed roughness (k value) could be taken as 0.12 m. The water temperature during sampling was 20°C. Calculate the suspended load transport for this flow in the creek.

Chapter 14

Chapter 15
Hydraulic Structures

15.1 Introduction

A variety of hydraulic structures are available to control water levels and regulate discharges for purposes of water supply, water storage, flood alleviation, irrigation and so on. These range from the weirs/sluices of small channels to the overflow spillways of large dams. Many of these structures may also be used as discharge-measuring devices, as described in standards such as BS3680 Part 4 (1981, 1990). This chapter mainly deals with the spillways and their associated energy dissipators as examples of common hydraulic structures.

15.2 Spillways

A **spillway** is the overflow device of a dam project which is essential to evacuate excess of water, which otherwise may cause upstream flooding, dam overtopping and eventually dam failure. Basically there are three types of spillways: overfall (ogee or side channel), shaft (morning glory) and siphon. The overfall types are the most frequently encountered, and they are described here in detail. The reader may refer to Novak *et al.* (2007) for other types and further information.

15.2.1 Overfall spillways

The basic shape of the overfall spillway (ogee spillway) is derived from the lower envelope of the overfall nappe (see Figure 15.1) flowing over a high vertical rectangular notch with an approaching velocity V_a close to zero and a fully aerated space beneath the nappe (see also Section 3.17.5). For a notch of width b, head h and discharge coefficient C'_d, the

Nalluri & Featherstone's Civil Engineering Hydraulics: Essential Theory with Worked Examples,
Sixth Edition. Martin Marriott.
© 2016 John Wiley & Sons, Ltd. Published 2016 by John Wiley & Sons, Ltd.
Companion Website: www.wiley.com/go/Marriott

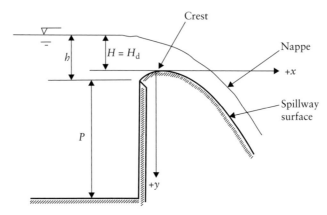

Figure 15.1 Crest of an ogee overfall spillway.

discharge equation (see also Equation 3.34) is

$$Q = \frac{2}{3}\sqrt{2gb}\,C_d' \left[\left(b + \frac{\alpha V_a^2}{2g}\right)^{3/2} - \left(\frac{\alpha V_a^2}{2g}\right)^{3/2}\right] \qquad [15.1]$$

which, for $V_a = 0$, reduces to

$$Q = \frac{2}{3}\sqrt{2gb}\,C_d'\,b^{3/2} \qquad [15.2]$$

This coefficient C_d' ($=0.62$) is valid for a rectangular sharp-crested weir.

Scimeni (see USBR, 1987) expressed the shape of the nappe in coordinates x and y, measured from the highest point (at a distance $0.282H_d$ away from the sharp-crested sill) of the water jet, for $H/H_d = 1.0$ as

$$y = Kx^n \qquad [15.3]$$

with

$$K = 0.5; \quad n = 1.85$$

For other values of H the nappes are similar and the equation can be rewritten as

$$\frac{y}{H} = K\left(\frac{x}{H}\right)^n$$

or

$$y = Kx^n H^{1-n} = 0.5x^{1.85}H^{-0.85}$$

Towards upstream of the summit point the nappe has the shape of two circular arcs, one with a radius $R = 0.5H_d$ up to a distance of $0.175H_d$ and the other with a radius of $R = 0.2H_d$ up to the sharp sill (Novak et al., 2007).

In Figure 15.1 it is clear that the head H above the new crest is smaller than the head b above the crest of the sharp-edged notch from which the shape of the overfall spillway was derived. For an overfall spillway one can rewrite the discharge equation as

$$Q = \frac{2}{3}\sqrt{2gb}\,C_{d_0}\,H_{d_e}^{3/2} \qquad [15.4]$$

or, if $V_a = 0$,

$$Q = \frac{2}{3}\sqrt{2g}bC_{d_0}H_d^{3/2}$$

In Equation 15.4, H_{d_e} is the design energy head given by

$$H_{d_e} = H_d + \alpha\frac{V_a^2}{2g} \qquad [15.5]$$

H_d being the design head and C_{d_0} the design discharge coefficient equal to 0.745 for spillways of $P/H_{d_e} > 3.0$.

For any other head (H) the discharge coefficient varies, suggesting that for $H/H_d < 1.0, 0.580 < C_d < 0.745$, and for $H/H_d > 1.0, C_d > 0.745$ (see Figures 3.19 and 3.20).

15.2.1.1 Negative pressures and cavitation

When $H/H_d > 1.0$, negative pressure exists on the underside of the nappe which may lead to cavitation problems. In order to avoid this problem it is suggested that $H < 1.65H_d$. To find the pressure under the downstream nappe one can use Cassidy's (see USBR, 1987) relation:

$$\frac{p_m}{\rho g} = -1.17H\left(\frac{H}{H_d} - 1\right)$$

in which p_m is the gauge pressure under the nappe.

15.2.1.2 Gated spillways

Gates are used in spillways to increase the reservoir capacity. For gated spillways, the placing of the sill by $0.2H_d$ (see Novak *et al.*, 2007) downstream of the crest substantially reduces the tendency towards negative pressures for outflow under partially raised gates. The discharge through partially raised gates can be calculated from

$$Q = \frac{2}{3}\sqrt{2g}bC_{d_1}\left(H^{3/2} - H_1^{3/2}\right) \qquad [15.6]$$

with $C_{d_1} = 0.6$, in which H is the distance from the spillway crest until the upstream (reservoir) water level and H_1 the distance from the lower gate lip until the same water level. Alternatively, an equation (orifice) of the type

$$Q = C_{d_2}ba(2gH_e)^{1/2} \qquad [15.7]$$

where a is the distance of the gate lip from the spillway surface and H_e the effective head on the gated spillway (which is very similar to H) could also be used.

15.2.1.3 Offset spillways

For slender dam sections as in the case of arch dams, it may be necessary to offset the upstream spillway face into the reservoir in order to gain enough space to develop its shape. This upstream offset has no effects on the discharge coefficient.

15.2.1.4 Effective spillway length

In all the previous equations, b refers to the spillway length. If the crest has piers (to support gates) this length must be reduced to (see Example 3.18)

$$b_e = b - 2(nk_p + k_a)H_e \qquad [15.8]$$

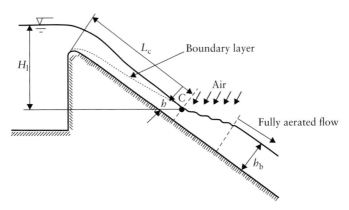

Figure 15.2   Air entrainment on overfall spillway surface.

where n is the number of piers, k_p the pier contraction coefficient and k_a the abutment contraction coefficient. The pier contraction coefficient is a function of the pier shape; for example,

Point nose pier: $k_p = 0$
Square nose pier: $k_p = 0.02$
Round nose pier: $k_p = 0.01$

The abutment contraction coefficient is $0 < k_a < 0.2$.

15.2.1.5 Self-aeration

An important design feature is the point at which self-aeration of the overfall nappe (in contact with the spillway) starts. A steep slope implies high velocities which in turn entrain air from the atmosphere into the flow. The appearance of the water is white with a violently agitated free surface. The air entrapment occurs at a location (say C, Figure 15.2) where

$$\frac{\rho h u^2}{\sigma} \left(\frac{\nu}{u_* h} \right)^{1/2} \geq 56 \qquad [15.9]$$

The location of point C (incipient point) is a function of the discharge per unit width (q), the surface roughness (k_s) and the slope (S) of the downstream channel:

$$L_C = f(q, k_s, S) \cong 15 \sqrt{q} \quad \text{(as an approximation)}$$

The entrainment of air has an effect of increasing (bulking effect) the flow depth downstream of the incipient point. To calculate the new depth one can use the following process:

1. The theoretical velocity at point C is

$$u_t = \sqrt{2gH_1}$$

   and the actual velocity is

$$u = 0.875 u_t$$

2. With this velocity, the depth at point C is

$$h = \frac{q}{u}$$

3. The air concentration in the fully aerated flow is given by

$$\bar{C} = \frac{1.35 n Fr_C^{3/2}}{1 + 1.35 n Fr_C^{3/2}}$$

where n is Manning's coefficient for the channel (spillway surface) and Fr_C the Froude number at C

$$Fr_C = \frac{u}{\sqrt{gh}}$$

4. Now the bulked depth (fully aerated) is

$$h_b = \frac{h_{n_a}}{1 - \bar{C}}$$

where h_{n_a} is the depth without air entrainment given by the Manning equation

$$h_{n_a} = \left(\frac{nq}{\sqrt{S}}\right)^{3/5}$$

15.2.2 Side channel spillways

In the case of a side channel spillway (in case part of the main body of the dam cannot be used as a spillway), the proper spillway is designed as a normal overfall spillway. The side channel itself must be designed in such a way that the maximum flood discharge passes without affecting the free flow operation of the spillway crest.

The flow in the side channel is an example of a spatially varied flow that is best solved by the application of the momentum principle, assuming that the lateral inflow into the channel (see Figure 15.3) has no momentum in the direction of the main flow towards the chute or tunnel (see also Section 8.16).

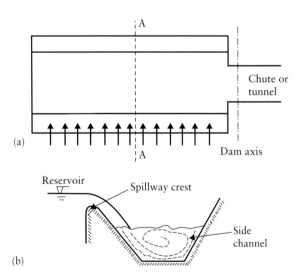

Figure 15.3 Side channel spillway: (a) plan view and (b) section A–A.

Equation 8.23 suggests the location of the control point (critical depth) in the side channel, and in a given length of the channel, the control section exists only if its slope satisfies Equation 8.24. Equations 8.23 and 8.24 may be modified for the use of Manning's n instead of Chezy's C by the relation

$$C = \frac{R^{1/6}}{n} \qquad [15.10]$$

To determine the spatially varied flow profiles, one can use a numerical integration technique with a trial-and-error process (see Chow, 1959; Henderson, 1966). The change in water surface elevation can be written as

$$\Delta h' = -\Delta h + S_0 \Delta x \qquad [15.11]$$

We can also write (using the finite difference approach) from the momentum equation

$$\Delta h' = \frac{Q_1}{g} \frac{V_1 + V_2}{Q_1 + Q_2} \left(\Delta V + \frac{V_2}{Q_1} \Delta Q \right) + S_f \Delta x \qquad [15.12]$$

Equations 15.11 and 15.12 can be used to determine the spatially varied water surface profile for the case of increasing (longitudinal direction) discharge that occurs in the side channel of the spillway (see Example 15.3).

15.3 Energy dissipators and downstream scour protection

Creation of a hydraulic jump downstream of the spillway toe is the most effective way of dissipating the high energy (supercritical flow) of the incoming water (see Section 8.8 and Example 8.16). In addition, numerous devices such as stilling basins, aprons and vortex shafts are known in general as energy dissipators. The most common type of energy dissipator is through the formation of a hydraulic jump (free or forced) in a stilling basin (horizontal apron downstream of the toe of spillway), which at times may consist of impact blocks, ramps, steps, baffles and so on (see USBR, 1987; Vischer and Hager, 1998).

Equation 8.17 suggests a sequential (conjugate) downstream subcritical depth (y_2) to incoming supercritical depth (y_1). The necessary measures to control erosion and dissipate energy (Equation 8.18) of the incoming flow depend on the available tailwater depth (y_T) and the conjugate depth (y_2) (see Figures 15.4–15.6 illustrating various scenarios of downstream conditions of energy dissipation). Figure 15.4a shows the classic formation of a hydraulic jump downstream of the toe when $y_2 = y_T$ (a condition rarely encountered), and in this case a simple 'horizontal apron' is adequate for downstream scour protection. Figure 15.4b suggests a remedial method to achieve this condition, if necessary.

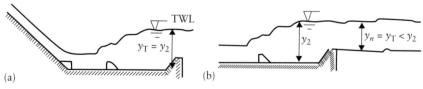

Figure 15.4 Hydraulic jump with conjugate depth equal to tailwater level: (a) simple horizontal (stilling basin) apron and (b) stilling basin with lowered bed to achieve $y_2 = y_T$.

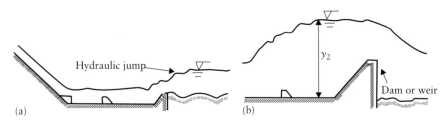

Figure 15.5 Hydraulic jump with conjugate depth greater than tailwater level: (a) hydraulic jump outside of the stilling basin when $y_2 > y_T$ and (b) downstream prominent weir pushing hydraulic jump into the stilling basin.

Figure 15.5 shows the scenario when the tailwater level is less than the conjugate depth. In this case the hydraulic jump would occur downstream of the stilling basin (see Figure 15.5a) without achieving the energy dissipation and scour protection. To avoid this, one can use a stilling basin with lowered level, as before, or a secondary dam or weir to increase the tailwater depth causing the jump to form at the toe of the main dam.

A third possible scenario is shown in Figure 15.6, the case when the tailwater level is greater than the conjugate hydraulic jump depth. This case is not desirable as a submerged jet (highly inefficient for energy dissipation) is likely to form downstream of the spillway face (see Figure 15.6a). Figure 15.6b suggests an alternative solution when $y_2 < y_T$, thus creating adequate energy dissipation.

The USBR (*Design of Small Dams*, 1987) suggests elaborate layouts of these dissipating structures, and Example 15.4 illustrates various design elements of such structures (stilling basin).

In order to choose the appropriate lowered bed level of a stilling basin, to achieve $y_2 > y_T$, it is necessary to develop the following design procedure, which uses the hydraulic jump and energy conservation equations and the geometry shown in Figure 15.7. The hydraulic jump equation is derived in Section 8.8, with y_1 being the incoming supercritical flow depth for a given (design) discharge.

$$y_2 = \frac{y_1}{2} \left(\sqrt{1 + 8\,Fr_1^2} - 1 \right) \qquad [8.17]$$

where $Fr_1^2 = V_1^2/gy_1$.

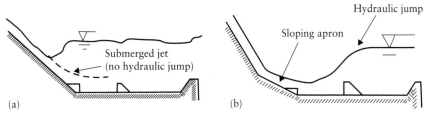

Figure 15.6 Hydraulic jump with conjugate depth smaller than the tailwater level: (a) submerged jump at the dam toe when $y_2 < y_T$ and (b) extension of the toe by a sloping apron (free jump in stilling basin).

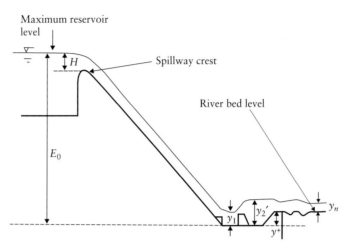

Figure 15.7 Design of a stilling basin bed level.

Making use of Figure 15.7, the following equations can be established:

$$y'_2 > y_2 \;\Rightarrow\; y'_2 = \sigma y_2 \;\Leftarrow\; \sigma > 1.0$$
$$y^+ = y'_2 - y_n = \sigma y_2 - y_n \qquad\qquad [15.13]$$

The energy conservation equation between the reservoir upstream and the point just down-stream of the basin entrance, including all the losses, is given by (see Figure 15.7)

$$E_0 = y_1 + \frac{V_1^2}{2g} + \xi \frac{V_1^2}{2g}$$

with ξ being the spillway (control and chute) loss coefficient. By writing

$$\zeta^2 = \frac{1}{1+\xi}$$

we can have

$$E_0 = y_1 + \frac{q^2}{2gy_1^2\zeta^2} \qquad\qquad [15.14]$$

The value of σ, which represents the safety factor for the conjugate jump depth, must be between the limits of $1.05 < \sigma < 1.20$, with 1.10 being the best value. The value of ξ increases with the crest height and roughness of the spillway, which implies more losses. The range of ζ, $1 > \zeta > 0.5$, is suggested for the design considerations. Example 15.4 illustrates the design process of a stilling basin.

Worked examples

Example 15.1

An overflow spillway is to be designed to pass a discharge of 2000 m³/s of flood flow at an upstream water surface elevation of 200 m. The crest length (spillway width) is 75 m and the elevation of the average bed is 165 m. Determine

(a) the design head
(b) the spillway profiles, upstream and downstream
(c) the discharge through the spillway if the water surface elevation reaches 202 m.

Solution:

(a) The discharge over the spillway per unit width is

$$q_d = \frac{Q}{b} = \frac{2000 \, \text{m}^3/\text{s}}{75 \, \text{m}} = 26.67 \, \text{m}^2/\text{s}$$

For the first calculation, one can assume a discharge coefficient corresponding to a deep upstream approaching channel. For such a case,

$$\frac{P}{H_{d_e}} \geq 3.0$$

$$\Rightarrow C_{d_0} = 0.74$$

Now the design head can be computed using the discharge equation:

$$q_d = \frac{2}{3} C_{d_0} \sqrt{2g} H_{d_e}^{3/2}$$

$$26.67 = \frac{2}{3} 0.74 \sqrt{2 \times 9.81} H_{d_e}^{3/2}$$

$$\Rightarrow H_{d_e} = 5.30 \, \text{m}$$

The approach velocity (channel upstream of the spillway crest) is

$$V_a = \frac{q}{P + H_d} = \frac{26.67}{200 - 165} \, \text{m/s} = 0.762 \, \text{m/s}$$

and the velocity head is

$$\frac{V_a^2}{2g} = \frac{0.762^2}{2 \times 9.81} \, \text{m} \approx 0.03 \, \text{m}$$

Therefore the spillway crest elevation is

$$200 \, \text{m} + \frac{V_a^2}{2g} - H_{d_e} = 194.73 \, \text{m}$$

Now the previous assumption of a deep approaching channel can be checked:

$$P = 194.73 - 165 = 29.73 \, \text{m}$$

$$\Rightarrow \frac{P}{H_{d_e}} = \frac{29.73}{5.30} = 5.6 > 3$$

$$\Rightarrow \text{Assumption} \rightarrow \text{OK}$$

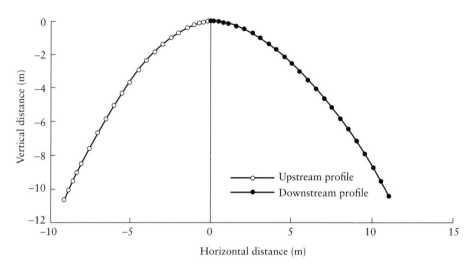

Figure 15.8 Upstream and downstream profiles for the proposed spillway.

The design head is

$$H_d = H_{d_e} - \frac{V_a^2}{2g} = 5.30 \text{ m} - 0.03 \text{ m} = 5.27 \text{ m}$$

(b) The upstream spillway profile is given by Equation (i) that can replace the two circular curves mentioned earlier. The result of this profile must be extended until it reaches the same slope of the upstream face of the dam.

$$\frac{y}{H_d} = 0.724 \left(\frac{x}{H_d} + 0.27 \right)^{1.85} - 0.432 \left(\frac{x}{H_d} + 0.27 \right)^{0.625} + 0.126 \tag{i}$$

i.e. $\dfrac{y}{5.27} = 0.724 \left(\dfrac{x}{5.27} + 0.27 \right)^{1.85} - 0.432 \left(\dfrac{x}{5.27} + 0.27 \right)^{0.625} + 0.126$

And the downstream spillway profile is given by Scimeni's equation (ii):

$$\frac{y}{H_d} = 0.50 \left(\frac{x}{H_d} \right)^{1.85} \tag{ii}$$

i.e. $\dfrac{y}{5.27} = 0.50 \left(\dfrac{x}{5.27} \right)^{1.85}$

The resulting upstream and downstream profiles are shown in Figure 15.8.
(c) The new head causing the flow is

$$H = 202.0 \text{ m} - 194.73 \text{ m} = 7.27 \text{ m}$$

As one does not know the approach velocity, it can be assumed, for first iteration, say

$$\frac{V_a^2}{2g} = 0.05 \text{ m}$$

$$\Rightarrow H_e = 7.27 \text{ m} + 0.05 \text{ m} = 7.32 \text{ m}$$

The discharge coefficient and the new discharge per unit width can now be found (see Figure 3.20):

$$\frac{H_e}{H_{d_e}} = \frac{7.32}{5.30} = 1.38$$

$$\Rightarrow \frac{C_d}{C_{d_0}} = 1.04$$

$$\Rightarrow C_d = 0.768$$

$$\Rightarrow q = \frac{2}{3} C_d \sqrt{2g} H_e^{3/2}$$

$$q = \frac{2}{3} 0.768 \sqrt{2 \times 9.81} \times 7.32^{1.5}$$

$$q = 45 \text{ m}^2/\text{s}$$

Using this new discharge per unit width, the approach velocity and the corresponding velocity head are

$$V_a = \frac{q}{P+H} = \frac{45}{29.73 + 7.27} = 1.216 \text{ m/s}$$

$$\Rightarrow \frac{V_a^2}{2g} = \frac{1.216^2}{2 \times 9.81} \text{ m} = 0.08 \text{ m}$$

Second iteration: Using the new velocity head, it can be found that

$$H_e = 7.27 \text{ m} + 0.08 \text{ m} = 7.35 \text{ m}$$

$$\Rightarrow \frac{H_e}{H_{d_e}} = 1.386$$

$$\Rightarrow C_d = 0.7683$$

Therefore

$$q = 45.28 \text{ m}^2/\text{s}$$

and finally

$$Q = qb = 45.28 \text{ m}^2/\text{s} \times 75 \text{ m} = 3396 \text{ m}^3/\text{s}$$

Example 15.2

A vertical-faced overfall spillway of large height (deep approaching channel) is to be spanned by a bridge deck. The design discharge is 450 m^3/s under a maximum design head of 3.45 m. The pier contraction coefficient is 0.012 and the abutment contraction coefficient is 0.1.

The following table suggests the variation of C_d with H for this spillway:

H/H_d	C_d/C_{d_0}
1.000	1.000
0.908	0.988
0.682	0.952
0.455	0.920
0.227	0.855

(a) Find the number of spans required if the maximum span length is 7 m.
(b) Establish the stage–discharge relationship for the proposed design.

Solution:

(a) To find the number of spans required, the spillway discharge equation, with the design discharge coefficient, can be used:

$$Q = \frac{2}{3}C_{d_0}b_e\sqrt{2g}H_d^{3/2}$$

$$450 = \frac{2}{3}0.74b_e\sqrt{2 \times 9.81} \times 3.45^{3/2}$$

$$\Rightarrow b_e = 32.136 \text{ m}$$

As the maximum span between piers is 7 m, 4 piers (5 spans of 7 m) are needed. The real effective width can now be found:

$$b_e = b - 2(nK_p + K_a)H$$
$$b_{e_R} = 35 \text{ m} - 2(4 \times 0.012 + 0.1) \times 3.45 \text{ m}$$
$$b_{e_R} = 33.98 \text{ m} > b_e$$
$$\Rightarrow \text{Safe}$$

(b) Stage–discharge relationship for the proposed design.

Using the data given for the variation of the discharge coefficient and an Excel spreadsheet, it is possible to construct the graph shown in Figure 15.9.

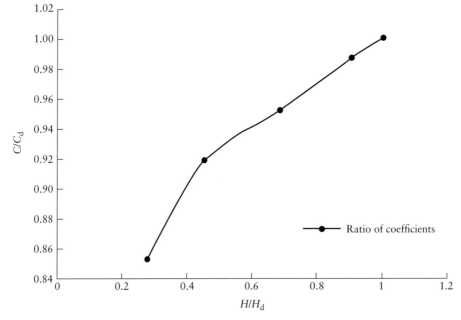

Figure 15.9 Variation of the discharge coefficient as a function of H/H_d.

Table 15.1   Stage–discharge data of the proposed spillway.

H (m)	H/H_d (–)	C_d/C_{d_0} (–)	C_d (–)	b_e (m)	Q (m³/s)
0.00	0.00000	0.7903050	0.58483	35.0000	0.0000
0.30	0.08696	0.8150876	0.60316	34.9112	10.2174
0.60	0.17391	0.8398702	0.62150	34.8224	29.7021
0.90	0.26087	0.8913478	0.65960	34.7336	57.7631
1.20	0.34783	0.9041304	0.66906	34.6448	89.9769
1.50	0.43478	0.9169130	0.67852	34.5560	127.1970
1 80	0.52174	0.9296957	0.68797	34.4672	169.1010
2.10	0.60870	0.9424783	0.69743	34.3784	215.4650
2.40	0.69565	0.9552609	0.70689	34.2896	266.1280
2.70	0.78261	0.9680435	0.71635	34.2008	320.9720
3.00	0.86957	0.9808261	0.72581	34.1120	379.9020
3.45	1.00000	1.0000000	0.74000	33.9788	475.8020
3.80	1.10145	1.0149130	0.75104	33.8752	556.5140
4.10	1.18841	1.0276957	0.76049	33.7864	629.9010
4.40	1.27536	1.0404783	0.76995	33.6976	707.1330
4.70	1.36232	1.0532609	0.77941	33.6088	788.1810
5.00	1.44928	1.0660435	0.78887	33.5200	873.0190
5.30	1.53623	1.0788261	0.79833	33.4312	961.6280
5.70	1.65217	1.0958696	0.81094	33.3128	1085.6100

Using Figure 15.9 and the discharge equation, it is possible to establish the stage–discharge relationship for the proposed design (again an Excel spreadsheet was used to construct the graph – see Table 15.1 and Figure 15.10).

The proposed design, finally, has the following results:

$$H = 3.45 \text{ m} = H_d \quad \Rightarrow C_d = 0.74 \Rightarrow Q = 475.8 \text{ m}^3/\text{s} > Q_d$$
$$H = 5.7 \text{ m} = 1.65H_d \Rightarrow C_d = 0.81 \Rightarrow Q_{max} = 1085.6 \text{ m}^3/\text{s}$$
(to avoid cavitation problems)

Example 15.3

A rectangular lateral spillway channel 77 m long is designed to carry a discharge which increases at a rate of 6 m³/(s m). The cross section has a width of 20 m. The longitudinal slope of the channel is 0.12 and begins at an upstream elevation of 100 m. If Manning's n is equal to 0.013, determine the water surface profile.

Solution:

The first step is to determine if a critical cross section exists, and if so, the second step is to determine its longitudinal position. This requires a trial-and-error method to solve Equation 8.23. Table 15.2 suggests the final results with the following procedure:

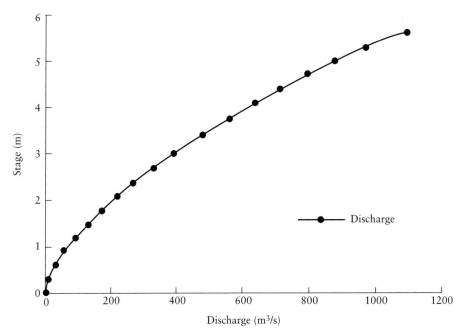

Figure 15.10   Stage–discharge relationship for the proposed spillway.

Column 1: An assumed longitudinal position
Column 2: The total discharge at the assumed longitudinal position (in this case it is
    equal to Column 1 times 6 m^3/(s m))
Column 3: Critical depth of flow corresponding to the discharge of Column 2,

$$y_c = \sqrt[3]{\frac{Q^2}{gb^2}}$$

Column 4: Wetted area = $20y_c$
Column 5: Wetted perimeter = $20 + 2y_c$

Table 15.2   Location of the control section.

1	2	3	4	5	6	7
Trial X_c (m)	Q (m^3/s)	y_c (m)	A (m^2)	P (m)	R (m)	X_c (m)
60.0000	360.000	3.2084	64.1685	26.417	2.4291	44.2515
50.0000	300.000	2.8412	56.8244	25.682	2.2126	44.2568
40.0000	240.000	2.4485	48.9697	24.897	1.9669	44.2719
44.0000	264.000	2.6091	52.1823	25.218	2.0692	44.2643
44.2600	265.560	2.6194	52.3876	25.239	2.0757	44.2639
44.2640	265.584	2.6195	52.3908	25.239	2.0758	44.2639
44.2639	265.583	2.6195	52.3907	25.239	2.0758	44.2639

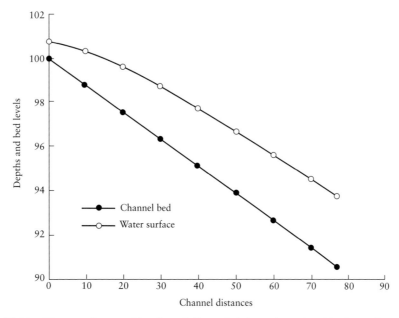

Figure 15.11   Water surface profile of spatially varied flow along the side channel.

Column 6: Hydraulic radius = Column 4 divided by Column 5
Column 7: The calculated distance to the critical section (Equation 8.23, with $\beta \approx 1.0$)

$$x_c = \frac{8q^2}{gb^2 \left[S_0 - (gP/C^2b)\right]^3} \qquad [8.23]$$

C being Chezy's coefficient given by Equation 15.10.

With the critical flow section located 44.264 m downstream of the beginning of the channel, the water surface profile both upstream (subcritical) and downstream (supercritical) of this point (control point) can be estimated by Equation 5.11. The calculations for the water surface profile upstream of the critical flow section are contained in Table 15.3, while those for the profile downstream are contained in Table 15.4 (see also Figure 15.11). In these tables,

Column 1: Distance between the point of computation and the beginning of the channel
Column 2: Incremental distance between adjacent points of calculation
Column 3: Elevation of the channel bottom
Column 4: Assumed depth of flow
Column 5: Elevation of water surface (Column 3 + Column 4)
Column 6: Change in water surface elevation

$$\Delta h' = -\Delta h + S_0 \Delta x \qquad [15.11]$$

Column 7: Wetted area
Column 8: Discharge = Column 1 times the discharge per unit length of the channel

Chapter 15

Table 15.3   Water surface profile calculations upstream of the control section.

1	2	3	4	5	6	7	8	9	10	11	12	13	14	15	16	17
x (m)	Δx (m)	Z_0 (m)	b (m)	Z (m)	$\Delta b'$ (m)	A (m²)	Q (m³/s)	V (m/s)	Q_1+Q_2 (m³/s)	V_1+V_2 (m/s)	ΔQ (m³/s)	ΔV (m/s)	$\Delta b'_m$ (m)	R (m)	S_f (−)	$\Delta b'$ (m)
44.26	–	94.688	2.620	97.3079	–	52.39	265.58	5.069	–	–	–	–	–	–	–	–
40.00	4.26	95.200	2.550	97.7500	0.442	51.00	240.00	4.706	505.583	9.7752	25.583	0.3634	0.4275	2.0319	0.006201	0.4539
40.00	4.26	95.200	2.570	97.7700	0.462	51.40	240.00	4.669	505.583	9.7385	25.583	0.4000	0.4432	2.0446	0.006054	0.4690
40.00	4.26	95.200	2.580	97.7800	0.472	51.60	240.00	4.651	505.583	9.7204	25.583	0.4181	0.4508	2.0509	0.005983	0.4764
40.00	4.26	95.200	2.590	97.7900	0.482	51.80	240.00	4.633	505.583	9.7025	25.583	0.4361	0.4584	2.0572	0.005912	0.4837
40.00	4.26	95.200	2.595	97.7950	0.487	51.90	240.00	4.624	505.583	9.6936	25.583	0.4450	0.4622	2.0603	0.005878	0.4873
30.00	10.00	96.400	2.550	98.9500	1.155	51.00	180.00	3.529	420.000	8.1537	60.000	1.0949	0.9391	2.0319	0.008180	1.0209
30.00	10.00	96.400	2.500	98.9000	1.105	50.00	180.00	3.600	420.000	8.2243	60.000	1.0243	0.9218	2.0000	0.008692	1.0088
30.00	10.00	96.400	2.400	98.8000	1.005	48.00	180.00	3.750	420.000	8.3743	60.000	0.8743	0.8838	1.9355	0.009853	0.9823
30.00	10.00	96.400	2.370	98.7700	0.975	47.40	180.00	3.797	420.000	8.4217	60.000	0.8268	0.8713	1.9159	0.010242	0.9737
30.00	10.00	96.400	2.367	98.7670	0.972	47.34	180.00	3.802	420.000	8.4266	60.000	0.8220	0.8701	1.9140	0.010282	0.9729
20.00	10.00	97.600	2.200	99.8000	1.033	44.00	120.00	2.727	300.000	6.5296	60.000	1.0750	0.7924	1.8033	0.005727	0.8496
20.00	10.00	97.600	2.000	99.6000	0.833	40.00	120.00	3.000	300.000	6.8023	60.000	0.8023	0.7498	1.6667	0.007697	0.8268
20.00	10.00	97.600	1.980	99.5800	0.813	39.60	120.00	3.030	300.000	6.8326	60.000	0.7720	0.7447	1.6528	0.007942	0.8241
20.00	10.00	97.600	1.990	99.5900	0.823	39.80	120.00	3.015	300.000	6.8174	60.000	0.7872	0.7473	1.6597	0.007818	0.8255
20.00	10.00	97.600	1.993	99.5930	0.826	39.86	120.00	3.011	300.000	6.8128	60.000	0.7917	0.7481	1.6618	0.007782	0.8259
10.00	10.00	98.800	1.800	100.6000	1.007	36.00	60.00	1.667	180.000	4.6772	60.000	1.3439	0.6920	1.5254	0.002673	0.7188
10.00	10.00	98.800	1.700	100.5000	0.907	34.00	60.00	1.765	180.000	4.7752	60.000	1.2458	0.6906	1.4530	0.003198	0.7226
10.00	10.00	98.800	1.600	100.4000	0.807	32.00	60.00	1.875	180.000	4.8855	60.000	1.1355	0.6883	1.3793	0.003870	0.7270
10.00	10.00	98.800	1.530	100.3300	0.737	30.60	60.00	1.961	180.000	4.9713	60.000	1.0498	0.6859	1.3270	0.004456	0.7304
10.00	10.00	98.800	1.524	100.3240	0.731	30.48	60.00	1.969	180.000	4.9790	60.000	1.0420	0.6856	1.3225	0.004511	0.7307
0.00	10.00	100.000	1.300	101.3000	0.976	26.00	00.00	0.000	60.000	1.9685	60.000	1.9685	0.3950	1.1504	0.000000	0.3950
0.00	10.00	100.000	1.100	101.1000	0.776	22.00	00.00	0.000	60.000	1.9685	60.000	1.9685	0.3950	0.9910	0.000000	0.3950
0.00	10.00	100.000	0.800	100.8000	0.476	16.00	00.00	0.000	60.000	1.9685	60.000	1.9685	0.3950	0.7407	0.000000	0.3950
0.00	10.00	100.000	0.720	100.7200	0.396	14.40	00.00	0.000	60.000	1.9685	60.000	1.9685	0.3950	0.6716	0.000000	0.3950
0.00	10.00	100.000	0.719	100.7190	0.395	14.38	00.00	0.000	60.000	1.9685	60.000	1.9685	0.3950	0.6708	0.000000	0.3950

Table 15.4 Water surface profile calculations downstream of the control section.

1	2	3	4	5	6	7	8	9	10	11	12	13	14	15	16	17
x (m)	Δx (m)	Z_0 (m)	b (m)	Z (m)	$\Delta b'$ (m)	A (m²)	Q (m³/s)	V (m/s)	Q_1+Q_2 (m³/s)	V_1+V_2 (m/s)	ΔQ (m³/s)	ΔV (m/s)	$\Delta b'_m$ (m)	R (m)	S_f (−)	$\Delta b'$ (m)
44.26	–	94.688	2.620	97.308	–	52.391	265.58	5.069	–	–	–	–	–	–	–	–
50.00	5.74	94.000	2.650	96.650	0.7188	53.000	300.00	5.660	565.58	10.730	34.42	0.59109	0.6803	2.09	0.01159	0.74678
50.00	5.74	94.000	2.660	96.660	0.7298	53.200	300.00	5.639	565.58	10.708	34.42	0.56981	0.6666	2.10	0.01145	0.73235
50.00	5.74	94.000	2.661	96.661	0.7298	53.220	300.00	5.637	565.58	10.706	34.42	0.56769	0.6653	2.10	0.01144	0.73092
50.00	5.74	94.000	2.662	96.662	0.7308	53.240	300.00	5.635	565.58	10.704	34.42	0.56558	0.6639	2.10	0.01143	0.72949
50.00	5.74	94.000	2.662	96.662	0.7303	53.230	300.00	5.636	565.58	10.705	34.42	0.56663	0.6646	2.10	0.01144	0.73020
60.00	10.00	92.800	2.700	95.500	1.2385	54.000	360.00	6.667	660.00	12.303	60.00	1.03075	1.3476	2.13	0.02748	1.62238
60.00	10.00	92.800	2.800	95.600	1.3385	56.000	360.00	6.429	660.00	12.064	60.00	0.79265	1.1618	2.19	0.02460	1.40777
60.00	10.00	92.800	2.810	95.610	1.3485	56.200	360.00	6.406	660.00	12.042	60.00	0.76977	1.1443	2.19	0.02433	1.38760
60.00	10.00	92.800	2.820	95.620	1.3585	56.400	360.00	6.383	660.00	12.019	60.00	0.74706	1.1270	2.20	0.02407	1.36765
60.00	10.00	92.800	2.823	95.623	1.3615	56.460	360.00	6.376	660.00	12.012	60.00	0.74028	1.1218	2.20	0.02399	1.36171
70.00	10.00	91.600	3.000	94.600	1.3770	60.000	420.00	7.000	780.00	13.376	60.00	0.62380	1.1268	2.31	0.02715	1.39833
70.00	10.00	91.600	3.100	94.700	1.4770	62.000	420.00	6.774	780.00	13.150	60.00	0.39800	0.9448	2.37	0.02459	1.19070
70.00	10.00	91.600	3.010	94.610	1.3870	60.200	420.00	6.977	780.00	13.353	60.00	0.60055	1.1078	2.31	0.02688	1.37660
70.00	10.00	91.600	3.005	94.605	1.3820	60.100	420.00	6.988	780.00	13.365	60.00	0.61216	1.1173	2.31	0.02702	1.38744
70.00	10.00	91.600	3.007	94.607	1.3840	60.140	420.00	6.984	780.00	13.300	60.00	0.60751	1.1135	2.31	0.02696	1.38310
77.00	7.00	90.760	3.200	93.960	1.0330	64.000	462.00	7.219	882.00	14.202	42.00	0.23505	0.6597	2.42	0.01893	0.79221
77.00	7.00	90.760	3.100	93.860	0.9330	62.000	462.00	7.452	882.00	14.435	42.00	0.46791	0.8500	2.37	0.02083	0.99582
77.00	7.00	90.760	3.120	93.880	0.9530	62.400	462.00	7.404	882.00	14.388	42.00	0.42014	0.8105	2.38	0.02043	0.95351

Chapter 15

Column 9: Velocity = Column 8 divided by Column 7
Column 10: Addition of discharges
Column 11: Addition of velocities
Column 12: Change in discharge
Column 13: Change in velocity
Column 14: Drop in the water surface due to impact loss,

$$\Delta h'_m = \frac{Q_1}{g} \frac{V_1 + V_2}{Q_1 + Q_2} \left(\Delta V + \frac{V_2}{Q_1} \Delta Q \right) \qquad [15.12a]$$

Column 15: Hydraulic radius associated with the assumed depth of flow
Column 16: Head loss due to friction computed from

$$h_f = S_f \Delta x = \left(\frac{nQ}{AR^{2/3}} \right)^2 \Delta x$$

Column 17: Drop in the water surface between two adjacent sections calculated by

$$\Delta h' = \frac{Q_1}{g} \frac{V_1 + V_2}{Q_1 + Q_2} \left(\Delta V + \frac{V_2}{Q_1} \Delta Q \right) + S_f \Delta x \qquad [15.12b]$$

At each station a trial-and-error process is used until the values of Columns 6 and 17 agree.

Example 15.4

In a small dam for water supply purposes it is necessary to design a baffle stilling basin. The level of the reservoir is located 30 m above the original river bed. The chute has a width of 15 m and a rectangular section, and moves a maximum discharge of 170 m³/s. Select the basin floor level and carry out its design. At this maximum flow rate the normal depth in the river is 4.5 m and for this chute assume $\sigma = 1.10$ and $\zeta = 0.7$.

Solution:

For the first iteration, y^+ (see Figure 15.12) may be assumed as 8 m. Thus the energy head and the discharge per unit width are

$$E_0 = 30\,m + 8\,m = 38\,m$$

$$q = \frac{Q}{b} = \frac{170}{15}\,m^2/s = 11.33\,m^2/s$$

Using Equation 15.14 we get

$$E_0 = y_1 + \frac{11.33^2}{2gy_1^2 0.7^2} = y_1 + \frac{13.36}{y_1^2} \qquad (i)$$

Also Equation 8.17 results in

$$y_2 = \frac{y_1}{2} \left(\sqrt{1 + \frac{104.74}{y_1^3}} - 1 \right) \qquad (ii)$$

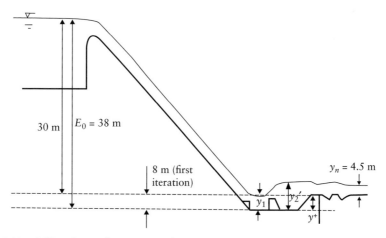

Figure 15.12   Stilling basin design example.

and Equation 15.13 gives

$$y^+ = y_2' - 4.5 \tag{iii}$$

where $y_2' = \sigma y_2$.

Using Equations (i), (ii) and (iii), it is possible to follow the design procedure, which is iterative. Table 15.5 shows the result for the final basin floor level; note that the method converges very fast.

Thus the stilling basin floor level is 2.12 m below the river bed. Using USBR (1987), it is possible to choose one of the standard stilling basin designs. For this case, the Froude number and the velocity of the incoming supercritical flow are

$$Fr_1 = 6.88 > 4.5$$
$$V_1 = 17.39 \, \text{m/s} < 20 \, \text{m/s}$$

Using these results, it is clear that USBR type III stilling basin must be used (see USBR, 1987). The details of the stilling basin are shown in Figure 15.13.

Table 15.5   Stilling basin floor level design.

Trial y^+ (m)	E_0 (m)	y_1 [Eq. (i)] (m)	V_1 (m/s)	Fr_1 (−)	y_2 [Eq. (ii)] (m)	y^+ [Eq. (iii)] (m)
8.000	38.000	0.5976	18.96475	7.832624	6.327557	2.460313
2.460	32.460	0.6480	17.48971	6.936824	6.041229	2.145351
2.145	32.145	0.6513	17.40110	6.884169	6.023559	2.125915
2.126	32.126	0.6515	17.39575	6.881000	6.022492	2.124741

References and recommended reading

British Standards Institution (1981) *Methods of Measurement of Liquids in Open Channels – Thin Plate Weirs*, BS 3680, Part 4A, British Standards Institution, London.

Chapter 15

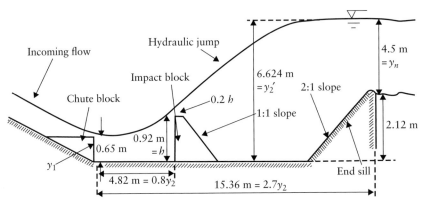

Chute block: width and spacing = y_1
Impact block: width and spacing = $0.75h$, h being its height given by

$$h = \frac{y_1(4 + \mathrm{Fr}_1)}{6}$$

Figure 15.13 Scheme of the stilling basin final design including its floor level and general dimensions (USBR, 1987; Novak *et al.*, 2007).

British Standards Institution (1990) *Methods of Measurement of Liquids in Open Channels – Rectangular Broad Crested Weirs*, BS 3680, Part 4E, British Standards Institution, London.
Chow, V. T. (1959) *Open Channel Hydraulics*, McGraw-Hill, New York.
Henderson, F. M. (1966) *Open Channel Flow*, MacMillan, New York.
Novak, P., Moffat, A. I. B., Nalluri, C. and Narayanan, R. (2007) *Hydraulic Structures*, 4th edn, Spon, London.
United States Bureau of Reclamation – USBR (1987) *Design of Small Dams*, 3rd edn, US Department of the Interior, Washington, DC.
Vischer, D. L. and Hager, W. H. (1998) *Dam Hydraulics*, John Wiley & Sons, Chichester, UK.

Problems

1. An overflow spillway is to be designed to pass a discharge of 1700 m³/s of flood flow with an upstream water surface elevation of 230 m. The crest length is 22 m and the elevation of the average river bed is 183 m.
    (a) Determine the design head.
    (b) Determine the spillway profiles, upstream and downstream.
    (c) Calculate the discharge through the spillway if the water surface elevation reaches the point of maximum allowable head.
    (d) What would be the minimum pressure downstream (underside of the nappe) of the spillway crest under part (c) discharge condition? Use Cassidy's relation given in the text.
    For the discharge coefficients:

C_d/C_{d_0}	H/H_d
1.06	1.50
1.07	1.60

2. If the spillway of Problem 1 is required to have piers to support a bridge deck over it and if the maximum free span between them is 6.3 m, what would be the discharge corresponding to the design head? What is the new allowable maximum discharge? The pier coefficient is 0.011, and the abutment contraction coefficient is 0.12.

3. A trapezoidal lateral spillway channel 127 m long is designed to carry a discharge which increases at a rate of 11 m³/(s m). The cross section has a bottom width of 18 m and lateral slopes of 2 (vertical) : 1 (horizontal). The longitudinal slope of the channel is 0.08 and starts at an upstream elevation of 520 m. If Manning's n is equal to 0.014 (concrete), determine if there is a critical section and the corresponding water surface profile.

4. A hydraulic jump stilling basin is to be designed to dissipate the energy of a maximum discharge of 750 m³/s. The water surface in the upstream reservoir is located at a level of 350 m. The chute channel has a slope of 45°, rectangular in cross section with a width of 22.5 m. At this discharge the river downstream of the basin has a normal depth of 4.4 m, and the bed is located at an elevation of 307 m. For this chute $\sigma = 1.05$ and $\zeta = 0.75$.
   (a) Determine the sequent depth of the hydraulic jump and the bed level of the stilling basin.
   (b) Choose the stilling basin design according to the USBR (1987) standards. Make a sketch of the basin.
   (c) Determine the percentage of energy dissipation in the basin.

Chapter 16
Environmental Hydraulics and Engineering Hydrology

16.1 Introduction

This chapter extends the discussion from previous chapters to cover wider issues concerning water in the environment and presents some further worked examples relevant to environmental engineering topics, with a particular theme of flood alleviation.

The well-known natural hydrological cycle traces the path of rainfall as it results in infiltration to the ground, surface runoff, evaporation and transpiration from trees and plants, returning to clouds and resulting in further rainfall. Alongside this, there is also a cycle of human use of water. Water in the environment is impounded by dams or pumped from the ground, treated for human consumption, distributed through pipes and channels for drinking and irrigation purposes, collected after use by sewerage systems and returned to the environment usually after suitable treatment. Many of the theories and calculation techniques from previous chapters find useful applications at different points in these cycles and enable engineers to analyse and design works to manage water in the environment.

Management of water has increasingly been recognised as involving environmental and sustainability issues, with particular concerns arising from urbanisation and climate change which have a bearing on flood risk management. Recommended reading includes more extended treatment of such topics, whilst the following sections serve to introduce the worked examples.

16.2 Analysis of gauged river flow data

Much of hydrology is concerned with analysis of data. Where river discharges are required, it is considered preferable where possible to work with actual discharge data, rather than to deduce results from rainfall and catchment characteristics. The latter approach may be needed where reliable discharge data are not available, and reference for this can be made to more specialised texts in the recommended reading. The analysis that follows illustrates the former approach where gauged records of discharge values are available.

Nalluri & Featherstone's Civil Engineering Hydraulics: Essential Theory with Worked Examples,
Sixth Edition. Martin Marriott.
© 2016 John Wiley & Sons, Ltd. Published 2016 by John Wiley & Sons, Ltd.
Companion Website: www.wiley.com/go/Marriott

For gauged catchments, a time series of measured annual maximum discharge values may be analysed to provide values with particular probabilities for design purposes. The measured data may come from a weir or flume, as described in Chapter 8, or from an ultrasonic gauging station or other velocity–area method. The annual exceedance probability P is related to the annual probability of non-exceedance F and to the return period T years by the following expression:

$$P = 1 - F = \frac{1}{T} \qquad [16.1]$$

Return period T has been widely used to express the average time interval between events greater than or equal to the value specified. However, this term has been criticised for implying a periodic regularity, which is not the case in reality (see Fleming, 2002).

Having ranked the annual maximum data for N years, in descending order from greatest $i = 1$ to least $i = N$, each point may be assigned a value of P from a formula of the form

$$P = \frac{i - a}{N + b} \qquad [16.2]$$

where a and b are constants depending on the probability distribution of the data. Alternatively, if ranked ascending from least with $j = 1$ to greatest with $j = N$, such a formula produces the annual probability of non-exceedance:

$$F = \frac{j - a}{N + b} \qquad [16.3]$$

From the two approaches to ranking, either in descending order i or in ascending order j, it is noted that

$$i + j = N + 1 \qquad [16.4]$$

A combination of Equations 16.1–16.4 then yields the result

$$b = 1 - 2a \qquad [16.5]$$

Widely used values include $a = 0$ (Weibull) and $a = 0.44$ (Gringorten), with the latter generally preferred for extreme value distributions. The Gringorten plotting position formula for exceedance probability P is thus

$$P = \frac{i - 0.44}{N + 0.12} \qquad [16.6]$$

and the non-exceedance probability F may be calculated either from Equation 16.1 or from ranked data:

$$F = \frac{j - 0.44}{N + 0.12} \qquad [16.7]$$

For a two-parameter analysis, a straight line may be fitted to the points plotted against an appropriate reduced variate. One such is the Gumbel reduced variate:

$$y_G = -\ln(-\ln(F)) \qquad [16.8]$$

This relates to the extreme value type 1 (EV1) distribution, for which the cumulative distribution function $F(x)$ is

$$F(x) = e^{-e^{-[(x-u)/a]}} \qquad [16.9]$$

Hence, when values of flow rate Q are plotted as the variable x against the Gumbel reduced variate y_G, we have the straight-line relationship

$$Q = u + \alpha y_G \qquad [16.10]$$

The parameters u and α may be found approximately by a best-fit line through the displayed data or estimated by a method of moments drawing on the properties of the EV1 distribution, using the mean (μ) and sample standard deviation (s) of the data as follows:

$$u = \mu - 0.45s \qquad [16.11]$$

and

$$\alpha = 0.78s \qquad [16.12]$$

This is shown in Example 16.1 using the River Thames data.

An alternative approach recommended in the *Flood Estimation Handbook* (Institute of Hydrology, 1999) advocates the use of the generalised logistic distribution, and data are displayed using the logistic reduced variate

$$y_L = -\ln\left(\frac{1-F}{F}\right) \qquad [16.13]$$

In this approach, the flow rates Q are non-dimensionalised using the median annual maximum flood Q_{MED}, and related to the following growth curve:

$$\frac{Q}{Q_{MED}} = 1 + \frac{\beta}{k}\left[1 - \left(\frac{1-F}{F}\right)^k\right] \qquad [16.14]$$

Values of k and β are estimated by a method using L moments, illustrated in Example 16.2 using the River Thames data, and this method is considered more appropriate for skewed data. In the special case of $k = 0$, the two-parameter logistic distribution is used with

$$\frac{Q}{Q_{MED}} = 1 + \beta y_L \qquad [16.15]$$

Space does not permit a fuller discussion here of the details of this method, or other recommended procedures such as the pooling of data from hydrologically similar catchments to extend the amount of data (see Institute of Hydrology, 1999, volume 3). The comparison of Examples 16.1 and 16.2 usefully illustrates how the results depend on the method of analysis, as well as demonstrates the methods involved.

16.3 River Thames discharge data

For the worked examples that follow, 120 years of published Thames discharge data are used for illustration, and some comments follow on the data which serve to indicate some of the issues associated with such data series. These data for the water years 1884–2003 are presented by Herschy (2003, 2004), with the Thames data updated and corrected in the later supplement dated 2004. These data are the highest daily mean flows (for the hydrological day 0900–0900 hours) for each water year (October to September) and have been naturalised (adjusted to account for major upstream abstractions). For large river basins worldwide, the mean daily value is usually not significantly different from the instantaneous maximum value, and peak flows quoted are often daily mean flows. However in the

UK, catchment response times are usually such that instantaneous or 15 minute peak flow data are generally used (Institute of Hydrology, 1999), hence the reason for noting this point. Herschy (2003) advises that for these Thames data, instantaneous peak flows can by estimated by increasing the quoted daily mean values by 2–5%. One further comment on the data used is that the earlier records are from Teddington Weir, whereas since 1977, data have been recorded at an ultrasonic gauging station at Kingston; as noted in Marsh *et al.* (2005), this is still considered as one continuous series, due to the negligible difference in catchment areas. The inherent assumption of hydrological stationarity remains that the data do not contain a significant trend over time from a cause such as climate change. The examples that follow are analysed on that basis, and the present pragmatic advice in this context concerning effects of climate change is to investigate the sensitivity of future designs to the addition of a percentage to predicted peak river flows to allow for such future changes.

16.4 Flood alleviation, sustainability and environmental channels

The main components of constructed works to alleviate the effects of flooding are the provision of storage, the increase in channel conveyance capacity and the provision of protection by means of banks or walls. Other measures include better warning and emergency planning.

One of the main components of sustainable urban drainage systems is to provide storage near to the source, by use of swales, permeable pavements and tank sewers. Advice for developers on such issues is now included in such publications as Water UK (2012). Storage can also be beneficial and cost-effective on larger rural catchments, as provided, for example, by the use of check dams on tributaries. The flood routing in Example 11.8 has demonstrated the beneficial effect of storage on reducing the magnitude of the peak of a flood hydrograph, as illustrated in Figure 11.6.

Sustainability in a broad sense involves balancing the effects of economic, social and environmental constraints, and places the traditional engineering concerns of quality, cost and time within a complex framework of other issues (see Fenner *et al.*, 2006). The trend in river engineering works has been to maintain to a greater extent the natural features of channels, with sediment transport issues as discussed in Chapter 14 being of particular significance.

Section 8.4 and Example 8.6 demonstrated the properties of a two-stage channel, and this combination of river and flood plain provides the combined conveyance for a valley. Often the main channel meanders within the flood plain, and there has been a growing concern to maintain such features. The degree of meandering is measured by the sinuosity s, which is defined in relation to the length L or slope S of the flood plain (subscript 'F') and the main channel (subscript 'C') as follows:

$$s = \frac{L_C}{L_F} = \frac{S_F}{S_C} \qquad [16.16]$$

In a simple one-dimensional approach, the discharge equation may be adjusted for a meandering channel to allow for the different hydraulic gradients on the channel and flood plain components of the flow, for example using the Manning formula,

$$Q = \frac{A_F}{n_F} R_F^{2/3} S_F^{1/2} + \frac{A_C}{n_C} R_C^{2/3} S_C^{1/2} \qquad [16.17]$$

but the previous vertical division between flood plain and main channel shown in Figure 8.3 becomes less appropriate as the degree of meandering increases.

The complex nature of interaction between the components is the subject of ongoing research beyond the scope of this text, concerning not only the conveyance but also the sediment transport (see e.g. Loveless *et al.*, 2000).

16.5 Project appraisal

A final section is included on the appraisal of schemes such as those for flood allevia-tion, using cost–benefit analysis. The discounted cash flow approach is used to express the present values of costs and benefits that arise at different points in time. Benefits need to be expressed in cost terms, and represent desirable consequences of the project, and tech-niques exist to help to quantify and include non-tangible and environmental benefits, as well as those that simply relate to the prevention of damage in monetary terms. All costs are expressed at present price levels, and a discount rate r is selected, which reflects the extent to which positive benefits are preferred sooner rather than later, and adverse or undesirable costs later rather than sooner. For a single sum arising at the end of year n in the future, the appropriate factor is the inverse of the well-known formula for compound interest.

$$\text{Single sum factor} = (1 + r)^{-n} \qquad [16.18]$$

Hence the present value of sum X arising at the end of the year n

$$= X(1 + r)^{-n} \qquad [16.19]$$

For a regular stream of annual sums at the end of each year, the cumulative present value factor is given by

$$\text{Cumulative factor} = \sum_{i=1}^{n}(1 + r)^{-i} = \frac{1 - (1 + r)^{-n}}{r} \qquad [16.20]$$

Hence the present value of annual sums Y for n years $= Y\left(\dfrac{1 - (1 + r)^{-n}}{r}\right) \qquad [16.21]$

The formula for the cumulative present value factor is derived using the mathematical technique for summing a geometric progression, and may be seen to tend to the limit of $1/r$ as n tends to infinity. Sometimes that simplification is taken, but more usually a value of n is set for the design life of the scheme, which it should be noted may well differ from the level of protection provided and from the asset lives of the various components of the scheme.

Costs are often a mix of capital and annual running or maintenance costs, to which Equations 16.19 and 16.21 would apply, respectively. Benefits arising from the damage prevented are assessed over the design life of the scheme as the stream of annual expected benefits, using the cumulative present value factor approach as in Equation 16.21. The annual expected benefit is calculated from combining the damage and probability data, as illustrated in Example 16.3. It should be noted that as well as the damage that has been prevented, there may also be other benefits, such as the opening up of areas for recreation such as riverside footpaths, which could be valued using willingness to pay or contingent valuation methods (see e.g. Penning-Rowsell and Green, 2000, for a discussion of these and other issues, such as above-design-standard benefits).

Having established the present value of benefits B and the present value of costs C, these may be compared as the benefit-to-cost ratio B/C, and values exceeding 1 show that the scheme may be justified in this way. The calculation approach above is illustrated in Example 16.3.

Worked examples

Example 16.1

Display the Thames data from Table 16.1 using the Gringorten plotting position formula and the Gumbel reduced variate. Estimate the magnitude of the flood discharge which has an annual exceedance probability of 2% (average recurrence interval or return period of 50 years).

Solution:

From ranked data on a computer spreadsheet, produce Figure 16.1.

For example, if ranked in ascending order, the smallest annual maximum flood discharge of 94 m^3/s has rank $j = 1$ and $N = 120$.

$$\text{Gringorten } F = \frac{1 - 0.44}{120 + 0.12} = 0.00466$$

$$\text{Gumbel reduced variate, } y_G = -\ln(-\ln(0.00466)) \approx -1.68$$

and the largest flood of 806 m^3/s has

$$\text{Gringorten } F = \frac{120 - 0.04}{120 + 0.12} = 0.995338$$

(do not round this off too soon)

$$\text{Gumbel reduced variate, } y_G = -\ln(-\ln(0.995338)) \approx 5.37$$

The target flood, often known as Q_{50} in view of the 50-year return period, has $P = 0.02$ and $F = 0.98$, so $y_G = -\ln(-\ln(0.98)) = 3.90$.

From a best-fit straight line through the displayed data on Figure 16.1, it may be seen that the flood discharge corresponding to this value of the Gumbel reduced variate is approximately 630 m^3/s.

By calculation from the sample data set,

$$\text{mean, } \mu = 324.025 \, m^3/s$$

$$\text{standard deviation, } s = 117.6 \, m^3/s$$

$$u = 324.0 - 0.45 \times 117.6 = 271.1$$

$$\alpha = 0.78 \times 117.6 = 91.7$$

$$Q_{50} = 271.1 + 91.7 \times 3.90 = 629 \, m^3/s$$

Example 16.2

Analyse the Thames data from Table 16.1 using the generalised logistic approach recommended in the *Flood Estimation Handbook*. Hence estimate the magnitude of the flood discharge with an annual exceedance probability of 2%, for comparison with the solution from Example 16.1.

Table 16.1   River Thames maximum mean daily flows.

Water year	Discharge (m^3/s)	Rank i (descending)	Rank j (ascending)	Annual exceedance probability (Gringorten)
1884	231	93	28	0.771
1885	229	96	25	0.796
1886	244	88	33	0.729
1887	284	74	47	0.612
1888	207	105	16	0.870
1889	237	90	31	0.746
1890	204	106	15	0.879
1891	171	113	8	0.937
1892	339	46	75	0.379
1893	300	66	55	0.546
1894	173	112	9	0.929
1895	806	1	120	0.005
1896	202	108	13	0.895
1897	351	42	79	0.346
1898	171	114	7	0.945
1899	262	79	42	0.654
1900	533	7	114	0.055
1901	200	109	12	0.904
1902	162	116	5	0.962
1903	386	25	96	0.204
1904	516	10	111	0.080
1905	229	97	24	0.804
1906	249	86	35	0.712
1907	220	102	19	0.845
1908	375	31	90	0.254
1909	204	107	14	0.887
1910	231	94	27	0.779
1911	428	18	103	0.146
1912	367	39	82	0.321
1913	255	84	37	0.696
1914	256	82	39	0.679
1915	585	4	117	0.030
1916	373	34	87	0.279
1917	327	52	69	0.429
1918	350	43	78	0.354
1919	334	49	72	0.404
1920	251	85	36	0.704
1921	240	89	32	0.737
1922	197	110	11	0.912
1923	231	95	26	0.787
1924	297	69	52	0.571
1925	522	9	112	0.071
1926	370	35	86	0.288
1927	374	32	89	0.263
1928	526	8	113	0.063

(Continued)

Table 16.1   (*Continued*)

Water year	Discharge (m³/s)	Rank *i* (descending)	Rank *j* (ascending)	Annual exceedance probability (Gringorten)
1929	235	92	29	0.762
1930	552	6	115	0.046
1931	228	98	23	0.812
1932	274	75	46	0.621
1933	478	12	109	0.096
1934	94	120	1	0.995
1935	227	99	22	0.821
1936	478	13	108	0.105
1937	237	91	30	0.754
1938	247	87	34	0.721
1939	369	36	85	0.296
1940	409	20	101	0.163
1941	384	28	93	0.229
1942	298	68	53	0.562
1943	457	15	106	0.121
1944	115	119	2	0.987
1945	261	81	40	0.671
1946	256	83	38	0.687
1947	714	2	119	0.013
1948	227	100	21	0.829
1949	299	67	54	0.554
1950	324	55	66	0.454
1951	385	26	95	0.213
1952	377	30	91	0.246
1953	263	78	43	0.646
1954	214	104	17	0.862
1955	452	17	104	0.138
1956	315	59	62	0.488
1957	314	60	61	0.496
1958	316	58	63	0.479
1959	374	33	88	0.271
1960	308	63	58	0.521
1961	456	16	105	0.130
1962	344	45	76	0.371
1963	285	73	48	0.604
1964	369	37	84	0.304
1965	131	118	3	0.979
1966	323	57	64	0.471
1967	312	62	59	0.512
1968	600	3	118	0.021
1969	369	38	83	0.313
1970	224	101	20	0.837
1971	362	40	81	0.329
1972	330	51	70	0.421
1973	266	77	44	0.637
1974	353	41	80	0.338

(*Continued*)

Table 16.1 (*Continued*)

Water year	Discharge (m³/s)	Rank i (descending)	Rank j (ascending)	Annual exceedance probability (Gringorten)
1975	559	5	116	0.038
1976	157	117	4	0.970
1977	334	50	71	0.413
1978	326	53	68	0.438
1979	324	56	65	0.463
1980	393	24	97	0.196
1981	289	71	50	0.587
1982	314	61	60	0.504
1983	345	44	77	0.363
1984	286	72	49	0.596
1985	270	76	45	0.629
1986	408	21	100	0.171
1987	304	64	57	0.529
1988	402	22	99	0.179
1989	262	80	41	0.662
1990	427	19	102	0.155
1991	220	103	18	0.854
1992	165	115	6	0.954
1993	378	29	92	0.238
1994	400	23	98	0.188
1995	385	27	94	0.221
1996	301	65	56	0.537
1997	192	111	10	0.920
1998	295	70	51	0.579
1999	325	54	67	0.446
2000	335	48	73	0.396
2001	463	14	107	0.113
2002	338	47	74	0.388
2003	482	11	110	0.088

Solution:

Enter data onto a computer spreadsheet, rank in ascending order and calculate the probability of weighted moments b (all with units of m³/s) as follows:

$$b_0 = \frac{1}{N}\sum_{j=1}^{N} Q_j = 324.025$$

$$b_1 = \frac{1}{N}\sum_{j=2}^{N} \frac{j-1}{N-1} Q_j = 193.750$$

$$b_2 = \frac{1}{N}\sum_{j=3}^{N} \frac{(j-1)(j-2)}{(N-1)(N-2)} Q_j = 141.416$$

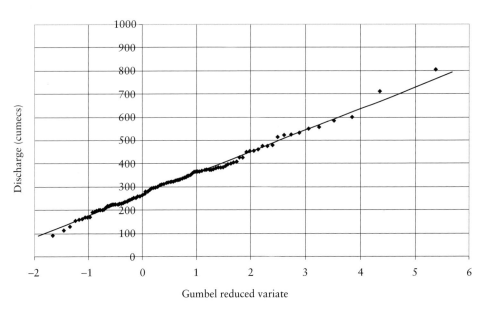

Figure 16.1 River Thames Gumbel extreme value type 1 distribution, 1884–2003; points displayed using Gringorten formula and trend line fitted by method of moments.

Now calculate the L moments as follows (again all with units of m³/s):

$$L_1 = b_0 = 324.025 \quad \text{(which is the mean of the data)}$$
$$L_2 = 2b_1 - b_0 = 63.475$$
$$L_3 = 6b_2 - 6b_1 + b_0 = 10.02$$

Now calculate the dimensionless L moment ratios t as follows:

$$t_2 = \frac{L_2}{L_1} = \frac{63.475}{324.025} = 0.196$$

$$t_3 = \frac{L_3}{L_2} = \frac{10.02}{63.475} = 0.158$$

These ratio values are used to calculate the growth curve parameters as follows:

$$k = -t_3 = -0.158$$

$$\beta = \frac{t_2 k \sin \pi k}{k\pi(k + t_2) - t_2 \sin \pi k} = 0.198 \quad \text{(noting that angles are in radians)}$$

The growth curve is then found from Equation 16.14.

For Q_{50}, we have the annual exceedance probability $P = 0.02$ and $F = 0.98$, and we note that the median value from the sample data set is 314 m³/s, so

$$Q_{50} = 314 \left\{ 1 + \frac{0.198}{-0.158} \left[1 - \left(\frac{0.02}{0.98} \right)^{-0.158} \right] \right\} = 648 \, \text{m}^3/\text{s}$$

which is approximately 3% higher than the result from Example 16.1. Further comparison between the two approaches is discussed in Marriott and Hames (2007).

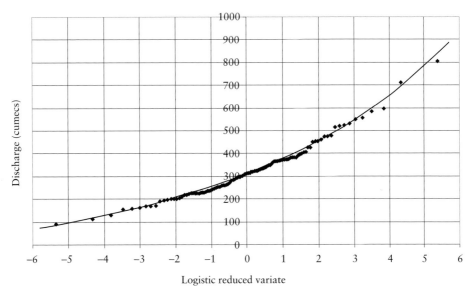

Figure 16.2   River Thames generalised logistic distribution, 1884–2003; points displayed using Gringorten formula and trend line fitted using L moments.

The growth curve is displayed in Figure 16.2, showing that this data set is unbounded above. Note that had k been zero, a straight line would have resulted using $\beta = t_2$ in Equation 16.15. Note also that for this approach it is recommended that if the record length N is less than twice the target return period T, further data are sought from other similar catchments to form a pooling group, with the aim of achieving $5T$ station years of data (see Institute of Hydrology, 1999).

Example 16.3

A proposed project, costing £4.5 million to construct, will protect against flooding up to a 100-year return period, but may be assumed not to affect higher floods. In addition to the initial capital cost of the project, there are annual operation costs of £0.05 million. The estimated costs of damage caused by floods of various magnitudes are given in the table below, with all costs being at present price levels.

Return period (years)	Damage (£ million)
5	0
10	2
20	4
50	6
100	7
200	8

Calculate the benefit-to-cost ratio of the proposed project, assessed over a 50-year period, using an annual discount rate of 6%.

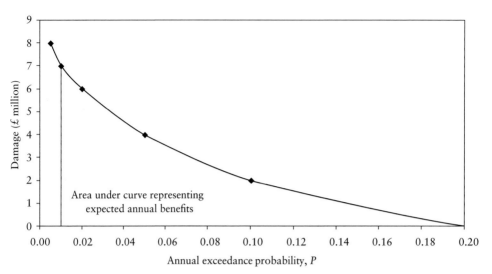

Figure 16.3   Loss versus probability relationship.

Solution:

Either plot the damage data against annual exceedance probability and find the relevant area under this curve as shown in Figure 16.3, or do this calculation in a tabular way as follows:

T	$P = 1/T$	Damage ($£$ million)	ΔP	Average damage $\times \Delta P$
5	0.20	0		
			0.10	0.100
10	0.10	2		
			0.05	0.150
20	0.05	4		
			0.03	0.150
50	0.02	6		
			0.01	0.065
100	0.01	7		

The annual expected benefit from prevention of damage up to $T = 100$ years is given by the sum of the final column above, $0.100 + 0.150 + 0.150 + 0.065 = £0.465$ million.

With n given as 50 years, and the discount rate $r = 6\% = 0.06$, the relevant cumulative present value factor

$$= \frac{1 - (1.06)^{-50}}{0.06} = 15.762$$

So the present value of benefits $= 0.045 \times 15.762 = £7.33$ million

The present value of costs $= 4.50 + 0.05 \times 15.762 = £5.29$ million

Hence the benefit-to-cost ratio $= 7.33/5.29 = 1.4$

Taking a vertical cut-off line at $P = 0.01$ in Figure 16.3 is consistent with the instruction to assume that the proposed works will not affect rarer floods, but it is likely not to be completely realistic. If the project will have some beneficial impact on higher return period floods, the value presented above may be regarded as a lower bound to the true benefit-to-cost ratio. With further data, the vertical line could be replaced by a more realistic post-project loss against probability curve, to evaluate what are referred to as above-design-standard benefits (Penning-Rowsell and Green, 2000).

References and recommended reading

Butler, D. and Davies, J. W. (2011) *Urban Drainage*, 3rd edn, Spon, Abingdon, UK.

Chadwick, A., Morfett, J. and Borthwick, M. (2013) *Hydraulics in Civil and Environmental Engineering*, 5th edn, Taylor & Francis, Abingdon, UK.

Fenner, R. A., Ainger, C. M., Cruickshank, H. J. and Guthrie, P. M. (2006) Widening engineering horizons: addressing the complexity of sustainable development, *Proceedings of the Institution of Civil Engineers Engineering Sustainability*, 159, ES4 December, 145–154.

Fleming, G. (Ed) (2002) *Flood Risk Management, Learning to Live with Rivers*, Thomas Telford, London.

Gringorten, I. I. (1963) A plotting rule for extreme probability paper, *Journal of Geophysical Research*, 68, 813–814.

Herschy, R. (2003) *World Catalogue of Maximum Observed Floods*, IAHS Publication, Wallingford, UK.

Herschy, R. (2004) *World Catalogue of Maximum Observed Floods (Errata and Supplementary Data)*, IAHS Publication, Wallingford, UK.

Institute of Hydrology (1999) *Flood Estimation Handbook*, Institute of Hydrology, now the Centre for Ecology and Hydrology, Wallingford, UK.

Knight, D. W. and Shamsheldin, A. Y. (Eds) (2006) *River Basin Modelling for Flood Risk Mitigation*, Taylor & Francis, London.

Loveless, J. H., Sellin, R. H. J., Bryant, T. B., Wormleaton, P. R., Catmur, S. and Hey, R. (2000) The effect of overbank flow in a meandering river on its conveyance and the transport of graded sediments, *Journal of the Chartered Institution of Water and Environmental Management*, 14, 447–455.

Mansell, M. G. (2003) *Rural and Urban Hydrology*, Thomas Telford, London.

Marriott, M. J. and Hames, D. P. (2007) Comparison of reduced variates for flood frequency estimation from gauged river flow records, *Second IMA International Conference on Flood Risk Assessment,* The Institute of Mathematics and its Applications, Plymouth, UK.

Marsh, T. J., Greenfield, B. J. and Hannaford, J. A. (2005) The 1894 Thames flood – a reappraisal, *Proceedings of the Institution of Civil Engineers Water Management*, 158, WM3 September, 103–110.

McGahey, C., Samuels, P. G., Knight, D. W. and O'Hare, M. T. (2008) Estimating river flow capacity in practice, *Journal of Flood Risk Management*, 1, 1, 23–33.

Penning-Rowsell, E. C. and Green, C. (2000) New insights into the appraisal of flood-alleviation benefits: (1) Flood damage and flood loss information, and (2) the broader context, *Journal of the Chartered Institution of Water and Environmental Management*, 14, 347–362.

Shaw, E. M., Beven, K. J., Chappell, N. A. and Lamb, R. (2011) *Hydrology in Practice*, 4th edn, Taylor & Francis, Abingdon, UK.

Tebbut, T. H. Y. (1998) *Principles of Water Quality Control*, 5th edn, Butterworth-Heinemann, Oxford, UK.

Water UK/WRc (2012) *Sewers for Adoption – A Design and Construction Guide for Developers*, 7th edn, WRc, Swindon, UK.

Chapter 16

Problems

1. Use the following annual maximum river discharge data to estimate the value of discharge that has a return period of 10 years (annual exceedance probability of 0.1). Make use of the Gumbel distribution. Comment on the likely accuracy of the result.

$$48, 42, 35, 33, 30, 28, 25, 22, 20, 18, 15, 12, 10 \, \text{m}^3/\text{s}$$

2. The onset of flooding at a location takes place at a river flow rate of 40 m³/s. Twenty-one years of data for annual maximum river flows are available as given below (in metres cubed per second) in ascending order. Estimate the annual probability of flooding, using the generalised logistic distribution. Compare the result obtained from using the Gumbel distribution.

$$9, 10, 12, 13, 13, 15, 17, 18, 20, 22, 24, 25, 28, 30, 33, 33, 35, 42, 48, 51, 56 \, \text{m}^3/\text{s}$$

3. A flood hydrograph as tabulated below is to be routed through a reservoir as part of a proposed flood alleviation scheme. Determine the magnitude and time of the peak outflow from the reservoir, and comment on the amount of attenuation. The reservoir may be assumed to have an effectively constant surface area of 0.5 km². Outflow Q (m³/s) from the reservoir is determined from the water level h (m) above the crest of the outflow control structure, by the discharge equation $Q = 30 \, h^{3/2}$. Assume that the reservoir is full before the start of the storm and that a steady flow of 1.0 m³/s has been passing through the reservoir for some hours.

Time (h)	0	1	2	3	4	5	6	7	8
Flow (m³/s)	1	15	30	17	10	6	4	2	1

4. For the compound channel described in Chapter 8, Problem 9, take the cross section and roughness data as given, but assume that the main channel has a sinuosity of 1.1 within the flood plain of gradient 0.00125. Calculate (a) the bank-full discharge, and (b) the total flood discharge when the depth of flow over the flood plain is 2.5 m.

5. A river has bank-full flow capacity of a 5-year return period, and the estimated amounts of damage caused by various floods are as shown in the table below. The second table below shows three proposed flood alleviation schemes, with their levels of protection and construction costs. Calculate which scheme shows the most favourable benefit-to-cost ratio, considered over a project design life of 50 years using an annual discount rate of 6%. Ignore any above-design-standard benefits, and other environmental or intangible benefits, but discuss the likely effect of these and other assumptions inherent in the calculations.

Flood return period (years)	Damage caused by flood (£ million)
5	0
10	16
20	30
50	41
100	49

Scheme	Level of protection, expressed as a return period (years)	Construction cost (£ million)
A	20	16
B	50	22
C	100	28

Chapter 17
Introduction to Coastal Engineering

17.1 Introduction

This chapter provides an introduction to theory that underpins coastal engineering, an extremely wide-ranging specialism with major challenges to design coastal structures and manage shorelines. Starting with waves and wave theories, the text leads through wave processes and properties, tides, surges and mean sea level including a section on tsunami waves. Worked examples are included, with references for further study, as well as additional problems to tackle.

17.2 Waves and wave theories

Coastal waves are mainly generated by the action of wind blowing across the sea surface. This causes the surface of the sea to exert a frictional drag on the lower layer of the wind, with the top layer (with the least drag) moving faster than the lower layer. This sets up a circular motion of wind energy that acts on the sea and thus creates waves. These waves travel vast differences in various directions, reducing gradually in height in the absence of further winds, generally only stopping when they meet an obstacle such as land or a coastal structure. Figure 17.1 illustrates two wave types in general: (a) regular long-crested waves following a sinusoidal pattern, and (b) random short-crested waves consisting of a number of sinusoidal waves travelling in different directions superimposed on top of each other. In general long-crested waves (see also Section 17.2.1) are observed in the nearshore region, whereas random short-crested waves (also see Section 17.2.4) are more commonly observed offshore.

In general, there are two types of wave theories: linear wave theory, sometimes known as Airy's waves, is the earliest, simplest and most widely used by coastal engineers; and, secondly, non-linear wave theory (e.g. Cnoidal and Stokes waves) is more complex and gives accurate predictions in shallow waters.

Nalluri & Featherstone's Civil Engineering Hydraulics: Essential Theory with Worked Examples, Sixth Edition. Martin Marriott.
© 2016 John Wiley & Sons, Ltd. Published 2016 by John Wiley & Sons, Ltd.
Companion Website: www.wiley.com/go/Marriott

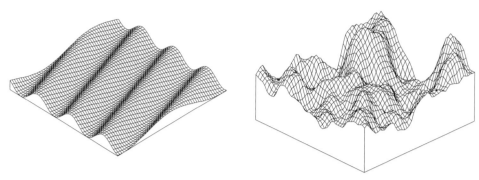

Figure 17.1 Wave patterns of (a) long-crested waves (regular waves), and (b) short-crested waves (irregular waves).

17.2.1 Linear wave theory

Linear wave theory is derived with the help of assumptions given in this section, and in addition, it is assumed that the waves to be described have heights (H) which are small compared with the wavelength (L) and the water depth (d) ($H/L < 0.04$ and $HL^2/d^3 < 40$). Under these conditions, the wave profile is essentially sinusoidal (regular wave).

Linear wave theory is exact for zero wave height. However, its range of applicability is surprisingly large, and for most purposes it will provide a very good estimate of values. Despite its relative simplicity, the derivation of linear wave theory is fairly complex, and depends on a solution to the Laplace equation that is satisfied throughout the body of the flow. Further details on the derivation of linear wave theory can be obtained for example from Dean and Dalrymple (1991), and Sorensen (2010).

The assumptions made for the derivation are:

- Water is homogeneous (of the same kind, uniform) and incompressible, and so it has a uniform density.
- Water lacks viscosity and surface tension.
- Waves are long crested (their analysis may be considered as a two-dimensional problem).
- Waves are of constant form (they do not change shape as they travel across the water surface).
- The seabed is horizontal and impermeable.
- Waves are propagating on quiescent (dormant) water (there is no motion of the water apart from that induced by the waves).

Under this set of assumptions, the wave height (H), the wave period (T) and the water depth (d) uniquely define a train of regular waves. All other characteristics of the wave train, including the wavelength (L), are functions of these three independent parameters.

17.2.2 Non-linear wave theory

In circumstances where wave heights are not small compared with wavelengths and water depths, then the wave profile is no longer approximately sinusoidal – wave crests become more peaked, and troughs are longer and flatter. There are corresponding changes in other wave properties. For example, the wavelength and wave speed (c) are no longer independent of the wave height. In these cases, it may be necessary to employ a non-linear wave theory.

These wave theories are very much more complex than linear (or Airy) wave theory. The main difference between them and linear wave theory is that they produce more accurate answers in the shallow water region. The most commonly referred to non-linear wave theories are Stokes' (first, third and fifth order) and Cnoidal wave theories.

17.2.3 Solutions from linear wave theory

The surface profile η of a linear wave relative to sea water level (SWL) is given by:

$$\eta = \frac{H}{2} \cos 2\pi \left(\frac{x}{L} - \frac{t}{T} \right) = a \cos(kx - \omega t) \tag{17.1}$$

A definition sketch for this sinusoidal wave pattern is given in Figure 17.2.

The wavelength L is related to the wave period by the equation:

$$L = \frac{gT^2}{2\pi} \tanh \left(\frac{2\pi d}{L} \right) = \frac{gT^2}{2\pi} \tanh kd \tag{17.2}$$

and the wave celerity c is:

$$c = \frac{L}{T} = \frac{gT}{2\pi} \tanh kd = \sqrt{\frac{gL}{2\pi} \tanh kd} \tag{17.3}$$

These are the two most widely used equations in linear wave theory. These equations depend not only on the wave period but also on the water depth. However, when $d/L > 0.5$, $\tanh kd \approx 1.0$, the waves can be considered to be in 'deep water', and the expressions for wavelength and celerity are given as Equations 17.4 and 17.5. In these equations, subscript $_0$ refers to deep water.

$$L_0 = \frac{gT^2}{2\pi} \tag{17.4}$$

$$c_0 = \frac{gT}{2\pi} = \sqrt{\frac{gL}{2\pi}} \tag{17.5}$$

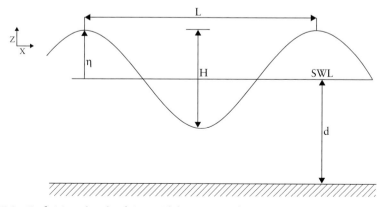

Figure 17.2 Definition sketch of sinusoidal wave (regular wave).

When $d/L < 0.04$, $\tanh kd \approx kd$, the waves can be considered to be in 'shallow water', and the expressions for wavelength and celerity are given as Equations 17.6 and 17.7. In these equations, subscript $_s$ refers to shallow water.

$$L_s = T\sqrt{gd} \qquad [17.6]$$

$$c_s = \sqrt{gd} \qquad [17.7]$$

The terms 'deep' and 'shallow' water are relative terms that depend on the lengths of the waves being considered. For example, a 5 m water depth is considered deep water for wave periods shorter than about 2.5 s, but shallow for wave periods in excess of about 18 s.

Equations 17.4 and 17.6 provide simple solutions for the wavelength under these conditions. However, in general L must be evaluated from an expression for which there is no explicit solution. Alternatively, an approximation can be made such as that given by Fenton and McKee (1990):

$$L = L_0\left\{ \tanh\left[\left(2\pi\sqrt{d/g}/T\right)^{3/2}\right] \right\}^{2/3} \qquad [17.8]$$

Figures 17.3 and 17.4 show how the wavelength and wave celerity vary as a function of wave period and water depth.

The corresponding equations for the horizontal and vertical velocities, accelerations and displacements of a water particle at a mean depth $-z$ below the still water level are given below. Note that as stated, in these equations, z is measured positively upwards. Therefore, at the water surface $z = 0$, and at the seabed $z = -d$. x_m and z_m denote the mean positions of water particles.

horizontal water particle velocity $(u) = a\omega\dfrac{\cosh k(d + z)}{\sinh kd}\cos(kx - \omega t) \qquad [17.9]$

vertical water particle velocity $(w) = a\omega\dfrac{\sinh k(d + z)}{\sinh kd}\sin(kx - \omega t) \qquad [17.10]$

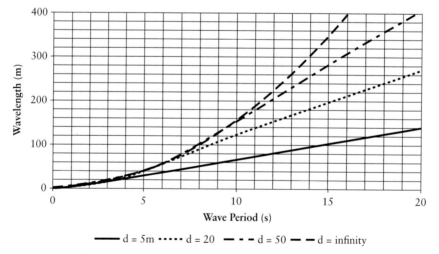

Figure 17.3   Wavelength as a function of wave period and water depth.

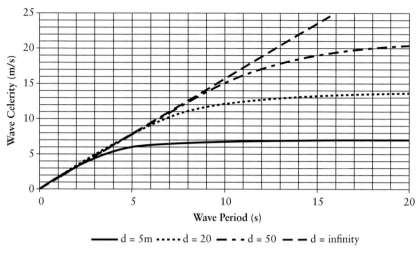

Figure 17.4   Wave celerity as a function of wave period and water depth.

horizontal water particle acceleration $\left(\dfrac{du}{dt}\right) = a\omega^2\dfrac{\cosh k(d+z)}{\sinh kd}\sin(kx - \omega t)$ [17.11]

vertical water particle acceleration $\left(\dfrac{dw}{dt}\right) = -a\omega^2\dfrac{\sinh k(d+z)}{\sinh kd}\cos(kx - \omega t)$ [17.12]

horizontal water particle displacement $(\xi_x) = -a\dfrac{\cosh k(d+z_m)}{\sinh kd}\sin(kx_m - \omega t)$ [17.13]

vertical water particle displacement $(\xi_y) = a\dfrac{\sinh k(d+z_m)}{\sinh kd}\cos(kx_m - \omega t)$ [17.14]

These equations are elliptical. Figure 17.5 illustrates graphically the ellipse formed by the water particle displacement through a sinusoidal wave pattern. You will notice that for

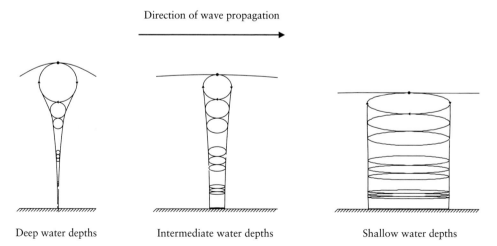

Figure 17.5   Water particle displacement through a sinusoidal wave pattern.

deep water waves, the ellipses are almost circular at the water surface. The further you get below the surface of the water level, the ratio of the vertical amplitude of the ellipse to the horizontal amplitude of the ellipse reduces. At the seabed, there is no vertical motion of the particles. The horizontal motion at the seabed is proportionally larger at smaller d/L ratios. This has important effects in respect to sediment transport.

As can be seen in Figure 17.5, the effects of the seabed on the deep water waves, and conversely those of the deep water waves on the seabed, are negligible. The deep water wavelength and celerity equations can therefore be considered to be solely a function of wave period (see Equations 17.4 and 17.5).

The seabed is affected by shallow water waves, and conversely shallow water waves are affected by the seabed. The shallow water wavelength can therefore be considered to be a function of wave period and depth, whereas the wave celerity can be considered to be solely a function of wave period (see Equations 17.6 and 17.7).

The energy of a wave consists of potential, kinetic and surface energies of all the water particles per unit wavelength, and it is given as energy per unit area of the sea (ignoring surface tension energy, which is negligible):

$$\text{wave energy (E)} = \frac{\rho g H^2}{8} \qquad [17.15]$$

This is a considerable amount of energy. For example, wave heights around the coasts of Britain generally exceed 5–6 m on average at least once a year. This will produce wave energy in excess of 30–45 kJ/m^2. Due to the shortage of hydrocarbons and the demand for renewable energy such as wave energy, this is an important area of interest at present.

The wave power P, or rate of transmission of wave energy, is given by:

$$P = Ec_g \qquad [17.16]$$

where c_g is the group wave celerity and is given by the product of group velocity parameter (n) and wave celerity (c):

$$c_g = nc = \frac{1}{2}\left(1 + \frac{2kd}{\sinh 2kd}\right)c \qquad [17.17]$$

In deep water ($d/L > 0.5$), the group velocity $c_g \to c/2$. In shallow water ($d/L < 0.04$), the group velocity $c_g \to c$. For intermediate depths, the group velocity is between these two values. Group velocity is the velocity with which a train of waves moves through an area. It is less than that of individual waves.

17.2.4 Irregular waves

In reality, waves are random, with many different waves of different directions, amplitudes and phases travelling in many different directions. These kinds of waves are known as 'short-crested waves'.

Offshore waves are generally short-crested. As the waves travel towards the shore, waves from certain directions are intercepted by landmasses, and the waves tend to become more long-crested. Also, the effects of refraction (see Section 17.3) will tend to align the waves closer with the seabed contours.

In the analysis of wave data, two kinds of analysis can be considered, a short-term analysis or a long-term analysis. A short-term analysis is an analysis of waves over a short duration (i.e. hours or minutes), such as a storm. A long-term analysis is the analysis of

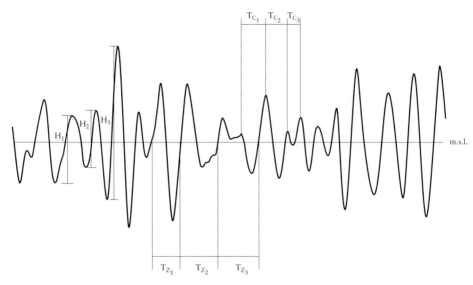

Figure 17.6   Time domain analysis of a wave record.

waves over a long duration such as several years. In coastal engineering, a long-term analysis is the analysis that is usually of most importance, and it is the analysis that you will encounter the most.

Based on random short-crested waves, two types of analysis may be performed: a time domain analysis or a frequency domain analysis. Only the time domain analysis is outlined in this chapter. Details on the frequency domain analysis can be found in Kamphuis (2000).

17.2.4.1 Time domain analysis
The following wave parameters can be defined with respect to time domain analysis of irregular wave records as illustrated in Figure 17.6.

The zero-crossing period (T_Z) is the average time between successive upward crossings of the mean sea level (m.s.l.), and it is given by:

$$T_Z = \frac{1}{N} \sum_{i=1}^{N} T_{Z_i} \qquad [17.18]$$

The crest period (T_C) is the average time between successive crests (not all of which are above the m.s.l.), and it is given in Equation 17.19. Note that $T_C \leq T_Z$.

$$T_C = \frac{1}{N} \sum_{i=1}^{N} T_{C_i} \qquad [17.19]$$

The mean wave height ($\overline{H}$) is the vertical distance between the highest crest and the lowest trough bounded by successive up-crossings of the m.s.l.

$$\overline{H} = \frac{1}{N} \sum_{i=1}^{N} H_i \qquad [17.20]$$

Further wave properties are as follows:

- The spectral wave period (T_P) is determined in the frequency domain, and it is included in this chapter for completeness. This is the mode of the 'spectral' period (i.e. the most commonly occurring wave period).
- The root-mean square wave height (H_{rms}) is the square root of the average of the square of the individual wave heights, and it is given by:

$$H_{rms} = \sqrt{\frac{1}{N} \sum_{i=1}^{N} H_i^2}$$
[17.21]

- The significant wave height (H_s) is defined, and is sometimes referred to, as the average of the highest one-third of the wave heights.
- Other wave properties are given by $H_{1/n}$, which is the average of the highest $1/n$'th of the wave heights. Aside from when $n = 3$, the most commonly used wave height is when $n = 10$.

If the surface elevation of waves, $\eta(x, t)$, is a Gaussian process, then it can be shown that wave heights closely follow a Rayleigh distribution, given by Equation 17.22. Figure 17.7 shows a Rayleigh distribution for a variance of 1.

$$p(H) = \frac{H}{4\sigma_\eta^2} \exp\left[-\frac{H^2}{8\sigma_\eta^2}\right]$$
[17.22]

The cumulative distribution function (cdf) is given by

$$P(H \leq H_*) = 1 - \exp\left[-\frac{H_*^2}{8\sigma_\eta^2}\right]$$
[17.23]

and therefore:

$$P(H > H_*) = \exp\left[-\frac{H_*^2}{8\sigma_\eta^2}\right]$$
[17.24]

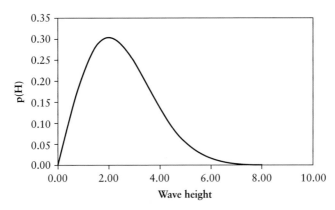

Figure 17.7 Rayleigh distribution.

From Equation 17.22, the following properties for random waves can be determined:

$$\overline{H} = \int_0^\infty H\,p(H)dH = 2.51\sigma_\eta \qquad [17.25]$$

$$H_{rms} = \sqrt{\int_0^\infty H^2 p(H)dH} = 2\sqrt{2}\sigma_\eta \qquad [17.26]$$

From Equation 17.26, therefore, Equation 17.24 may be rewritten as:

$$P(H > H_*) = \exp\left[-\left(\frac{H_*}{H_{rms}}\right)^2\right] \qquad [17.27]$$

The average height of the highest $1/n$'th of the waves may be found by considering the height H_* above which the highest $1/n$'th of the waves lie, given by:

$$P(H > H_*) = \int_{H_*}^\infty p(H)dH = \frac{1}{n} \qquad [17.28]$$

Therefore, $H_{1/n}$ may be written:

$$H_{1/n} = \frac{\int_{H_*}^\infty Hp(H)dH}{\int_{H_*}^\infty p(H)dH} = n\int_{H_*}^\infty Hp(H)dH \qquad [17.29]$$

Equating Equations 17.27 and 17.28 gives

$$\frac{H_*}{H_{rms}} = \sqrt{\ln n} \qquad [17.30]$$

Using these relationships, it can be shown that:

$$H_{1/3} = 1.416H_{rms} = 1.60\overline{H} = 4.0\sigma_\eta \qquad [17.31]$$

$$H_{1/10} = 1.800H_{rms} = 2.03\overline{H} \qquad [17.32]$$

$$H_{1/100} = 2.359H_{rms} = 2.66\overline{H} \qquad [17.33]$$

If m waves out of a total of N waves are larger than H_*, then:

$$P(H > H_*) = \exp\left[-\left(\frac{H_*}{H_{rms}}\right)^2\right] = \frac{m}{N} \qquad [17.34]$$

Therefore,

$$\frac{H_*}{H_{rms}} = \sqrt{-\ln\left(\frac{m}{N}\right)} \qquad [17.35]$$

This is the same as Equation 17.30, since $m/N = 1/n$.

Putting $n = 1$ into Equation 17.35 gives H_{max}, the wave height expected to be exceeded only once in a sample of N waves. Therefore, H_{max} can be written as:

$$H_{max} = H_{rms} \sqrt{\ln N} = H_s \sqrt{\left(\frac{\ln N}{2}\right)} \qquad [17.36]$$

Therefore, for a 20-minute wave trace of waves with a mean zero-crossing period of 7 s, the maximum wave height that could be expected in that period would be about 1.6 H_s.

17.2.4.2 Wave prediction from wind records

For a given wind speed, the waves produced will depend on the duration of the wind D and the fetch F. The bigger the fetch or the longer the duration, the bigger the waves produced. However, as the wind contains only a limited amount of energy, there is a limit to the height of waves produced for a given duration and/or fetch. This is the point where the rate of transfer of energy to the waves equals the energy dissipation by wave breaking and friction.

Wind is recorded for meteorological purposes at many sites. This and the fact that wave measurement can be very difficult for several reasons, including the frequent inaccessibility of the recording apparatus and maintenance problems, mean that wind speeds are often more frequently recorded than waves. As storm waves are dependent solely on wind, duration and fetch length, the most common technique used to determine the wave climate is using wind records. This technique is commonly referred to as 'hindcasting'.

17.2.4.3 Pierson Moskowitz wave spectrum

The most common technique used to determine wind speeds offshore is by means of the Pierson Moskowitz spectrum (PMS). For a fully arisen sea, this is given as:

$$S(f, \theta) = \frac{\alpha g^2}{(2\pi)^4 f^5} \exp\left\{-\frac{5}{4}\left(\frac{f_m}{f}\right)^4\right\} \qquad [17.37]$$

The PMS was derived from measurements of ocean waves taken by weather ships in the North Atlantic. It does not describe conditions in fetch-limited seas. It is not going to be considered any further in this chapter.

17.2.4.4 JONSWAP wave spectrum

The JONSWAP (Joint North Sea Wave Project) spectrum was derived for fetch-limited seas. This spectrum was determined based on observations of wave heights in the North Sea. It is more complicated than the equation for the PMS, because it is a function of both wind and fetch. This is the spectrum that is used for locally generated seas across (for example) estuaries. This is also the spectrum most likely to be encountered in coastal engineering.

The JONSWAP spectrum (Hasselmann et al., 1973) is given by the formula:

$$S(f, \theta) = \frac{\alpha g^2}{(2\pi)^4 f^5} \exp\left\{-\frac{5}{4}\left(\frac{f_m}{f}\right)^4\right\} \gamma^a \qquad [17.38]$$

The PMS and JONSWAP spectra are deep water spectra. Where a spectrum is required in transitional water depths, the TMA spectrum may be used, which is given in Section 17.2.4.5; this is a modified JONSWAP spectrum.

17.2.4.5 TMA spectrum

The JONSWAP spectrum is applicable for deep water conditions. To correct for shallow water conditions, the TMA spectrum is used (Bouws *et al.*, 1985, 1987). Based on linear wave theory, this can be written as:

$$S(f,\theta)_{TMA} = S(f,\theta)_{JONSWAP}\,\Phi(f) \qquad [17.39]$$

where $\Phi(f)$ can be expressed in the form:

$$\Phi(f) = \frac{\tanh^3 kd}{\tanh kd + kd - kd\tanh^2 kd} = \frac{\tanh^2 kd}{1 + 2kd/\sinh 2kd} \qquad [17.40]$$

Considering $\Phi(f)$ as a function of $\Phi(f_d)$, then the following non-dimensional relationship can be drawn:

$$f_d = 2\pi f\sqrt{\frac{d}{g}} = \sqrt{[kd\tanh(kd)]} \qquad [17.41]$$

Alternatively, an approximation accurate to 4% was given by Thompson and Vincent (1983) as:

$$\Phi(f) = \begin{cases} \dfrac{f_d^2}{2} & \text{for } f_d \leq 1 \\[2mm] 1 - \dfrac{1}{2}(2 - f_d)^2 & \text{for } f_d > 1 \end{cases} \qquad [17.42]$$

This is shown graphically in Figure 17.8.

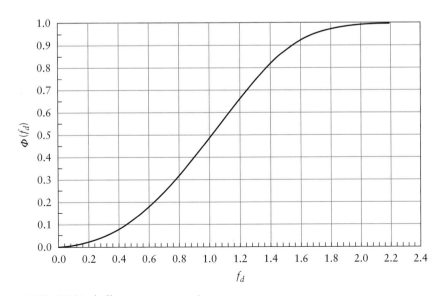

Figure 17.8   TMA shallow water correction.

17.2.4.6 Effective fetch

In cases where the fetch width is small in comparison with the fetch length, then waves will not be able to reach their full height as dictated by the JONSWAP spectrum. Therefore, a technique has to be used which considers the effective fetch length for the wind speed being looked at. Several methods have been proposed; however, the one considered here uses what is known as 'directional spreading'. This assumes that the wave energy is distributed evenly according to a cosine[6] function over a 180° arc, and is given by:

$$S(f) = \frac{3.2\delta\theta}{\pi} \sum_{i=1}^{n} \left\{ S(f, \theta_i) \cos^6 (\theta_i - \theta_0) \right\} \tag{17.43}$$

The significant wave height (H_s) and the zero crossing period (T_m) from the TMA spectrum are given as:

$$H_s = 4\sqrt{m_0} \tag{17.44}$$

$$T_m = \sqrt{\frac{m_0}{m_2}} \tag{17.45}$$

where m_0 and m_2 are, respectively, the zeroth and second moments of the spectrum $S(f, \theta)_{TMA}$. The mean wave direction is given by:

$$\theta_w = \theta_0 + \frac{\iint S(f,\theta)(\theta - \theta_0)df\, d\theta}{\iint S(f,\theta)df\, d\theta} \tag{17.46}$$

17.3 Wave processes

17.3.1 Refraction

As waves move into shallower water, they shorten and generally steepen. This process is known as wave shoaling. If a wave approaches the coast in such a way so that the wave front is at an angle to the seabed contours, then each part of the front travels at a different speed. Those parts in deep water travel faster than those in shallow water; therefore, the wave direction tends to change so that the wave crests become more nearly aligned with the seabed contours. This process is known as wave refraction. This will tend to cause waves to spread out, and therefore reduce their height in bays and bunch together and increase in height at headlands.

Linear wave theory for wave refraction gives:

$$Ec_g b = \text{constant} \tag{17.47}$$

where b is the distance between adjacent orthogonals. The wave height H at any particular inshore location relative to the value H_o in deep water is given by:

$$H = K_R K_S H_0 \tag{17.48}$$

where $K_R = (b_o/b)^{1/2}$ is the refraction coefficient which accounts for changes in wave height associated with changes in orthogonal spacing, and b_0 is the spacing between orthogonals in deep water. $K_S = (2n \tanh kd)^{-1/2}$ is the shoaling coefficient which accounts for changes in wave height resulting from changes in wave group velocity induced by variations in water depth.

To evaluate the refraction coefficient, wave orthogonals may be constructed from deep water to the point of interest using Snell's law, an approach originally considered by O'Brien (1942).

Snell's law for wave refraction is given as:

$$\frac{\sin \alpha_1}{\sin \alpha_2} = \frac{L_1}{L_2} = \frac{c_1}{c_2} \qquad [17.49]$$

The change of direction of an orthogonal as it passes over the seabed is therefore given by:

$$\sin \alpha_1 = \frac{c_2}{c_1} \sin \alpha_2 \qquad [17.50]$$

The assumptions of this theory are as follows:

- No wave energy is transmitted across orthogonals (this is reasonable provided that there is a small variation in wave height and no currents of any significance in the area).
- The direction of the wave advance is perpendicular to the wave crest (i.e. in the direction of the orthogonals).
- The speed of a wave with a given period at a particular location depends only on the depth at that location.
- Changes in bottom topography are gradual.
- Waves are long-crested, of constant period, small-amplitude and monochromatic.
- The effects of currents, winds and the reflections from beaches and underwater topographic variations are considered negligible.

If α_1 is taken to apply in deep water, then the local wave direction α_2 at water depth d is given by:

$$\alpha_2 = \sin^{-1}\left(\sin \alpha_1 \tanh kd\right) \qquad [17.51]$$

and the corresponding refraction coefficient K_R (for straight parallel bed contours) by:

$$K_R = \sqrt{\frac{\cos \alpha_1}{\cos \alpha_2}} \qquad [17.52]$$

Wave refraction can be considered graphically as shown in Figure 17.9.

Consider the wave front XX travelling over the seabed from the deep water region to the shallower water region as shown in Figure 17.9. The part of the wave front that travels from A to B travels a distance $gT^2/2\pi$ over a wave period. However, the part of the wave front that travels from C to D travels a shorter distance over the same time, as it is in a shallower water depth. Hence, the wave front that travelled from AC is at BD one wave period later, which has rotated to be more in line with the seabed contours.

Repeating this process over successive time steps means that the changing direction of a wave can be traced as it travels across a seabed. This is demonstrated in Figure 17.10, which shows an example of predicted wave refraction patterns as waves move into a shallow water area over an offshore shoal. The part of the wave 'nearer' the shallow areas will slow down more, and therefore 'bend' towards these shallower areas. This is more pronounced as the water becomes shallower, as indicated by the waves as they travel over the edge of the shoal.

When wave rays cross, caustics are formed. At these points, $b \to 0$, and wave heights according to this theory are infinite. This is indicated in Figure 17.10, where the wave rays passing over the edge of the shoal would cross if continued. In reality, however, diffraction

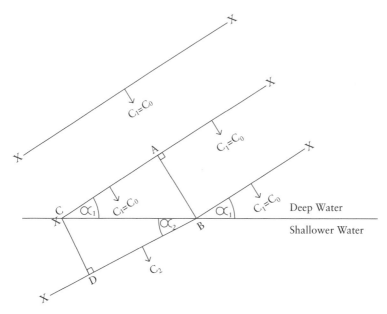

Figure 17.9   Wave refraction by Snell's law.

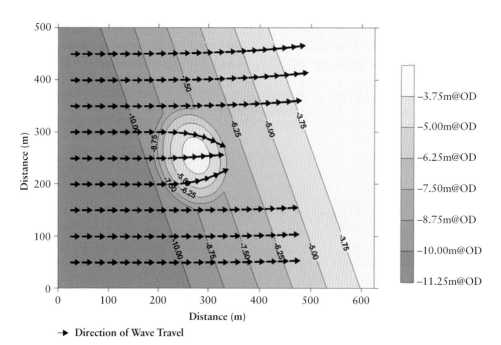

→ **Direction of Wave Travel**

Figure 17.10   Predicted wave refraction over an offshore shoal (wave height 1 m, wave period 4 s, water level −1 m relative to ordnance datum (OD)).

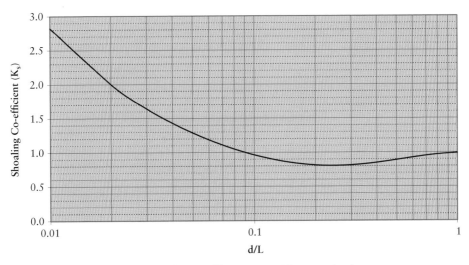

Figure 17.11   Variation of shoaling coefficient (K_s) with water depth.

effects (see Section 17.3.3) will cause wave energy to be transferred laterally, and a caustic would not be formed.

Deeply dredged entrance channels to ports can cause very strong refraction effects. If d_1 is the water depth outside the channel and d_2 is the depth inside, then $d_1 < d_2$ and thus $c_1 < c_2$. Since $\sin \alpha_2$ cannot be greater than unity, the above equation can only be satisfied provided that:

$$\sin \alpha_1 \leq \frac{c_1}{c_2} \qquad [17.53]$$

If this is not the case, then the waves cannot enter the channel and they will be reflected from it. The presence of currents as well as changes in water depth can also cause wave refraction. When currents are present, Equation 17.47 is no longer generally valid, and instead use is made of the principle of wave action conservation.

17.3.2 Shoaling

The variation of the shoaling coefficient K_s with the d/L ratio is shown in Figure 17.11.

17.3.3 Diffraction

Wave diffraction is the process by which energy is transferred in a direction perpendicular to that in which the waves are propagating. Figure 17.12 gives an example of wave diffraction, with waves passing the tip of a breakwater. If the transfer of energy behind the breakwater did not occur, then the water behind the breakwater would be perfectly calm.
Figure 17.12 shows three distinct regions:

1. the region shadowed from the main wave direction, where diffraction takes place;
2. the region where incident waves meet reflected waves, causing a short-crested sea to be set up; and
3. the undisturbed region where incident waves pass by.

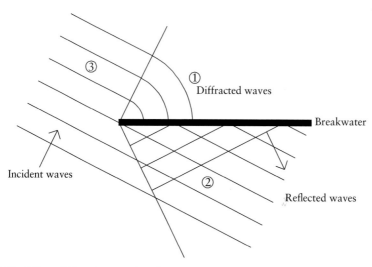

Figure 17.12   Wave diffraction at a breakwater.

17.3.4 Wave breaking

Wave breaking is one of the most commonly observed features in the nearshore zone. Wave breaking is a non-linear process that is difficult to describe analytically. Considering the wave-breaking process, if you return to Equation 17.54, the speed of the wave in shallow water is given by:

$$c_s = \sqrt{gd} \qquad [17.54]$$

For this demonstration, we assume that this equation is correct locally at every point in the wave. The velocity of the water particles in the top portion of the wave will therefore be significantly greater than the velocity of the water particles in the bottom portion of the wave. This will cause the wave to steepen towards the front, and eventually topple over (i.e. break) (Figure 17.13).

Waves may break in a number of different ways. Steep waves on mild slopes tend to break by spilling water gently from their crests, and there is little reflection of the incident

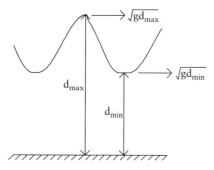

Figure 17.13   Wave-breaking phenomenon.

wave energy. In contrast, long low waves on steep slopes tend not to break at all. Instead, they surge up and down the slope with most of the wave energy being reflected.

The Iribarren number, N_I, also sometimes known as the 'surf similarity parameter', gives the most useful parameter that is used for describing wave behaviour on a slope:

$$N_I = \frac{\tan \beta}{\sqrt{H/L_0}} \qquad [17.55]$$

where H is measured at the toe of the slope, and the slope is β.

Equation 17.55 is as applicable for random waves replacing H with H_s as it is for linear waves.

The breaker types can be classified as follows, and they are represented graphically in Figure 17.14.

- Spilling breakers break gradually and are characterised by white water at the crest ($N_I < 0.4$).
- Plunging breakers curl over at the crest with a plunging forward of the mass of water at the crest ($0.4 < N_I < 2.3$).

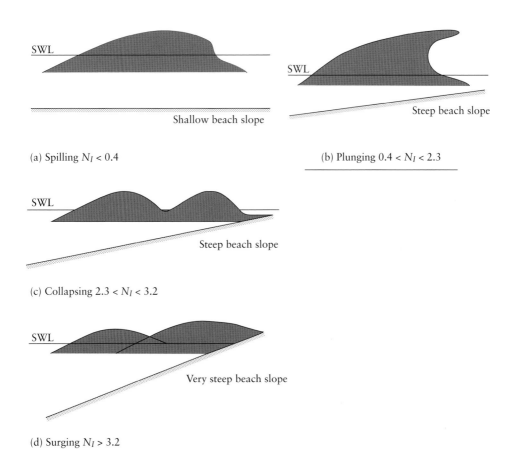

(a) Spilling $N_I < 0.4$

(b) Plunging $0.4 < N_I < 2.3$

(c) Collapsing $2.3 < N_I < 3.2$

(d) Surging $N_I > 3.2$

Figure 17.14   Wave breaker types.

- Collapsing breakers are waves in the transition between plunging and surging ($2.3 < N_I < 3.2$).
- Surging breakers build up as if to form a plunging breaker, but the base of the wave surges up the beach before the crest can plunge forward ($3.2 < N_I$).

In general, the point at which waves break depends on the wavelength, the water depth and the seabed slope. Much research has been done into wave breaking at a coastline, and no definitive equations are available which are generally considered the most appropriate to define the wave-breaking process.

Goda's (2000) wave-breaking index is the most commonly used formula to estimate the limiting height of individual breaking waves within the surf zone.

$$\frac{H_b}{L_0} = A\left\{1 - \exp\left[-1.5\frac{\pi h}{L_0}(1 + 15\tan^{4/3}\theta)\right]\right\} \qquad [17.56]$$

The coefficient A is set to 0.17 for regular waves, whilst it is 0.12 and 0.18 for random wave breaking of the lower and upper limits of the surf zone, respectively.

Moreover, Goda (2000) derived a set of formulae for wave height distribution within the surf zone. In general, Equation 17.57 is based on random waves that exist in real life and that the coastal engineer will encounter.

$$H_s = \begin{cases} K_s H_0' & : h/L_0 \geq 0.2 \\ \min\left\{\left(\beta_0 H_0' + \beta_1 h\right), \beta_{max} H_0', K_s H_0'\right\} & : h/L_0 < 0.2 \end{cases} \qquad [17.57]$$

where the coefficients $\beta_0, \beta_1, \beta_{max}$ are formulated as follows:

$$\beta_0 = 0.028\left(H_0'/L_0\right)^{-0.38} \exp[20\tan^{1.5}\theta]$$
$$\beta_1 = 0.52\exp[4.2\tan\theta]$$
$$\beta_{max} = \max\left\{0.92, 0.32\left(H_0'/L_0\right)^{-0.29} \times \exp[2.4\tan\theta]\right\}$$

where θ is the beach slope.

The ratio of wave height (H) to wavelength (L) is called wave steepness. Wave steepness is a commonly used term in coastal engineering. It is usually given the notation n, and it is usually referenced to deep water conditions (e.g. L_0). Under these conditions, wave steepness is given as in Equation 17.58:

$$n = \frac{2\pi H}{gT^2} \qquad [17.58]$$

For locally generated large wind waves (e.g. within a bay, or across a narrow stretch of water such as the Severn Estuary), the wave steepness value is usually high (e.g. >0.05). For swell waves, the wave steepness condition is usually very low (e.g. <0.04). Swell waves are waves arriving at a coastline from a storm occurring some distance away. For example, waves generated by a storm in the Atlantic Ocean could arrive in Britain from a westerly direction possibly several days later. The effects of travelling long distances will have reduced their wave height and therefore the wave steepness. Waves generated by storms off an exposed coast within a large body of water usually have wave steepness values in the range of 0.04–0.05.

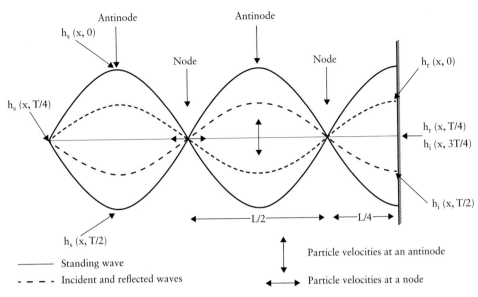

Figure 17.15   Standing wave pattern off a vertical wall.

17.3.5 Wave reflection

Waves normally incident on solid vertical or near vertical walls such as a harbour or sea-wall are reflected. The reflected wave has the same phase but the opposite direction as the incident wave. For vertical or near vertical sea walls, the reflected wall is almost of the same amplitude as the incident wave.

The resulting wave pattern set-up off a vertical or near vertical wall is called a stand-ing wave, or clapotis. A typical standing wave pattern is shown in Figure 17.15. At the nodal points, there is no vertical movement with time, but the horizontal velocities are at a maximum. By contrast, at the antinodes, crests and troughs appear alternately, but the horizontal velocities are zero.

Standing waves at seawalls can cause considerable damage to structures and can lead to large movements of sediment at the seabed, and hence increased erosion at the base of the wall. This is especially noticeable at the base of seawalls built during Victorian times, where beach levels have almost disappeared as a result of the construction of vertical seawalls.

The equation of the standing wave pattern is found by adding the waveform of the incident wave to that of the reflected wave. The formula of the incident wave is given by Equation 17.59, where the subscript i stands for the 'incident' wave.

$$\eta_i = \frac{H}{2} \cos \left[2\pi \left(\frac{x}{L} - \frac{t}{T} \right) \right] \qquad [17.59]$$

The reflected wave is given by Equation 17.60, where in this instance, the subscript r stands for the 'reflected' wave.

$$\eta_r = \frac{H}{2} \cos \left[2\pi \left(\frac{x}{L} + \frac{t}{T} \right) \right] \qquad [17.60]$$

This gives the resultant standing wave pattern as:

$$\eta_s = \eta_i + \eta_r \qquad [17.61]$$

where the subscript s is in this instance for a standing wave.

If the amplitude of the incident and the reflected wave are taken to be the same, then Equation 17.61 can be written as:

$$\eta_s = H \cos\left(\frac{2\pi x}{L}\right) \cos\left(\frac{2\pi t}{T}\right) \qquad [17.62]$$

17.4 Wave set-down and set-up

As waves propagate towards the shore, not only is there a flow of wave energy (see Equation 17.15), but also there is a flow of momentum. Changes in the wave height as waves travel shoreward cause changes in the flow of momentum that must be balanced by variations in the mean water level. When the wave height increases, the mean water level must fall; and when the wave height decreases, the mean water level must rise. Consequently, as waves approach a slope there is a gradual lowering of the mean water level as the wave height increases until the breaking point is reached. After breaking begins, the wave height continues to reduce as a result of the decreasing water depth, and the mean water level rises. At the shoreline, the mean water level is above the level of the still water. The negative and positive changes of water level due to the presence of a train of water waves are known as the wave set-down and wave set-up, respectively (Figure 17.16).

Denoting the difference between the still water level and the mean water level as S, its value at the breaking point, S_b, is given as:

$$S_b = \frac{-k H_b^2}{8 \sinh 2k \, d_b} \qquad [17.63]$$

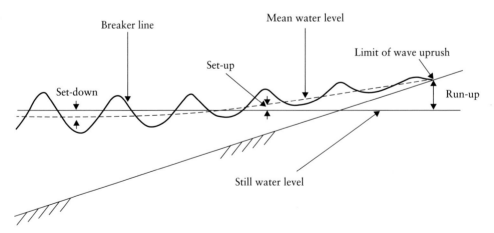

Figure 17.16   Wave set-down and set-up.

in which d_b is the depth of water at the breaking point measured to the mean water level (negative since the mean water level is below the still water level).

If we assume that shallow water conditions exist at the breaking point, then $\sinh 2kd_b \approx 2kd_b$, giving:

$$\frac{S_b}{d_b} \approx \left(\frac{H_b}{4d_b}\right)^2 \qquad [17.64]$$

Now, the maximum value of the ratio H_b/d_b is about 1.3, suggesting that the greatest set-down is about 10% of the mean water depth. However, the theory is not really valid for such highly non-linear conditions, and in practice, the maximum set-down is only about half of this theoretical value.

The maximum set-up, S_{max}, occurs at the shoreline. Ignoring the small set-down at the breaking point, then,

$$S_{max} \approx \frac{3\gamma H_b}{8} \qquad [17.65]$$

Since the value of γ is in the range of 0.7 to 1.3, S_{max} is between about 25% and 50% of the breaking wave height.

It is clear from Equation 17.64 that a change in the breaking wave height causes a change in the set-up. Thus, when groups of waves in a natural 'sea state' arrive at a beach, there is a slow fluctuation in the mean water level on the beach. This is known as surf beat. This rise and fall in the water level causes very long, low waves to radiate seaward.

17.5 Wave impact, run-up and overtopping

17.5.1 Wave impact

Wave impact occurs when an incoming wave meets a coastal defence structure, particularly such as a vertical seawall, in the nearshore region. Forces caused by wave impacts are the most intense and dangerous that a structure can suffer. Wave impact is highly dependent on both the wave conditions and the type of structure being considered. The typical wave conditions of the sea state are unbroken, breaking and broken waves. When a small amount of air is trapped between the structure and the breaking wave at the point of impact, the impact pressure increases considerably. Recent research studies show that air pocket dynamics play a paramount role in the evaluation of impact wave pressures (see Bullock *et al.*, 2007; Bredmose *et al.*, 2015).

17.5.2 Wave run-up

After a wave has broken, much of the wave energy that remains when the mass of water reaches the shoreline is used in driving it up the face of the beach or structure. This movement of water up the slope is called wave run-up. The run-up, R, which is measured vertically above the mean surface of the sea, is the maximum height above still water level to which the water rises, and it includes the effect of wave set-up.

The run-up on a smooth impermeable slope may be estimated with the aid of Hunt's formula (1959) as:

$$\frac{R_{si}}{H} \approx N_i \frac{\tan \beta}{\sqrt{\left(\frac{H}{L_0}\right)}} \qquad (for\ N_I \leq 2.3 - spilling\ plunging\ waves) \qquad [17.66]$$

$$\frac{R_{si}}{H} \approx \sqrt{\left(\frac{\pi}{2\beta}\right)} + \frac{\pi H}{L} \cot\left(\frac{2\pi d}{L}\right) \qquad (for\ N_I > 3.2 - surging\ waves) \quad [17.67]$$

In the range of $2.3 < N_I < 3.2$ (collapsing waves), R_{si} may be estimated using both equations, with the lower value chosen.

To allow for slope roughness and permeability, R_{si} may be multiplied by a roughness factor r. Note, however, that R_{si} cannot be less than S_{max}.

17.5.3 Wave overtopping

Wave overtopping occurs when a wave impacts a coastal structure, and the resulting momentum of the wave as it breaks on the structure causes some of the water to rise above and over the top of the structure. Based on the structure type (either permeable or impermeable), there will be two processes taking place: wave transmission and/or passing of water (green water) over the structure. If the structure freeboard exceeds the maximum run-up height, the overtopping occurs due to splashing (white water).

Prediction of wave overtopping is usually based on empirical formulae fitted with the help of experimental data. In general, overtopping is measured as mean discharge of water per linear metre of width of the structure; for example, $m^3/s/m$ or $l/s/m$. The EurOtop Manual (2007) highlights the latest overtopping models based on wave conditions and variety of structure types. The manual also gives tolerable discharge values for human and structural safety.

The principal formula employed to calculate wave overtopping (q) is of the form:

$$\frac{q}{\sqrt{gH_{m0}^3}} = a \exp(-bR_c/H_{m0}) \qquad [17.68]$$

where H_{m0} is the significant wave height estimated from spectral analysis; R_c is the freeboard of the structure; and a and b are overtopping coefficients. R_c will be replaced with R_c/γ, where γ is the roughness coefficient, when a structure consists of roughness elements or roughened surface (see EurOtop Manual, 2007).

17.6 Tides, surges and mean sea level

17.6.1 Sea level

Sea levels are constantly changing under the influence of both astronomical and meteorological effects. The astronomical effects result in a sinusoidal variation in the sea level surface, which is dominated by the semi-diurnal and diurnal tide components. The meteorological effects result in surges as a result of differential pressure changes and the wind, and waves as a result of the wind. A third component in the variation of the sea level surface, which is not easily noticeable without long-term measurements, is the tendency of the sea level surface to rise (or fall) over a long period of time. This is known as the 'trend' in the sea level, and it is usually referred to as a rate in the form of 'x mm/year'.

Sea levels relative to land can vary markedly from 'average' sea levels as a result of 'isostatic' effects (often referred to as 'glacial rebound'). During the last Ice Age, large areas of the Northern Hemisphere such as Northern Europe and Canada were covered with large ice sheets. This caused the land to be pushed down. With the disappearance of the ice, these regions are now rising, often at a rate greater than the increase in mean sea levels. This results in the phenomenon that exists in these areas of sea levels actually falling relative to a fixed datum on land (e.g. at Juneau in Alaska, sea levels relative to land are estimated to be falling by 13.8 mm/year) (National Research Council, 1987).

Britain is currently rotating along an axis running from the North Welsh coastline, through Liverpool to Hull, with land levels rising to the north of this line and falling to the south of it. The rise or fall is greater further away from this line.

17.6.2 Estimation of sea level rise

17.6.2.1 Use of tide gauges

Tide gauges are fixed relative to a land datum. Therefore, tide gauges only measure 'relative' changes in sea levels, and any calculation of trends from tide gauge data would need to take local land movements into account. However, the determination of rises in sea levels from tide gauge data is extremely difficult and notoriously inaccurate. There are several reasons for this, not least of which is the fact that long-term accurate records are only available from digital records, most of which have only been in place since the late 1980s at the earliest. To determine trends in sea levels from tide gauge records would probably need at least 50 years of digitally recorded data. The reason for needing such a large data set over such a long period of time is because tides repeat themselves over a period of 18.61 years. This means that trends determined over (say) the 9-year period between a peak and a trough of the 18.61-year cycle would almost certainly indicate a spurious negative trend. However, trends determined over (say) the 9-year period between a trough and a peak of the 18.61-year cycle would almost certainly indicate a spurious large positive trend.

17.6.2.2. Manually recorded sea level records

Before digital tide gauges were in operation, sea levels were measured manually (i.e. by watching the sea level go up and down on a tide board and estimating the sea level from this observation). Long-term records of these measurements have usually been transferred into a digital format only for the maximum recorded sea level in any particular year or, at best, the maximum recorded sea level in every month. Apart from the obvious human error that exists in these measurements, this technique has the added problem that the largest sea level recorded in any one year will almost certainly correspond to a very windy day, with large onshore waves. This therefore makes the determination of the sea level to the nearest (say) 10 cm extremely difficult, and these records are notoriously inaccurate (Hames *et al.*, 2004) However, many of these records exist back to the mid-late nineteenth century, and the effect of the 18.61-year cycle of tides is therefore mostly negated.

17.6.3 Current and future trends in sea levels

The most important body who publish information on changes in sea level are the Inter-governmental Panel on Climate Change (IPCC). They have published five assessment reports since 1988 on scientific, technical and socio-economic information related to climate change, its impact and future risks, and countermeasures for adaptation and mitigation. IPCC (2013) predicted the mean rate of sea level rise over the next 100 years between 26 and 82 cm based on four scenarios considered. CU Sea Level Research Group (2015) presented a global mean sea level rise of 3.3 ± 0.4 mm/year for the period of 1993–2015.

Many other organisations publish reports and give guidance on sea level rises throughout the world. Apart from the IPCC, the main body responsible for analysing sea levels are the Permanent Service for Mean Sea Level, based at the Proudman Oceanographic Laboratory (POL). National guidance is given by bodies such as the Environment Agency (UK).

17.7 Tsunami waves

Tsunamis are large ocean waves, or a series of waves with very large wavelengths and long periods of up to an hour, triggered by volcanic eruptions, explosions, landslides, earthquakes and planetary impacts. 'Tsunami' as a Japanese word is represented by two characters which translate into English as 'harbour wave'. Tsunamis are often incorrectly referred to as tidal waves or seismic sea waves.

In deep depths, tsunamis can be difficult to identify. There can be hundreds of kilometres between crests (typically 10–500 km), and the height of the wave may only be 1–2 m high at most. With several minutes passing between successive waves of a tsunami, they are hardly noticeable as they pass under ships in deep waters. When these waves approach a shoreline, however, they exhibit all the characteristics of waves covered in this chapter. Rapid draw-down of water levels can occur in harbours (i.e. set-down), often drawing people to the shoreline and subsequently to their deaths. Similar to water waves, when a tsunami wave approaches onshore, part of the wave energy is reflected back offshore, and the onshore propagating wave energy is dissipated through bottom friction and turbulence. Despite these losses, a tsunami waves inundates inland with a significant amount of energy, causing destruction to people, beaches and infrastructure such as coastal defences. The first part of the tsunami wave propagating inland can be either crest or trough-focused with a considerable amount of momentum and inertia. For example, recent studies revealed that most coastal defences such as coastal dikes and seawalls in northern Japan collapsed by scour failure due to the great erosive power of the 2011 Tohoku tsunami waves (see Jayaratne *et al.*, 2013, 2015; Esteban *et al.*, 2015).

Due to large wavelengths, tsunami waves act as shallow water waves. Under such conditions, the wavelength is very large compared to the water depth, and the wave speed depends only on water depth as seen in Equation 17.69:

$$c = \sqrt{gh} \qquad [17.69]$$

where h is the water depth.

For example, in offshore conditions, the water depth of around 5000 m generated a tsunami speed of around 220 m/s, or about 800 km/h.

The distance and height achieved by a tsunami depend on the natural contours of the land that it approaches. Waves flowing over flat land will reach a long way inshore (large inundation). Waves travelling up a river valley will increase in height, and potentially flood areas to a much greater height than adjacent areas at a lower land level, which may be perhaps unaffected by the effects of the tsunami.

On land, the velocity of the tsunami wave (horizontal velocity, or u) is affected by the topography of the land; therefore, Equation 17.69 is further modified as:

$$u = A\sqrt{g(R - h_G)} \qquad [17.70]$$

where h_G is the height of land above mean sea level; R is the tsunami run-up height; and A is a coefficient of order 1.0 (see Murata *et al.*, 2010).

The slowing down of the tsunami as it reaches the shore results in a shoaling wave, reaching many times its height offshore. It would not be unknown for a tsunami 1–2 m high in the deep depths of an ocean to reach 25–50 m by the time it reaches the shoreline (e.g. the 2004 Indian Ocean and 2011 Tohoku tsunamis). These waves will then surge unbroken onto the land, and one of the reasons for the significant damage and loss of life is the violent turbulent nature of the leading edge of the tsunami wave. The tsunami event also continues for a long time after the leading edge has passed, exacerbating the destruction caused.

The increase of the tsunami wave height as it approaches shallow water (as described in this section) is given by:

$$\frac{H_s}{H_d} = \left(\frac{h_d}{h_s}\right)^{0.25} \qquad [17.71]$$

where H_s and H_d are wave heights in shallow and deep water and h_s and h_d are the depths of the shallow and deep water, respectively. For example, a tsunami with a wave height of 1 m in the open ocean where the water depth is 5000 m would have a wave height approaching 5 m in water of depth 10 m.

Disaster prevention or preparedness for future tsunami events was brought into focus after the 2004 Indian Ocean tsunami. Since tsunami events are infrequent, people may forget about the implications for them and civil engineering infrastructure. Researchers attempted to classify different levels of tsunamis (Levels 1 and 2) and gave an introduction to classification of evacuation areas after the 2011 Tohoku tsunami (see Shibayama *et al.*, 2013). The key purpose of tsunami countermeasures is to mitigate casualties, injuries and physical damage to the infrastructure. There are two types of countermeasures being adopted in the coastal environment, namely hard and soft engineering methods. Defences such as mass concrete tsunami seawalls, breakwaters, coastal dikes and high-rise evacuation buildings are, to name a few, hard engineering countermeasures; whilst tsunami hazard maps with evacuation routes and signs, and coastal forests and wetlands, are examples of soft solutions.

Worked examples

Example 17.1

A wave in water depth of 100 m has a wave height of 3.0 m and a wave period of 12 s. Calculate wavelength, celerity and wave steepness of that wave. Check whether this wave is in shallow, transition or deep water.

Solution:

Given $d = 100m, H = 3m, T = 12s$,

Using $L_0 = \dfrac{gT^2}{2\pi}$

$L_0 = \dfrac{9.81 \times 12^2}{2\pi} = 225m$

Using the Fenton and McKee wavelength formula (Equation 17.8),

$$L = L_0 \left\{ \tanh \left[\left(2\pi \sqrt{d/g}/T \right)^{3/2} \right] \right\}^{2/3}$$

$$L = 225 \times \left\{ \tanh \left[\left(2\pi \times \sqrt{100/9.81}/12 \right)^{3/2} \right] \right\}^{2/3}$$

$\underline{L = 221m}$

Wave celerity is given by $c = \dfrac{L}{T}$

$\therefore c = \dfrac{221}{12} = \underline{18.4m/s}$

Wave steepness is given by $n = \dfrac{H}{L}$

$\therefore n = \dfrac{3.0}{221} = \underline{0.014}$

To check whether the wave is in shallow or deep water, using the d/L ratio,

$$d/L = 100/221 = \underline{0.45}$$

that is, $0.5 > d/L > 0.04$, so this is considered to be in the transition region between deep and shallow water.

Example 17.2

The distance between wave orthogonals at the breaker wave height is 15 m. The off-shore distance between wave orthogonals is 10 m. If the wave orthogonals at the breaker wave height are perpendicular to the seabed contours, calculate the wave height in deep water. Take nearshore wavelength = 40.1 m, wave height = 4.87 m and shallow water depth = 8 m.

Solution:

The refraction coefficient, $K_R = \sqrt{\dfrac{b_0}{b}} = \sqrt{\dfrac{10}{15}} = 0.816$

The group velocity parameter, n

$$n = \dfrac{1}{2}\left(1 + \dfrac{2kd}{\sinh 2kd}\right) = \dfrac{1}{2}\left(1 + \dfrac{2 \times (2\pi/40.1) \times 8}{\sinh[2 \times (2\pi/40.1) \times 8]}\right) = 0.706$$

The shoaling coefficient, K_S

$$K_S = (2n \tanh kd)^{-1/2} = (2 \times 0.706 \times \tanh[(2\pi/40.1) \times 8])^{-1/2} = 0.913$$

$\therefore$ Deepwater wave height, $H_0 = \dfrac{H}{K_R K_S} = \dfrac{4.87}{0.816 \times 0.913} = \underline{6.5m}$

Example 17.3

If a wave with a period of 12 s is travelling from offshore to a depth of 5 m with a seabed slope of 1 in 100, and it is on the point of breaking (upper limit), find the wave height at this point. Assume that the breaker wave height is given by Goda's breaker index formula.

Solution:

Given $h = 5m, T = 12s, \tan \beta = 1/100,$

From $\dfrac{H_b}{L_0} = A \left\{ 1 - \exp \left[-1.5\dfrac{\pi h}{L_0}(1 + 15 \tan^{4/3} \beta) \right] \right\}$

For upper limit of random wave breaking $A = 0.18$

$$L_0 = \frac{gT^2}{2\pi} = \frac{9.81 \times 12^2}{2\pi} = 225m$$

Re-arranging, $H_b = 0.18 \times L_0 \left\{ 1 - \exp \left[-1.5\dfrac{\pi h}{L_0}(1 + 15 \tan^{4/3} \beta) \right] \right\}$

$H_b = 0.18 \times 225 \times \left\{ 1 - \exp \left[-1.5\dfrac{\pi \times 5}{225}(1 + 15 \times (1/100)^{4/3}) \right] \right\}$

$H_b = \underline{4.15m}$

Example 17.4

Determine the wavelength, surface elevation and power generated by a wave with a period of 10 s travelling in a water depth of 20 m when it is approaching a sandy beach of 1:30. Assume that the waves are regular with plunging breaking characteristics (Iribarren number = 0.65), and follow linear wave theory.

Solution:

Given beach slope $\tan \beta = 1/30, T = 10s, d = 20m, N_I = 0.65,$

Using $L_0 = \dfrac{gT^2}{2\pi}$

$L_0 = \dfrac{9.81 \times 10^2}{2\pi} = 156m$

To calculate wavelength from the Fenton and McKee formula:

$$L = L_0 \left\{ \tanh \left[\left(2\pi \sqrt{d/g}/T \right)^{3/2} \right] \right\}^{2/3}$$

$$L = 156 \times \left\{ \tanh \left[\left(2\pi \times \sqrt{20/9.81}/10 \right)^{3/2} \right] \right\}^{2/3}$$

$$L = 122m$$

Using the Iribarren number to find H,

$$N_I = \frac{\tan \beta}{\sqrt{H/L_0}}$$

Re-arranging the above formula, $H = \dfrac{L_0 \tan^2 \beta}{N_I^2}$

$$H = \frac{156 \times (1/30)^2}{0.65^2} = 0.41m$$

General equation for a water wave is given by $\eta = a\cos(kx - \omega t)$

$$a = H/2 = 0.41/2 = 0.205$$
$$k = 2\pi/L = 2\pi/122 = 0.052$$
$$\omega = 2\pi/T = 2\pi/10 = 0.63$$

$\therefore$ Equation for the surface profile of a particular wave is given by

$$\eta = 0.205 \cos(0.052x - 0.63t)$$

To calculate power generated by the wave P,

$$\text{Energy density } E = \frac{1}{8}\rho g H^2$$

Taking seawater density, $\rho = 1025 \ kg/m^3$

$$E = \frac{1}{8} \times 1025 \times 9.81 \times 0.41^2$$
$$E = 211J/m^2 \tag{1}$$

Group velocity $c_g = \dfrac{1}{2}\left[1 + \dfrac{2kd}{\sinh 2kd} \right] c$

$$c_g = \frac{1}{2} \times \left[1 + \frac{2 \times 0.052 \times 20}{\sinh(2 \times 0.052 \times 20)} \right] \times \frac{122}{10} = 9.3m/s \tag{2}$$

From (1) and (2), $P = Ec_g = 211 \times 9.3$

$$P = 2.0KW/m$$

Example 17.5

The wave heights (in metres) given below were measured by a wave recorder over a period of 3 minutes. Determine the value of significant wave height (H_s). If the wave steepness of the coastal site is 0.045, calculate the maximum probable wave height in a storm of 3 hours.

0.68	1.21	0.58	0.36	1.60	1.18	0.47	0.96	1.02
1.70	0.28	0.92	1.75	1.20	0.93	0.41	1.75	0.79
0.55	1.75	0.59	1.01	1.00	0.54	0.24	0.86	1.36
1.73	0.48	0.38	1.72	1.52	1.69	0.23	0.99	1.12
1.77	0.86	0.42	0.72	1.73	1.53	1.49	1.68	0.53

Solution:

Rank the data from the highest to the lowest (45 values), and take the top one-third of the wave heights (15 values).

1.77	1.75	1.75	1.75	1.73	1.73	1.72	1.70	1.69
1.68	1.60	1.53	1.52	1.49	1.36			

Find the average, $H_s = \underline{1.65m}$

Using the wave steepness formula, $n = \dfrac{2\pi H_s}{gT_z^2}$

Re-arranging, $T_z = \sqrt{\dfrac{2\pi H_s}{gn}} = \sqrt{\dfrac{2\pi \times 1.65}{9.81 \times 0.045}} = 4.846s$

Number of storm waves, $N = \dfrac{3 \times 3600}{T_z} = \dfrac{3 \times 3600}{4.846} = 2229$

Maximum wave height, $H_{MAX} = H_s\sqrt{\dfrac{\ln N}{2}} = 1.65 \times \sqrt{\dfrac{\ln 2229}{2}} = \underline{3.24m}$

Example 17.6

A tsunami with a wave height of 1 m in the deep ocean where the water depth is 6000 m has moved to a nearshore depth of 2 m. Calculate the wave height at this water depth. If the distance between the epicentre of the tsunami and nearshore is about 1000 km, estimate how long it will take this wave to reach the nearshore region.

Solution:

Given $H_d = 1m, h_d = 6000m, h_s = 2m,$

Using $\dfrac{H_s}{H_d} = \left(\dfrac{h_d}{h_s}\right)^{0.25}$

Re-arranging above, the wave height at 2 m depth,

$$H_s = H_d \left(\frac{h_d}{h_s} \right)^{0.25} = 1.0 \times \left(\frac{6000}{2} \right)^{0.25} = \underline{7.4m}$$

Speed of the tsunami wave in deep water is approximated by $c = \sqrt{gh}$,

$$c = \sqrt{9.81 \times 6000} = 242.6 m/s$$

Time taken to reach the nearshore region, $T = \dfrac{Distance}{c} = \dfrac{1000 \times 1000}{242.6 \times 3600} hr$

$$\therefore T = \underline{1.15hr}$$

References and recommended reading

Bouws, E., Gunther, H., Rosenthal, W. and Vincent, C. L. (1985) Similarity of the wind wave spectrum in finite depth water. Part 1: spectral form. *Journal of Geophysical Research*, 90(C1), 975–986.

Bouws, E., Gunther, H., Rosenthal, W. and Vincent, C. L. (1987) Similarity of the wind wave spectrum in finite depth water. Part 2: statistical relationships between shape and growth stage parameters. *Deutsches Hydrographisches Zeitschrift*, 40, 1–24.

Bullock, G. N., Obhrai, C., Peregrine, D. H. and Bredmose, H. (2007) Violent breaking wave impacts. Part 1: results from large-scale regular wave tests on vertical and sloping walls. *Coastal Engineering*, 54, 602–617.

Bredmose, H., Bullock, G. N. and Hogg, A. J. (2015) Violent breaking wave impacts: part 3. Effects of scale and aeration. *Journal of Fluids Mechanics*, 765, 82–113.

CU Sea Level Research Group. (2015) University of Colorado, Boulder, CO, USA. http://sealevel.colorado.edu (accessed 8 September 2015).

Dean, R. G. and Dalrymple, R. A. (1991) *Water Wave Mechanics for Engineers and Scientists*, Advanced Series on Ocean Engineering, Vol. 2, World Scientific, Singapore.

Esteban, M., Takagi, H. and Shibayama, T. (2015) *Handbook of Coastal Disaster Mitigation for Engineers and Planners*, Elsevier, Amsterdam.

EurOtop Manual (2007) *Wave Overtopping of Sea Defences and Related Structures: Assessment Manual*, HR Wallingford, Wallingford, UK.

Fenton, J. D. and McKee, W. D. (1990) On calculating the lengths of water waves. *Coastal Engineering*, 14, 499–513.

Goda, Y. (2000) *Random Seas and Design of Maritime Structures*, Advanced Series on Ocean Engineering, Vol. 15, 2nd ed., World Scientific, Singapore.

Hames, D., Reeve, D., Marriott, M. and Chadwick, A. (2004) Effect of data quality on the analysis of water levels along the Cumbrian coastline, *IMA International Conference on Flood Risk Assessment*, Bath, UK.

Hasselmann, K., Barnett, T. P., Bouws, E., Carlsen, H., Cartwright, D. E., Enkee, K., Ewing, J. A., Gienapp, H., Hasselmann, D. E., Kruseman, P., Meerburg, A., Muller, P., Olbers, D. J., Richter, K., Sell, W. and Walden, H. (1973) Measurements of wind-wave growth and swell decay during the joint North Sea wave project (JONSWAP). *Deutsches Hydrographisches Zeitschrift*, 12, A8.

Hunt, I. J. (1959) Design of seawalls and breakwaters. *Proceedings of the American Society of Civil Engineers*, 85(WW3), 123–152.

Intergovernmental Panel on Climate Change (IPCC) (2013) *The Physical Scientific Basis of Climate Change*, The Working Group I, Cambridge University Press, Cambridge.

Jayaratne, R., Mikami, T., Esteban, M. and Shibayama, T. (2013) Investigation of coastal structure failure due to the 2011 Great Eastern Japan Earthquake Tsunami, in *Coasts, Marine Structures and Breakwaters Conference*, Institution of Civil Engineers, Edinburgh, doi:10.1680/fsts.59757.

Jayaratne, R., Premaratne, B., Abimbola, A., Mikami, T., Matsuba, S., Shibayama, T., Esteban, M. and Nistor, I. (2015) Failure mechanisms and local scour at coastal structures induced by tsunamis. *Coastal Engineering Journal*, World Scientific (accepted).

Kamphuis, J. W. (2000) *Introduction to Coastal Engineering and Management*, Advanced Series on Ocean Engineering, Vol. 16, World Scientific, Singapore.

Murata, S., Imamura, F., Katoh, K., Kawata, Y., Takahashi, S. and Takayama, T. (2010) *Tsunami: To Survive from Tsunami*, Advanced Series on Ocean Engineering, Vol. 32, World Scientific, Singapore.

National Research Council (1987) *Responding to Changes in Sea Level*, National Academy Press, Washington, DC.

O'Brien, M. P. (1942) *A Summary of the Theory of Oscillatory Waves*, TR 2, US Army Corps of Engineers, Beach Erosion Board, Washington DC.

Shibayama, T., Esteban, M., Nistor, I., Takagi, H., Thao, N. D., Matsumaru, R., Mikami, T., Arenguiz, R., Jayaratne, R. and Ohira, K. (2013) Classification of tsunami and evacuation areas. *Natural Hazards*, 67(2), 365–386, doi:10.1007/s11069-013-0567-4.

Sorensen, R. M. (2010) *Basic Wave Mechanics for Coastal and Ocean Engineers*, John Wiley & Sons, New York.

Thompson, E. F. and Vincent, C. L. (1983) Prediction of wave height in shallow water, in *Proceedings of Coastal Structures Conference*, American Society of Civil Engineers, 1000–1008.

Problems

1. A wave in water depth of 10 m has a wave height of 2.0 m and a wave period of 10 s. Calculate the wavelength, celerity and wave steepness of that wave. Check whether this wave is in shallow, transition or deep water.

2. Time series data measured at an offshore site are given in this table:

Incident wave height (m)	Wave period (s)
2.35	5.0
2.58	4.8
2.89	5.5
3.75	4.0
3.27	4.7
2.62	5.2
3.13	5.0
3.66	4.5
2.71	5.0
3.75	5.2
3.01	4.6
2.50	4.9

Determine the zero-crossing wave period (mean period) and the significant wave height. Given a wave steepness of 0.04, calculate the maximum probable wave height in a storm of 1 hour.

3. The water level at a coastal site is given relative to ordnance datum as 8.0 m, and prior to reaching the site, the waves pass over a sand bank at a level of 4.0 m.

    If a 7.5 m high offshore wave travels over the sand bank and the breaker wave height–to–water depth ratio is given as 0.78, determine the height of the waves after they have passed over this sand bank. Find the period of these broken waves, assuming that the steepness of the waves offshore is given as 0.045.

4. A tsunami has generated a wave height of 10 m at a nearshore depth of 2 m. Calculate the wave height at its offshore depth of 8000 m. If the time taken for the tsunami wave to travel from its epicentre to nearshore is about 2 hours, estimate the distance between the epicentre and the nearshore region.

Answers

1. Properties of Fluids

(1(b)) 5 N s/m^2; (2) 2 m/s, 0.4 N s/m^2; (3) 1.7 kW; (4) 21.3 m/s; (5) 5.0 mm, −1.6 mm; (6) 80 N/m^2.

2. Fluid Statics

(1(a)) 57.63 kN/m^2; (1(b)) 31.83 m; (2) 36.9 kN/m^2, 10.2 kN/m^2; (3) 32 mm, 31.6; (4) 2.94 kN/m^2; (5) 0.25 N/mm^2; (6) 0.67 m, 2.0 m, 3.53 m, 26.16 kN/m; (7) 0.19 m; (8) 100.7 kN, 3.49 m below water level, 53.3 kN; (9) 171.2 kN; 39.23° to the horizontal, 1.90 m below water surface; (10) 4.83 MN/m, 66°, 26.21 m from heel; (11) 0.85; (13) 12.82 kN; (14) $\frac{L^2}{4b} - \frac{b}{2}$ above water level; (15) 0, 11.92 sin θ kN m; (16) 1.41 m, 2.54°; (17) 244.6 mm; (18) 9.10°, 11.12°; (19) 17.17 kN.

3. Fluid Flow Concepts and Measurements

(1) 15 m/s^2, 150 m/s^2; (2) 21 m/s^2; (4) 3.62 kW, towards 450 mm diameter section; (5) −50.62 kN/m^2, 36.7 kN/m^2, 1.34 kW; (6) 3.98 m, 76.8°; (7) 23.7 kN, 45°, 16.75 kN; (8) 12.91 kN, 9.4° to the horizontal; (9) 811 kN; (10) 1.35 kN, 67.5° to the horizontal; (11) 856 kN, 12.54° to the vertical; (12) 130 mm; (13(a)) 46.36 mm; (13(b)) no change; (14) 25 km/h; (15) $C_v = 0.96$, $C_c = 0.62$, $C_d = 0.596$; (16) 7 min 51.7 s; (17) 25.5 mm, 39.5 mm; (18) m$^{1/2}$/s [L$^{1/2}$ T^{-1}] 62.3 mm, 0.6%; (19(b)) 1.6%; (20) 1.48, 2.5, 0.626; (21) 3 h 10 min.

4. Flow of Incompressible Fluids in Pipelines

(1(a)) 170.9 L/s; (1(b)) 20.174 m; (2) 8700 m, 22.78 kW; (3) 158 L/s; (4(a)) 350 mm, 214.7 L/s; (4(b)) 12.44 m; (5) 58.77 L/s, 0.28 mm; (6(a)) 215.5 L/s; (6(b)) 9350 m;

Nalluri & Featherstone's Civil Engineering Hydraulics: Essential Theory with Worked Examples, Sixth Edition. Martin Marriott.
© 2016 John Wiley & Sons, Ltd. Published 2016 by John Wiley & Sons, Ltd.
Companion Website: www.wiley.com/go/Marriott

(7) 0.1147 m, 13.37 kW; (8(a)) 9.5 L/s, laminar; (8(b)) 25 L/s, turbulent; (9(a)(i)) 62.25 L/s, (ii) 64.82 L/s; (9(b)(i)) 50.1 L/s, (ii) 49.12 L/s; (10) 0.0144.

5. Pipe Network Analysis

(1) $Z_B = 91.48$ m, $Q_{AB} = 107.56$ L/s, $Q_{BC_1} = 52.05$ L/s, $Q_{BC_2} = 55.51$ L/s; (2) $Z_B = 90.98$ m, $Q_{AB} = 110.82$ L/s, $Q_{BC_1} = 55.78$ L/s, $Q_{BC_2} = 55.03$ L/s; (3) $Z_B = 75.31$, $Q_{AB} = 156.43$ L/s, $Q_{BC} = 56.29$ L/s, $Q_{BD_1} = 59.20$ L/s, $Q_{BD_2} = 40.98$ L/s; (4) $Z_B = 132.3$ m, $Q_{AB} = 118$ L/s, $Q_{BC} = 40$ L/s, $Q_{BD} = 78$ L/s. Pump total head = 21.75 m, power consumption = 14.22 kW; (5) $Z_B = 126.73$ m, $Z_D = 109.41$ m, $Q_{AB} = 505.3$ L/s, $Q_{BC} = 133.2$ L/s, $Q_{BD} = 372.1$ L/s, $Q_{DE} = 221.2$ L/s, $Q_{DF} = 150.8$ L/s; (6) $Z_A = 200$ m, $Z_C = 100$ m, $Z_E = 60$ m, $Z_F = 50$ m, $Z_B = 100.19$ m, $Z_D = 71.71$ m, $Q_{AB} = 279$ L/s, $Q_{BC} = 126.7$ L/s, $Q_{BD} = 152.4$ L/s, $Q_{DE} = 52.1$ L/s, $Q_{DF} = 100.3$ L/s; (7) $Q_{AB} = 104.8$ L/s, $Q_{BC} = 45.4$ L/s, $Q_{DC} = 4.6$ L/s, $Q_{ED} = 44.6$ L/s, $Q_{AE} = 95.2$ L/s, $Q_{EB} = 0.6$ L/s, $Z_A = 60$ m, $Z_B = 39.03$, $Z_C = 17.37$, $Z_D = 18.14$ m, $Z_E = 39.06$; (8) $Q_{BCE} = 61.98$ L/s, $Q_{BE} = 44.91$ L/s, $Q_{BDE} = 93.10$ L/s, head loss in AF = 12.5 m; (9) $Q_{AB} = 106.4$ L/s, $Q_{BC} = 52.5$ L/s, $Q_{CD} = 2.5$ L/s, $Q_{ED} = 37.5$ L/s, $Q_{AE} = 93.6$ L/s, $Q_{EB} = 6.1$ L/s, $Z_A = 60$ m, $Z_B = 38.50$ m, $Z_C = 25.03$ m, $Z_D = 24.77$ m, $Z_E = 39.78$ m; (10).

Pipe	AB	BH	HF	GF	AG
Discharge (L/s)	136.41	56.57	2.51	53.59	93.59

Pipe	BC	CD	HD	DE	FE
Discharge (L/s)	29.84	9.84	24.06	13.90	26.10

Junction	A	B	C	D	E	F	G	H
Head elevation (m)	100.00	69.44	64.18	63.85	63.22	67.31	88.94	67.34

(11)

Pipe	AB	BC	CD	ED	FE	BE
Flow (L/s)	95.31	90.14	30.14	49.86	44.69	5.17

Junction	A	B	C	D	E	F
Head elevation (m)	100.00	87.70	71.67	65.48	87.66	90.00

(12) See (10); (13) $Q_{AB} = 131.56$ L/s, $Q_{BE} = 25.02$ L/s, $Q_{FE} = 48.44$ L/s, $Q_{AF} = 88.44$ L/s, $Q_{BC} = 46.54$ L/s, $Q_{CD} = 6.54$ L/s, $Q_{ED} = 23.46$ L/s; $H_A = 40.000$ m, $H_B = 31.294$ m, $H_C = 11.650$ m, $H_D = 10.124$ m, $H_E = 14.795$ m, $H_F = 38.418$ m.

6. Pump–Pipeline System Analysis and Design

(1) 136 L/s, 88.9 kW; (2(a)(i)) 182 L/s, (ii) 192 L/s; (2(b)(i)) 138.4 kW, (ii) 186.3 kW; (3) mixed flow; (4) 31 L/s, 20.8 L/s, 10.0 L/s; (5) $N_s = 5120$, axial flow, 125 L/s, 6.44 kW; (6(a)) 4.81 m, 0.12; (6(b)) 4.76 m; (7) 1386 rev/min; (8) 27.5 L/s; (9(a)) 137.6 L/s; (9(b)) 166 L/s; (10(a)) $Z_B = 90.46$ m, $Q_{AB} = 57.1$ L/s, $Q_{BC} = 38.5$ L/s, $Q_{BD} = 18.6$ L/s; (10(b))

$Z_B = 89.04$ m, $Q_{AB} = 61.4$ L/s, $Q_{BC} = 37.6$ L/s, $Q_{BD} = 23.8$ L/s, $H_p = 26.5$ m, power consumption $= 12.4$ kW.

7. Boundary Layers on Flat Plates and in Ducts

(1) 381.6 N, 0.186 m, 4.56 N/m²; (2(a)) 770.8 N; (2(b)) 748.9 N; (3) 22.23 m/s; (4) 1.95 mm, 3.7 m/s; (5) 16.38 N/m², 0.067, 23.65 L/s, 0.091 mm; (6) 2.97 mm, 0.0114, 0.094 m³/s, 0.0379.

8. Steady Flow in Open Channels

(1) 4.9 N/m²; (2(a)) $k = 2.027$ mm, $n = 0.0149$; (2(b)) $Q_{Darcy} = 56.21$ m³/s, $Q_{Manning} = 56.57$ m³/s; (3) 2.74 m; (4) 3.776 m³/s, 1.83 m/s, 5.48 N/m²; (5) 3.6 m³/s, 0.00324; (7) 17.6 m³/s, 1.67 m; (8(a)) 30.12 m³/s; (8(b)) 29.29 m³/s; (8(c)) 33.95 m³/s; (9) 2.5 m; (10) width $= 12.26$ m, depth $= 6.13$ m; (11) bed width $= 3.33$ m, depth $= 4.02$ m; (12) 1.45:1; (13) bed width $= 20.5$ m, depth $= 2.29$ m; (14) 2.09 m; (15(a))

z (m)	0.1	0.2	0.3	0.4	0.5
y_1 (m)	2.5	2.5	2.5	2.5	2.5
y_2 (m)	2.373	2.24	2.097	1.937	1.739

z (m)	0.6	0.7	0.8	0.9	1.0
y_1 (m)	2.54	2.66	2.78	2.89	3.01
y_2 (m)	1.45	1.45	1.45	1.45	1.45

(15(b)) $y_c = 1.45$ m; (15(c)) $z_c = 0.568$ m; (16) $y_1 = 1.544$ m, $y_2 = 1.047$ m, $Q = 0.877$ m³/s; (17) initial depth $= 0.639$ m, upstream depth $= 7.42$ m, force $= 786$ kN; (18) submerged flow at gate, 5.157 m, 2.176 m, 257 kN; (19) $y_n = 3.5$ m, 13 km, 4.44 m; (20) see table below; (21) $y_n = 1.48$ m, $y_c = 0.714$ m, $y = 1.32$ at $x = 350$ m; (22) > 10 km; (23(a)) 103.20 m, A OD; (23(b)) submerged inlet and reduced flow rate; (24(a)) 2.25 m; (24(b)) 2.000 m.
(20)

Depth (m)	4.0	3.9	3.8	3.7	3.6	3.5	3.4
Distance (m)	0	151	308	472	643	824	1018

Depth (m)	3.3	3.2	3.1	3.0	2.9	2.8
Distance (m)	1229	1463	1730	2049	2464	3106

9. Dimensional Analysis, Similitude and Hydraulic Models

(1) 85 L/s, 0.01445; (2) 35176 N/m²; (3) 2.42 L/s, 3.4; (4) (length scale)$^{3/2}$; (6) 0.4645 m, 118.1 L/s; (7) 93.75 m/s, 11.2 kN;
(8)

Q_p (L/s)	0.0	103.1	206.2	309.3	412.44
H_p (m)	83.06	78.89	66.45	43.61	10.38

(9(a)) $N_s = N\sqrt{P}/H^{5/4}$; **(9(b))** 1549.2 rev/min, 27.89 MW, 21.07 m³/s; **(10)** 2.85 s, 7.5 m, 0.5 m, 800 kN/m.

10. Ideal Fluid Flow and Curvilinear Flow

(1) $V(y\cos\alpha - x\sin\alpha)$, $Vr(\sin\theta\cos\alpha - \cos\theta\sin\alpha)$; **(2)** $\phi = (x^2 - y^2)/2$
(9) 0.075 m³/s; **(10)** 2.2 m, 2.658 m; **(11)** 105.8 m³/s, −5.36 m, −4.40 m;
(12) 0.2 L/s; **(13)** 0.20 m³.

11. Gradually Varied, Unsteady Flow from Reservoirs

(1) 175.9 h; **(2)** 39.69 h, 131.39 h; **(3)** 602.78 h; **(4)** peak outflow = 45 m³/s at $t = 11$ h.

12. Mass Oscillations and Pressure Transients in Pipelines

(1) $z_{max} = 40.5$ m at $t = 95$ s; **(2(a))** $z_{max} = 40.80$ m after 100 s;
(2(b)) $z_{max} = 37.12$ m after 150 s;
(3(a))

Time (s)	0	3	6	9	12	15
V (m/s)	4.233	4.221	4.171	4.031	3.559	0
h (m)	3.00	4.66	8.09	17.00	53.03	436.74

(3(b))

Time (s)	0	3	6	9	12	15	18	21	24	27	30
V (m/s)	4.2339	4.2278	4.2107	4.1768	4.1190	4.0253	3.8740	3.6230	3.1767	2.2733	0.0
h (m)	3.0	3.68	4.64	5.96	7.89	10.85	15.71	24.42	42.24	86.54	228.46

13. Unsteady Flow in Channels

(1(a)) $y_2 = 2.815$ m, $c = 3.54$ m/s; **(1(b))** $y_2 = 3.295$ m, $c = 4.31$ m/s; **(2)** $y_1 = 0.98$ m, $y_2 = 1.807$ m, $Q_2 = 46.15$ m³/s (upstream); **(3)** $y_2 = 2.98$ m, $y_1 = 3.54$ m, $c = 7.14$ m/s; **(4)** $y_1 = 2.631$ m, 1.515 min; **(5)** 36.45 min.

14. Uniform Flow in Loose-boundary Channels

(1) Safe; **(2)** 0.57 m; **(3)** 15 mm; **(4)** 9.2×10^{-5}; **(5)** 6.95 m/s, 0.6 m; **(6(i))** 0.69 m³/(s m), 8.2×10^{-6}; **(6(ii))** 2.5 m³/(s m), 1.73×10^{-4}; **(7(a))** 9×10^{-6} m³/(sm), 21 mm, dunes, rough; **(8)** 20 kg/(s m); **(9)** 83.4%; **(10)** 1060 N/s.

15. Hydraulic Structures

(1(a)) 10.599 m; **(1(c))** 3787 m³/s; **(1(d))** −11.9 m of water; **(2)** 1424 m³/s; 2764 m³/s; **(3)** $X_c = 105.38$ m; $y_c = 7.01$ m; **(4(a))** $y_1 = 1.423$ m; $y_2 = 11.924$ m; $y^+ = 8.12$ m; **(4(c))** 61%.

16. Environmental Hydraulics and Engineering Hydrology

(**1**) 41 m³/s; (**2**) 0.144; (**3**) 9 m³/s just after $t = 4$ h (**4(a)**) 1120 m³/s; (**4(b)**) 2360 m³/s; (**5**) Scheme B, 2.2.

17. Introduction to Coastal Engineering

(**1**) 94 m, 9.4 m/s, 0.021, transition; (**2**) 4.87 s, 3.61 m, 6.34 m; (**3**) 3.12 m, 10.3 s; (**4**) 1.26 m, 2017 km.

Index

absolute pressure, 10
acceleration
  centrifugal, 23
  convective, 50
  horizontal, 22
  local, 50
  normal, 50
  radial, 23
  tangential, 50
  uniform linear, 22
  vertical, 22
Ackers–White, 346
Airy, 409, 411
alluvial, 337
angular velocity, 23
antidune, 335
Archimedes, principle of, 17
atmospheric pressure, 9

backwater curves in channels, 187, 234
Barr, 91, 93
beach slope, 425–426, 435
bed forms, 335
bed hydraulic radius, 337
bed load, 340
bed shear, 335
benefits, 397
Bernoulli's equation, 53
Blasius, 90, 171

boundary layers
  boundary shear stress, 89, 172
  displacement thickness, 173
  drag, 172–173
  effect of plate roughness, 171
  flat plates, 172–173
  Kármán–Prandtl equations, 176
  laminar boundary layers, 171
  laminar sub-layer, 176
  mixing length, 174
  Nikuradse, 175–176
  Prandtl mixing theory, 174
  shear velocity, 174
  thickness, 171
  turbulent boundary layers, 172
  turbulent pipe flow, 174
Boussinesq, coefficient of, 57, 197
bulk modulus, 2, 248
  effect on wave speed, 311
buoyancy, centre of, 18
buoyant thrust, 17

canal delivery, 201
capillarity, 3
cavitation, 55, 59, 153
  number, 154
celerity, 309, 323, 411, 413–414
centrifugal pumps, 149
channels, *see* open channel flow

characteristic curves, pumps, 151–153
Chezy equation, 189
climate change, 396, 432
coastal engineering, 409
Colebrook–White, 91, 176, 188, 247, 251
compressibility, 2
concentration, 217, 343
continuity, equation of, 50, 120, 306, 311
    two-dimensional flow, 265
contraction, coefficient of, 57, 61
Coriolis coefficient, 55, 187
corresponding speed, 255
cost–benefit analysis, 397
critical depth, 196
critical shear, 335
critical velocity, 196
culvert, 202, 237–239
curved surface, hydrostatic thrust, 15
curvilinear flow, 273
    channel bend, 284
    duct bend, 278–281
    rotating cylinder, 281
    siphon spillway, 282–283

Darcy–Weisbach equation, 90, 188
density, mass, 2
    relative density, 2, 336
    specific volume, 4
    specific weight, 2
diffraction, 423–434
dimensional analysis, 247
    Buckingham π theorem, 249
    corresponding speed, 255
    dimensional forms, 248
    Froude number, 249–262
    model studies, 249, 257–259, 262
    non-dimensional groups, 248
    pipelines, 250
    prototype, 247
    rectangular weir, 254
    Reynolds number, 248–249, 254, 256–257
    rotodynamic pumps, 251–252, 260–262
    similitude, 249
    V-notch, 253, 256
    Weber number, 249, 254, 256–257
discharge, 50
    coefficient of, 59, 61, 371, 373
    under varying head, 63, 289
discounted cash flow, 397
displacement thickness, 173
drag coefficient, 340

drag, flow over flat plate, 172
dunes, 335
dynamic similarity, 249

eddy viscosity, 52, 343
Einstein's equation, 339, 342, 345
energy
    equation for ideal fluid flow, 52
    equation for real, incompressible flow, 54
    gradient, 89, 188
    kinetic, 54
    potential, 54
    pressure, 54
    total, 54
energy dissipators, 376
energy equation, 52, 54, 89, 187, 195
energy losses
    flowing fluid, 89, 198
    pipes, 54, 89
    sudden transitions, 57, 95
Engelund, 345
environment, 393
equilibrium
    neutral, 18
    relative, 22
    stable, 18
    unstable, 18
equipotential lines, 267
estuary models, 259
Euler's equation of motion, 53

fall velocity, 339
Fenton and McKee, 412
fetch, 247, 418, 420
floating bodies
    equilibrium of, 18
    liquid ballast, 20
    periodic time of oscillation, 20
    stability of, 17
flood alleviation, 396
flood routing, 291, 299–301, 396
flow
    channels, 187, 323
    curvilinear, 273
    dynamics of, 52
    Eulerian description of, 47
    gradually varied, steady, 198–200
    gradually varied, unsteady, 289, 323
    ideal, 52, 265
    incompressible fluids in pipelines, 89–118
    irrotational, 49

kinematics of, 47
Lagrangian description of, 47
laminar, 51, 90, 92
measurement of, *see* flow measurement
non-uniform, 48
one dimensional, 49
pipe networks, 119–148
rapidly varied, steady, 195
rapidly varied, unsteady, *see* pressure
    transients, waterhammer, surge
real, 54
rotational, 49
subcritical, 196
supercritical, 196
steady, 48, 195, 198–199
three dimensional, 49
turbulent, 51, 91–92
two dimensional, 49
uniform, 48
unsteady, 48, 287, 323
flow measurement, 58–68, 218–221, 227,
    376
flow nets, 268, 270, 275, 279
flow regimes, 335
fluid
    definition of, 1
    flow concepts and measurements, *see* flow,
        flow measurement
    ideal, definition, 52
    Newtonian, 1–2
fluid statics, 7
fluids in relative equilibrium, 22
    effect of acceleration, 22–23
fluids, real and ideal, 52, 54, 265
force exerted by a jet on a flat plate, 77–78
force on pipe bend, 76
forced vortex, 274, 281
form drag, 337
free vortex, 274
friction factor
    dependence on Reynolds number, 90–92
    laminar flow, 90, 92
    rough pipes, 91–94
    smooth pipes, 90–92
    turbulent flow, 90–94
friction losses in pipes, 89–90
Froude number, 196–197, 249–262, 323,
    335

Garde, 345
Garde–Ranga Raju's formula, 337

gas, definition of, 1
generalised logistic distribution, 395,
    398–403
geometric similarity, 247
Goda's index and formulae, 426
gradually varied flow, *see* flow
Graf, 345
grain resistance, 337
Gringorten, 394, 398, 402–403
group wave celerity, 414
Gumbel distribution, 394, 398, 402

Hagen–Poiseuille equation, 90, 247
Hardy-Cross, 120
head
    potential or elevation, 54
    pressure, 9, 54
    varying, 63
    velocity or kinetic, 54
HR Wallingford design chart, 93
hydraulic grade line, 89, 150
hydraulic jump, 196–198, 376–378
hydraulic radius, 90, 336
hydraulic structures, 371
hydrograph, 299–301, 396
hydrological cycle, 393
hydrology, 393
hydrostatic thrust (force)
    curved surface, 15
    plane surface, 11

ideal fluid flow, 52, 265
    boundary conditions, 272
    circulation, 267
    combination of basic flow patterns, 269
    curvilinear flow, 273, 281
    equipotential lines, 267
    flow nets, 268, 270, 275, 279
    forced vortex, 274, 281
    free vortex, 274
    graphical methods, 275
    Laplace equation, 270
    line sink, 269
    line source, 268
    numerical methods, 270
    pathline, 47, 265
    radial velocity component, 266
    siphon spillway, 282
    stream function, 265, 267
    streamlines, 47, 265
    streamtube, 49, 265

ideal fluid flow (*Continued*)
   tangential velocity component, 266
   uniform flow pattern, 268
impeller, 149, 260–262
incipient motion, 335
Iribarren number, 425
irrotational flow, 49, 265

JONSWAP wave spectrum, 418

Kalinske, 341, 344
Kármán–Prandtl equations, 91, 176
Kármán's constant, 343
kinetic energy correction factor, 55

Lacey, 337
laminar boundary layer, 171
laminar flow, 51
laminar sub-layer, 176
Lane and Kalinske, 344
Laplace's equation, 270
Laursen, 345
liquid, definition of, 1
logarithmic velocity distribution, 174
loose-boundary, 335
losses
   in pipe fittings, 94
   sudden contraction, 57
   sudden enlargement, 57

Manning formula, 189, 193, 336, 396
manometer
   differential, 10
   inclined, 10
   U-tube, 10
mass oscillations in pipelines
   finite difference methods, 308
   sudden discharge stoppage, 306–307
   surge chamber operation, 305
meandering, 396
metacentre, 18
metacentric height, 18, 20
Meyer–Peter and Muller's equation, 341
model studies, *see* dimensional analysis
model testing, *see* similitude
momentum
   correction factor, 56
   equation, 56, 77, 196
Moody diagram, 92
mouthpieces, 61
   hydraulic coefficients of, 61
movability parameter, 339, 343–344

negative surge, 327
net positive suction head, 154
Newtonian fluid, 1–2
Newton's law of motion, 56, 248
Newton's law of viscosity, 2
Nikuradse, 91, 175–176
notches, 64
   Bazin formula, 65
   end contractions, 65
   Francis formula, 65
   rectangular, 64
   Rehbock formula, 65
   trapezoidal, 66
   V or triangular, 65

ogee spillway, 67, 371–372
open channel flow (steady)
   best hydraulic section, 191
   channel design, 191
   Chezy equation, 189, 208
   Colebrook–White equation, 188
   composite roughness, 189
   compound section, 190, 210
   conveyance of river and flood plain, 396
   critical depth flume, 228
   critical tractive force, 192–193, 337
   Darcy–Weisbach equation, 188
   economic section, 191
   energy components, 195
   energy principles, 195
   flow measurement, 64–68, 218–221
   gradually varied flow, 198–200
   hydraulic jump, 196–198, 376–378
   Manning formula, 189, 193, 336, 396
   mobile boundary, 192
   momentum equation, 196
   part-full circular pipes, 194, 216
   rapidly varied flow, 195
   rigid boundary, 191
   sewers, 194
   spatially varied flow, 203
   specific energy, 195
   storm sewer, 194, 217
   uniform flow, 187
   uniform flow resistance, 188
   velocity distribution, 207
   venturi flume, 221, 228–229
   wastewater sewer, 194, 215
   water surface profile, 234–235
   wetted perimeter, 191, 339

open channel flow (unsteady)
  celerity (wave), 323
  dam break, 329
  downstream positive surge, 326
  gradually varied, 323
  negative surge waves, 327
  upstream positive surge, 325
orifice
  hydraulic coefficients of, 61
  large, 62
  small, 60
  submerged, 63
  varying head, 63, 289
orifice meter, 58

particle Reynolds number, 340
Pascal's law, 7
pathline, 47, 265
Pierson Moskowitz wave spectrum, 418
piezometer, 10
piezometric pressure head, 9
pipe networks, 119
  effect of booster pump, 128–129
  gradient method, 123
  head balance method, 120
  quantity balance method, 121
pipelines
  Colebrook–White equation, 91, 176
  Darcy–Weisbach equation, 90
  effective roughness size, 90, 176
  friction factor, 90
  Hagen–Poiseuille equation, 90
  incompressible steady flow resistance, 89
  Kármán–Prandtl equations, 91, 176
  laterally distributed outflow, 100
  local losses, 94
  Moody diagram, 92
  Moody formula, 91
  networks, see pipe networks
  pipes in series, 98–100
  resistance in non-circular sections, 94
pitot tube, 59
plane boundary, 315
Poiseuille, 90
Prandtl, 91, 174
pressure
  absolute, 10
  atmospheric, 9
  centre of, 11–14
  diagram, 14
  distribution, 14
  within a droplet, 3

gauge, 9
head, 9
hydrostatic pressure distribution, 9
measurement of, 9
piezometric pressure head, 9
at a point, 7–8
saturated vapour pressure, 3
stagnation, 59–60
vacuum, 9
vapour pressure, 2–3
variation with depth, 8
pressure transients in pipelines, 309
  Allievi equations, 312
  basic differential equations, 311
  waterhammer, 309, 311
  wave celerity, 309
probability, 394
project appraisal, 397
pumps
  cavitation in, 153–154
  characteristic curves, 151–152
  efficiency, 151
  impeller, 149, 260
  manometric head, 150
  manometric suction head, 154
  net positive suction head, 154
  parallel operation, 151, 160
  in pipe network, 128–129
  pipeline selection in pumping system, 158
  power input, 151
  pump–pipeline system, 150
  rotodynamic types, 149
  series operation, 152, 161
  specific speed, 149
  static lift, 150
  system curve, 158
  Thoma cavitation number, 154
  variable speed, 153, 162

quasi-steady flow, 289

rainfall, 217, 393
Rayleigh distribution, 416
reflection, 427
refraction, 420–423
regime channel design, 337, 346
  Blench's approach, 348
  Kennedy's approach, 346
  Lacey's approach, 337, 347
  non-scouring boundary, 350
  Simons–Albertson's method, 349
  stable erodible boundary, 350

Rehbock formula, 65
relative density, 2, 336
relative roughness, 90
reservoir routing, 291, 299–301
return period, 194, 394
Reynolds number, 52, 90, 249–262
Riemann, 313
rigid-bed channels
    friction factors, 351–352
    limit deposition, 350–352
    sediment transport, 350
ripples, 335
river models, 257–258
rotational flow, 49, 265, 273
roughness, 90
Rouse's distribution, 343
Runge–Kutta method, 240
runoff, 194, 217, 393

saltation, 340, 344
Schoklitsch's equation, 341
sea level, 430–432
sediment, 194, 335
sediment transport, 335, 350, 396
self-cleansing velocity, 194, 216, 351
separation, 55
sewers, 187, 194, 350
shear Reynolds number, 335
shear stress, 1, 7, 89, 171, 192–193,
    335
shear velocity, 174, 336
Shields criterion and diagram, 336
Shields transport equation, 341
side weir, 204
similitude, similarity, 247, 249
    dynamic, 249
    geometric, 249
    laws for river models, 257–260
    laws for rotodynamic machines,
        260
    laws for weirs and spillways, 262
sink, 269
sinuosity, 396
siphon, 75
sluice gate, 223–224
Snell's law, 421–422
source, 269
spatially varied flow, 203, 375
specific energy, in channels, 195
specific speed, rotodynamic machines, 149,
    261

specific volume, 4
specific weight, 2
spillway
    cavitation, 373
    effective spillway length, 373
    gated spillway, 373
    model, 262
    negative pressures, 373
    offset spillway, 373
    ogee spillway, 67, 371–372
    profile (ogee spillway), 372
    self-aeration, 374
    shaft (morning glory) spillway, 371
    side channel spillway, 375
    siphon spillway, 282–283
standing wave, 427
static lift, 150
steady flow energy equation, 53, 89
stilling basin, 376, 388–390
stream function, 265–266
streamline, 47, 265
    patterns of, 52
streamtube, 49, 265
Strickler's equation, 189, 336
surface tension, 248, 253, 259
surge, 305, 323–324, 430
surge chambers, 305–309
suspended load, 340, 343
sustainability, 393, 396
sustainable drainage systems, 396

Thames data, 395, 398–403
threshold, 335, 342
tidal modelling, 259
tidal period, 259, 431
tides, 431
time domain analysis, 415
time of concentration, 217
TMA spectrum, 419
Torricelli's theorem, 60
total load, 340, 345
tractive force, 192–193
tsunami, 432–433, 437–438
turbine, 110, 112
turbulent flow, 51–52, 90–91

units
    energy, 1
    force, 1
    power, 1
    pressure, 7

SI system, 1
  work, 1
unsteady flow
  negative surge wave, 327
  open channels, 324
  pipe flow, 289
  positive surge wave, 325–326

vapour pressure, 2
velocity
  angular, 23
  approach, 65
  average, 50
  coefficient of, 61
  components of, 47
  distribution, 171–172, 174, 207
  fluctuations of, 48
  potential, 262
  self-cleansing, 194, 216, 351
  temporal mean, 48
vena contracta, 57, 61
venturi flume, 187–188, 221, 228–229
venturi meter, 58
  coefficient of discharge, 59
viscosity, 2
  dynamic, 2, 248
  kinematic, 2, 248
  Newton's law of, 2
volume, specific, 4
vortex
  forced, 45, 274, 281
  free, 274

wash load, 340, 345
waterhammer, 309
wave breaking, 424–426
wave energy, 414
wave height, 410, 420
wave impact, 429
wavelength, 410–412
wave overtopping, 430
wave period, 410
wave propagation speed, 311, 411, 413
wave reflection, 313, 427
wave refraction, 420–423
wave run-up, 429
waves, 409, 420
  irregular, 410, 414
  linear, 410–411
  non-linear, 410
  regular, 410–411
wave set-down and set-up, 428
wave spectrum, 418–419
wave steepness, 426
Weber number, 249, 254
Weibull, 394
weirs
  broad-crested, 227
  Cipolletti, 66
  De Marchi coefficient, 204
  proportional or Sutro, 66
  Rehbock formula, 65
  side weir, 204
  submergence of, 68
wetted perimeter, 89, 94, 191, 339